AF588091

Trails in Modern Theoretical and Mathematical Physics

Giovanni Morchio

Andrea Cintio · Alessandro Michelangeli
Editors

Trails in Modern Theoretical and Mathematical Physics

A Volume in Tribute to Giovanni Morchio

Editors
Andrea Cintio
Institute for the Chemical-Physical Processes
National Council of Research
Pisa, Italy

Alessandro Michelangeli
Department of Mathematics and Natural Sciences
Prince Mohammad Bin Fahd University
Al Khobar, Saudi Arabia
Hausdorff Center for Mathematics
Bonn, Germany
Trieste Institute for Theoretical Quantum Technology
Trieste, Italy

ISBN 978-3-031-44987-1 (Hardcopy)
ISBN 978-3-031-44990-1 (Softcopy)
ISBN 978-3-031-44988-8 (eBook)
https://doi.org/10.1007/978-3-031-44988-8

This Springer imprint is published by the registered company Springer Nature Switzerland AG
The registered company address is: Gewerbestrasse 11, 6330 Cham, Switzerland

Paper in this product is recyclable.

Preface

This volume accomplishes a beautiful collective effort to celebrate and remember the late Giovanni (Gianni) Morchio on behalf of his friends, colleagues, collaborators, mentees, students, presenting at the same time updated overviews and perspectives on topical research themes in theoretical and mathematical physics which are intimately connected with Gianni's wide spectrum of scientific interests and activities.

When Gianni left us, way too early, on 10 November 2021, it did not take long for the dismay of that immense loss to be replaced by the desire to say goodbye to such an outstanding, humble, generous scientist through an editorial initiative in his memory and tribute.

As two of the many former students of Gianni, we took it upon ourselves to channel these intentions into a volume that immediately met with sincere interest and prompt approval by Springer.

Right from the start we were very positively impressed, but certainly not surprised, by the support and direct or indirect participation of a large number of people linked to Gianni by friendship and scientific ties of various kinds, including his many former students, several of his world-class collaborators and co-authors, and also international scholars involved in modern physical developments of classical results which Gianni had given a decisive contribution to over the years.

Indeed, Gianni's scientific, social, and human legacy trespasses the necessarily narrow boundaries of this volume. Today countless people can proudly bear witness to Gianni's righteousness, his humanity, his sensitivity towards the evolution, injustices, and hopes of contemporary society, his interests in human history and its lessons, his dedication to the cause of the labourers, the acuity of his political analysis, his scientific stature, the depth of his knowledge of modern physics and mathematics, his altruism and boundless generosity with friends, collaborators, and students, the kindness and moderation of his character, his permanent availability to help, support, encourage, always with his sweet smile.

It is with this spirit and in the light of such a warm response that we structured the volume into three parts.

The first presents the profile of Gianni from two complementary and inseparable points of view: on the one hand, the theoretical and mathematical physicist side, whose academic career spanned the last five decades, produced admirable results, and impacted all those who had the luck to know him at whatever stage of their scientific path; on the other hand, the social and political side of Gianni, often little known to those who knew him as a scientist, but surely central to his entire life and equally inspiring for so many. Remarkably indeed, Gianni had the utmost devotion to these two spheres of his life without those who knew him on the one side being well aware of what he did with equal dedication on the other. In the end, this first part of the volume is complemented by an intimate and touching memory by Gianni's children which opens up a discreet and intense look at the beauty of his family life.

The second part is the purely scientific one and, in a sense, it represents the core of the volume. Several outstanding scholars, many of which had been Gianni's co-authors, mentors, collaborators, kindly accepted to discuss themes that are topical in modern theoretical and mathematical physics and are very closely linked to Gianni's vast research field as it unfolded throughout his whole career. They all managed to produce valuable scientific chapters, each of which is self-contained and with its own bibliographic references, where the reader may find, in a straightforward and accessible form, a collection of retrospectives, modern reviews, recent developments, updated refinements, and future and open problems.

It is not by chance that we entitled this part "Trails in modern theoretical and mathematical physics", like the title of the whole volume: the readers—the experts as well as young graduates and interested researchers from neighbouring fields—may indeed enjoy surfing across those chapters and may well benefit from presentations that are clean, concise, updated, and surely inspirational.

At the same time we trust to have managed, by curating this second part, and in view of the calibre of the scholars involved, to convey a flavour of the indisputable resonance that Gianni Morchio had and has in theoretical physics and that part of mathematical physics concerned with the rigorous understanding of classical and quantum systems.

In the end, the third part of this volume collects a miscellaneous ensemble of short or not-so-short contributions by those who accepted to remember Gianni by revisiting and sharing their memories as former friends, colleagues, and of course former students of his—some of the latter kindly included instructive retrospectives of their scientific activity with Gianni as well. Gianni was, among other, a splendid teacher who inspired generations of students, a numerous group of which eventually chose him as supervisor and mentor. This resulted in a touching tribute across personal and private recollections and anecdotes which gives the volume a final brushstroke of intimate humanity, in the sadness of the loss accompanied by the warmth and happiness of the memory.

In fact, while curating the whole initiative many more had the pleasure to express to us their sincere adhesion while at the same time opting not to send a written contribution and instead cherish their own memory of Gianni privately. Analogously, many outstanding former co-authors of Gianni gently declined to send a scientific chapter, despite their appreciation for this initiative, due to compelling private reasons. We do consider all them equally part of the volume, as all the present authors, of course.

In addition to the genuine involvement of the many we already referred to, the completion of this volume was possible thanks to the decisive advice and encouragement from Franco Strocchi, who was Gianni's teacher first, and then was linked to him through a fortunate decades-long scientific and human tie, as well as thanks to the patient and professional technical support of Elena Griniari, Springer senior editor, and her team at Springer. We are also keen to acknowledge the discreet encouragement received from Gianni's family, who followed the genesis and implementation of this initiative.

May this volume be of inspiration and scientific reference, and keep Gianni's legacy alive.

Pisa and Bonn
July 2023

Andrea Cintio
Alessandro Michelangeli

Contents

Part I
Gianni: A Profile

Giovanni Morchio (1948–2021)

Sergio Albeverio and Alessandro Michelangeli

Giovanni ("Gianni") Morchio passed away on 10 November 2021.

He left with discretion and a smile, intelligence and generosity, culture and unique insight of physics, all features that have always distinguished him and through which for years he has trained his students and enriched his colleagues.

His name is linked to the Institute—later Department—of Physics of the University of Pisa, where he carried out his research, teaching, and training activity for almost fifty years, including those following his retirement. Even more is his name linked to his exceptional scientific profile, recognised by unanimous admiration and international esteem, including his enduring prolific collaboration with his former supervisor and later close collaborator Franco Strocchi.

For his thesis for the Diploma in Physics obtained in 1971, as a student at the University of Pisa and at the same time of the Scuola Normale Superiore of Pisa, Gianni was sent by Franco Strocchi to the ETH Zurich, to discuss the infrared problem with Klaus Hepp with non-perturbative techniques, also in low dimensions. As further explained later on, those were the years of the great successes of constructive field theory; Strocchi had previously returned to Pisa after his collaboration at Princeton with Arthur Wightman, and the ETH was the other world top node in this branch of theoretical physics. After graduating, Gianni left for the then-compulsory two-year army service, and subsequently joined the Physics Institute of Turin, from which he then moved to Pisa in 1974, which became his permanent home since then. Even after the forced "pre-retirement" imposed by law by the regulations of

S. Albeverio
Institute for Applied Mathematics, University of Bonn, Bonn, Germany
Hausdorff Center for Mathematics, University of Bonn, Bonn, Germany
e-mail: albeverio@iam-uni-bonn.de

A. Michelangeli (✉)
Department of Mathematics and Natural Sciences, Prince Mohammad Bin Fahd University, Al Khobar, Saudi Arabia
Hausdorff Center for Mathematics, Bonn, Germany
Trieste Institute for Theoretical Quantum Technology, Trieste, Germany
e-mail: amichelangeli@pmu.edu.sa

A. Cintio, A. Michelangeli (Eds.), *Trails in Modern Theoretical and Mathematical Physics*, https://doi.org/10.1007/978-3-031-44988-8_1

the time, Gianni remained most active in the Physics Department and in all exchanges with former students and collaborators.

The context in which Gianni started his research was the very lively activity on the theory of quantised fields in the second half of the 1960s.

After the pioneering works of P. A. Dirac, W. Heisenberg, and W. Pauli in the years 1927–1929, with the first formal treatment for passing from non-relativistic quantum mechanics to the theory of quantised fields, the problems that arose within perturbation theory—necessary to include the interactions, but bringing in physically and mathematically unacceptable divergences—had given rise to renormalisation techniques intended to "cure" such divergences. The dissatisfaction from the point of view of rigour led an amount of physicists and mathematicians, starting from the mid-1950s, to develop axiomatic frameworks that express in a mathematically precise way a few fundamental principles, which be sufficient for rigorous deductions (the Gårding–Wightman axioms, with deductions of physical relevance such as PCT invariance and the connection between spin and statistics, and the rigorous formulation by R. Haag and D. Ruelle of scattering theory for relativistic fields). It then became even more accessible and crucial to solve the problem of establishing whether quantum field theory axioms imply the non-triviality of the associated scattering amplitudes—the latter being the signature of actually interacting fields or particles, as indeed observed in Nature.

Around the mid-1960s, with the Princetonian theses of A. Jaffe and O. Lanford, researchers started analysing models (respectively, ϕ^4 and Yukawa) of local interactions with regularisation, intended as potential candidates, upon removing the regularisations in a mathematically consistent way, for the construction of fields satisfying all the axioms and with non-trivial scattering amplitudes. E. Nelson in 1964–66 had obtained, through probabilistic methods of Feynman–Kac type, a fundamental estimate (hypercontractivity) for the ϕ^4 model on a bounded spatial interval. In other seminal works, J. Glimm and A. Jaffe had begun a systematic study of the problem of regularisation removal. A course held by K. Hepp in 1969 at the École Polytechnique had indicated a 'road map' for the development of a constructive theory of fields in lower space-time dimensions (from 1 to 3, thereby requiring only a finite number of renormalisation counter-terms) and, later, in higher dimensions as well (from 4 on, with instead infinitely many counter-terms needed). The case of quantum electrodynamics was further complicated by the absence of mass in the photons, which implies a too slow decrease of the propagator at infinity (the necessary modifications to scattering theory for that scenario were only formulated at the end of the 1970s), and by gauge invariance, which introduces constraints between the potentials of the field variables.

All that had stimulated the study of simplified interacting models devised as approximations of electrodynamics accessible to known methods. After a visit at the ETH, E. Nelson had developed the 'Nelson model with mass' (non-relativistic scalar electrons interacting with massive scalar bosons, and consequent renormalisation), published in 1966, also discussed by K. Hepp in his 1969 course, and later generalised by J. P. Eckmann in his 1970 thesis by means of relativistic kinematics for scalar electrons. Scattering in this model was further studied by S. Albeverio at

Princeton in 1970–71. In his first stay at the ETH as a guest of K. Hepp, Gianni had the opportunity to discuss ideas and strategies for dealing with models of such type, including the scenario of massless bosons, hence closer to the case of electrodynamics, with J. Fröhlich, who in the course of 1971 was working on his doctoral thesis under the direction of K. Hepp and W. Hunziker.

It is in this scenario, so much fertile and in strong expansion, that in November 1971 Gianni discussed his thesis entitled ‘Infrared difficulties in field theory’ in Pisa. In it he analysed the problem of infrared divergence in quantum electrodynamics and in simple models such as the ‘models of currents’ (also time-dependent) of Bloch–Nordsieck (proposed in 1937) and of Pauli–Fierz (of 1938), the latter having been rigorously treated shortly before by Ph. Blanchard in 1969 in the doctoral thesis carried out under the direction of R. Jost and W. Hunziker. Gianni’s thesis ends with a discussion of Nelson’s model and J. Fröhlich’s results. Among other things, with this brilliant work Gianni begun a notable scientific closeness with J. Fröhlich, which would give rise to fundamental subsequent works.

Among all topics of all Gianni’s research activity, that of infrared divergences in quantum electrodynamics would in fact occupy a central position throughout, starting from the celebrated joint works with J. Fröhlich and F. Strocchi. In the first of them (Annals of Physics, 1979), it was demonstrated that under explicit hypotheses scattering states of quantum electrodynamics with non-zero charge have an associated algebra with disjoint representation from that of a free field—whereby the meaningfulness of ‘associated infra-particles’. The existence of a spontaneous symmetry breaking with respect to Lorentz group’s boosts was also highlighted, and the possibility of an extension of the Haag–Ruelle scattering theory to the case of charged infra-particles was outlined. In the subsequent works of this series (Physics Letters B, 1979 and 1980, and Nuclear Physics B, 1981) the Higgs phenomenon (of mass generation in bosonic gauge fields) is discussed both in the continuum (with a discussion in perturbative terms) and on lattice, and with emphasis on the role of gauge invariance in connecting the two approaches.

The success of those theoretical results on the formulation of the Higgs mechanism without an order parameter, later known as the ‘FMS mechanism’ (Fröhlich–Morchio–Strocchi), is the basis of a research line that is extremely active at present, in theoretical physics and phenomenology of fundamental interactions, as well in the direction of extending physics beyond the Standard Model and to quantum gravity.

Gianni kept investigating problems of the theories of quantised gauge fields also later in the course of his career, in particular the issues of: infrared singularities, choice of the Hilbert space of physical states, presence and multiplicity of invariant states, symmetries and their breaking, representation in spaces of indefinite metric. The results contained in various works by Gianni with F. Strocchi along these directions have become a reference for all subsequent studies of mathematical-physical nature. Significant examples were the general study of the structure of theories of local quantised fields with non-positive-metric Hilbert space, the formulation of both the relativistic and Euclidean axiomatic framework, the study of the algebras associated with the fields and their irreducibility.

Gianni's collaboration with F. Strocchi has produced many other remarkable studies of both general structures and specific models, across a multitude of areas of theoretical physics, which also includes, in addition to the axiomatic and constructive field theories and their models (by Schwinger, Kibble, and the others already mentioned), condensed matter theory, statistical mechanics, classical and quantum dynamical systems.

To these, he further added over time an intense research activity on problems from non-relativistic quantum theory (Bell's inequalities, quantisation on manifolds, topological effects, scattering theory, objectification of quantum mechanics, stationary and thermal states in quantum statistical mechanics, Bohmian mechanics and stochastic mechanics à la Nelson, rules of sum in frequency, Coulomb model of Jellium, partial Boolean structures, classical representability in quantum mechanics), as well as works of more algebraic-geometric nature (theory of the representation of algebras in spaces of Hilbert with non-positive definite product, Dirac operators on manifolds, C*-categories in relation to asymptotic abeliannnes, representation of commutation relations, Kreĭn representation of CCR-Heisenberg and non-regular algebras, algebraic bosonisation, spectral stochastic processes, Wigner crystals, Poisson–Rinehart spectral algebra on manifolds), and, recently, the study of the connection between general relativity and Newtonian gravitation.

In fact, the above overview is inevitably partial, even more impressively so in view of Gianni's extreme caution in publishing only works that be highly innovative, original, trailblazers.

Such a depth of contributions across a very broad spectrum of physical subjects reflects Gianni's characteristic, certainly atypical in the contemporary specialisation and fragmentation into disciplines and sub-disciplines, of dominating an immense physical and mathematical culture.

Owing to his extraordinary versatility, depth, and culture, not infrequently in the Institute of Physics Gianni was consulted as a sort of 'oracle' for difficult or impossible questions …

Gianni was, in the highest sense of the term, a theoretical physicist: for him the discussion of a model or a problem was primarily and profoundly physical, hinged on a coherent framework of first principles and specific assumptions and hypotheses, developed with rigorous methods where the mathematical rigour has the role of language and tool, and with ontological distinction between physical formalism and its interpretation.

Alongside the top-level physicist and scientist, internationally recognised, appreciated and esteemed far beyond the unfortunately limited national and local awards, Gianni was an extraordinary trainer of students and young collaborators, offering extraordinary and unique stimuli, opportunities, advice, and vision.

For years, generations of students of the physics study programme in Pisa came across him first for the class of mathematical methods for physics; then, a part of them, including students from the mathematics programme, decided to to attend his legendary final year elective course, centred on probabilistic, statistical, functional-analytic, operator-theoretic, as well as foundational methods and tools for quantum mechanics ('in the [penultimate year's] theoretical physics class you'd learn things, in [last year's] Morchio's class you'd understand them', recited a popular handbook for freshmen, written by senior students, to provide a presentation of all the courses of the study programme).

Fundamental was, for Gianni, the rigour that permeated all his pedagogical and didactic interaction with students: he shunned sterile abstractions or unnecessary generalisations; on the contrary, he adopted an 'experimental' approach in applying theoretical tools to identify the conceptually essential aspects of a problem, through an initial 'exploratory' phase consisting of identifying and examining particularly illuminating special cases, prior to leading to a general comprehension of the subject.

When even the arduous effort of Gianni's course was overcome, repaid in the end by the satisfaction of an unparalleled cultural experience in the whole physics study programme, the most daring students—indeed many, and in a regular and continuous flow over time—used to ask him to be their supervisor for the final thesis. Which, in turn, proved on the one hand to be an extraordinarily demanding undertaking, on problems already strongly characterised by the elements of current academic research, and on the other to be a path of profound understanding and superior breadth and vision. Gianni's former students, who today are researchers, university professors, journalists, professionals in Italy and abroad, all bring with them, in addition to the memory of the man's tenderness and generosity, the vivid recollections of their thesis' work, so intense, challenging, fulfilling.

Not to mention, it should not be forgotten, the innumerable theses for which Gianni acted as opponent in the final dissertation, always accepting the task with abnegation, and on the most disparate topics from a whole spectrum of subjects taught in the physics department, contributing with detailed and improving observations—yet one further sign of his cultural universality.

Interacting and discussing with Gianni, for those who knew him, went far beyond the scientific collaboration or the training relationship between the lecturer/supervisor and his students, and continued with broad and exciting dialogues on society, politics, history, literature, science's paths and trends, discussions on recent scientific or popular monographs, among many others.

Outside the academic sphere, Gianni has always maintained primary and extraordinary sensitivity and watchful attention to social and political problems of our time, to contemporary, local, and global injustices and distortions, to contradictions and inequalities in society and in the labour world, not only by sharing his own always deep and vigilant analysis, but also by personally devoting his own time and energy with great generosity to many labourers and 'non-scientists' of sort, who too now pay him their sincere and dear memory.

With profound sadness we are aware that there will no longer be another Gianni Morchio, an exquisite physicist, a researcher of exceptional calibre and boundless culture, a rigorous and scrupulous teacher and trainer, a person of unique humanity, sensitivity, altruism, and generosity. Yet, in many lucky ones remain all the affection, admiration, esteem, recognition, friendship for Gianni, and great many remember his example, his results, his scientific impact, and his high lesson in physics and life.

Chi ha compagni non morirà

Fulvio Cornolti, Pompeo Antonio De Biase, and Maurizio Rinaldi

CHI HA COMPAGNI NON MORIRÀ[1]

In a New Year email message, Gianni once wrote to us:

> *I've just found one thing in the Marxists Internet Archive, beautiful and simple. I had believed that the choice of both the adage and the motto had been mine, but someone did come first.*
> *The adage:* Humani nihil a me alienum puto.[2]
> *The motto:* De omnibus dubitandum.[3]

These were actually Karl Marx's answers to a questionnaire his daughters once submitted to him—it was then customary in England that girls elicited confessions from parents and friends.

We believe that there is the whole Gianni in such words.

Referring to Gianni, *Humani nihil* was meant in all its universality. Each and every single moment of human history raised condemnation and enthusiasm in Gianni. Condemnation for all the anguish and the material and moral devastation suffered by the vast majority of the humankind in the course of history; enthusiasm for all

[1] "*Those having comrades will never die*", from *The Internationale anthem*, Italian adaptation by Franco Fortini (1917–1994) in *Poesie Inedite*, curated by Pier Vincenzo Mengaldo (1997). Original Italian version of the present Chapter included in Appendix A.1.
[2] "*I consider nothing human is alien to me*", Terence (195–159 b.C.).
[3] "*Doubt everything*", René Descartes (1596–1650).

F. Cornolti
University of Pisa, Pisa, Italy
e-mail: Fulvio.Cornolti@gmail.com

P. A. De Biase (✉)
"Ulisse Dini" Scientific High School, Pisa, Italy
e-mail: Pompeoantonio.DeBiase@gmail.com

M. Rinaldi
Firenze, Italy

A. Cintio, A. Michelangeli (Eds.), *Trails in Modern Theoretical and Mathematical Physics*, https://doi.org/10.1007/978-3-031-44988-8_2

what man had created and would further create, against all odds, so as to bring the humankind more and more towards the dreamed world.

Gianni totally identified himself in these words by Marx: "*It will then become plain that the world has long since dreamed of something of which it needs only to become conscious for it to possess it in reality. It will then become plain that our task is not to draw a sharp mental line between past and future, but to complete the thought of the past. Lastly, it will become plain that mankind will not begin any new work, but will consciously bring about the completion of its old work.*"[4]

Not a blind faith, though. Rather, a faith based on reason.

In fact, quite the opposite of the fetishistic reason that worships the *magnifiche sorti e progressive*,[5] the blind positivist belief in an unlimited and extraordinary progress of the course of history. And quite the opposite also of the instrumental rationality of those who only exploit the use of reason to achieve the best personal advantages within a society given as a fact and hence accepted as a destiny. Neither was Gianni's standpoint the contemplation typical of those living detached from the society, isolated in the world of ideas or of science. On the contrary: his was the reason meant as the foundation of ethics and common good, as well as the ground of passion and political action. This is why Gianni used to love the Dutch philosopher Baruch Spinoza, whom he read until his last days, as well as the Neapolitan philosopher Antonio Labriola.

Gianni was adamant in his belief that each person may eventually reach the capability to comprehend even the most complex subjects. That was actually the main legacy of the Age of Enlightenment. One should have seen how unlimited his patience and passion were when he used to illustrate abstract categories of economics to the labourers whom he shared decades of political and union activity with. He was driven by the confidence that the exchange with them would elevate himself, for thought and knowledge are the outcome of the real life of real people.

The simple fact that every succeeding generation finds productive forces (including science) *acquired by the preceding generation and which serve it as the raw material of further production, engenders a relatedness in the history of man, engenders a history of mankind, which is all the more a history of mankind as man's productive forces, and hence his social relations, have expanded. From this it can only be concluded that the social history of man is never anything else than the history of his individual development, whether he is conscious of this or not. His material relations form the basis of all his relations. These material relations are but the necessary forms in which his material and individual activity is realised.*[6]

De omnibus dubitandum: Gianni firmly believed that "*there is nothing so far removed from us as to be beyond our reach, or so hidden that we cannot discover it*",[7] and that both the object of knowledge and the subject itself are inherently part of human activity as steps of the social production process. As a result, every truth

[4] Karl Marx (1818–1983), Letter to Arnold Ruge, September 1843.

[5] "*the magnificent and progressive fate*", from *La Ginestra*, by Giacomo Leopardi (1798–1837).

[6] Karl Marx, Letter to Pavel Vasilevič Annenkov, December 1846.

[7] René Descartes, *The Discourse on The Method* (1637).

must be considered provisional, continually subjected to criticism, both in the check of its internal consistency and in its accordance to historical and scientific facts.

The actual truth of a theory lies first and foremost in its ability to stimulate human actions and subsequent researches, and ultimately in its capacity to lead beyond itself. In this respect, questioning any proposition or consolidated result is precisely the opposite of relativism and scepticism: constantly reflecting on what is considered established is the driving principle of any development of thinking.

Gianni used to devour one book after another, in continuous search of historical facts that enhance experience and widen ideas to discuss on. Whenever he reached a conclusion that broadened his understanding of history and relationships between classes he promptly shared it with us with great zeal, impatient to test it and to subject it to objections and criticism. Not rarely, however, in the very moment when he was reporting and substantiating his deductions, his concern was palpable, as a thought unsatisfied of itself and of the findings reached until that moment. And then he restarted the analysis of the problem from yet another perspective.

It was wonderful to witness and follow his thoughts in their making.

Nonetheless, never did Gianni consider this process as an end in itself. It was the search for truth that drove him in every single moment of his life, intimately convinced, as he was to his very foundation, that the search for truth was one with the struggle for a future society as a community of free people, as is indeed feasible with the technical means available nowadays, with the goal of transforming those social relations that bound people to suffer and their souls to sadden.

The split between man and society has characterised all historical forms of social life. So far, the society has founded its very existence on an immediate oppression or it has emerged as the blind result of opposing forces: it is certainly not the outcome of a spontaneous and aware action of free individuals.

Gianni's indestructible assumption was that the truth is pushed forward to the extent that men who own it adamantly side with it, apply and impose it, acting in accordance with it. "*The process of cognition includes real historical will and action just as much as it does learning from experience and intellectual comprehension.*"[8] Therefore, the process of cognition is always the collective struggle for the transformation of the existing state of things.

On the historical and political level, in particular, Gianni, together with his fellow comrades, had arrived at the conclusion of the necessity to take stock of the communist movement and of the problems that had put it in difficulty. Such an analysis had to account, in a unified perspective, for the historical expansion of capitalism beyond the USA and Western Europe, and for the political crisis of the workers' movement between the last years of the Second International and the crisis of the Third International. Furthermore, one had to take stock of such processes with no shortcuts and within the framework of Marx's theses and their later developments.

This is not the forum to summarise and discuss the results of such an analysis; what matters and is to be emphasised is that the coherent reconstruction of the history of the last century and a half from the point of view of the communists has

[8] Max Horkheimer (1895–1973), from *On the Problem of Truth* (1935).

been Gianni's principal theoretical activity throughout. He was not moved by mere intellectual curiosity: it was the indispensable prerequisite for the political struggle of the workers' movement.

All of Gianni's life bears witness to what Anatole France says: "*We have nothing that belongs to us alone but ourselves; we truly give only when we give our work, our minds, our genius. And this splendid offering of one's whole self to all men enriches the giver as much as the community.*"[9]

And for sure Gianni would endorse each single word of Walter Benjamin's *Angelus Novus*: "*The class struggle, which is always present to a historian influenced by Marx, is a fight for the crude and material things without which no refined and spiritual things could exist. Nevertheless, it is not in the form of the spoils which fall to the victor that the latter make their presence felt in the class struggle. They manifest themselves in this struggle as courage, humour, cunning, and fortitude. They have retroactive force and will constantly call in question every victory, past and present, of the rulers. As flowers turn toward the sun, by dint of a secret heliotropism the past strives to turn toward that sun which is rising in the sky of history. A historical materialist must be aware of this most inconspicuous of all transformations.*"[10]

Even if all this today seems impossible and crazy to many.

We are left with the rare and immense privilege of having known him.

Pompeo, Maurizio, Fulvio

(*Translation and adaptation by Alessandro Michelangeli*)

[9] Anatole France (1844–1924), from *Monsieur Bergeret in Paris* (1901).
[10] Walter Benjamin (1892–1940), from *On the Concept of History* (1940).

Our Dad

Cecilia Morchio and Iacopo Morchio

We have long hesitated to write something about our father; the reason is that he was fundamentally convinced that nobody is ever capable of judging, describing, and narrating about other people. He sincerely thought this, not because he was disinterested in others, but because he deeply respected other people's lives, and because he was extremely humble and modest.

In all his relationships, he was the one who listened, who always maintained that listening to the opinions of those who think differently from you is one of the most interesting things in the world.

However, we would like to convey an impression of our father that could somehow show a less known side of his personality in the pages that follow.

The first images that come to our mind when we think about him are the moments in which he came back from work. In these occasions, we felt almost as if he was the one welcoming us, not vice versa: he entered our home with a gorgeous smile, eager to hear how our day had been and to exchange ideas and opinions, often so complex that they were hard for us to comprehend, although we could always feel his unlimited trust in intelligence and in the ability to understand, ours and everyone's.

He spoke little of his work and of physics, and when he did it was exclusively in an entertaining way, making jokes about how the laws of physics determine how water boils, how ice cream gets made, how frittatas are turned.

He spoke little of it also because he shared with us many of his other passions. He translated from latin and greek without needing a dictionary, he had a deep knowledge of philosophy, he knew entire sections of the Divine Comedy and many poems by heart, he was a great lover of classical music and opera, to the point that

C. Morchio (✉)
Pisa, Italy
e-mail: Cecilia.Morchio@gmail.com

I. Morchio
Pisa, Italy
e-mail: Iacopo.Morchio@gmail.com

A. Cintio, A. Michelangeli (Eds.), *Trails in Modern Theoretical and Mathematical Physics*, https://doi.org/10.1007/978-3-031-44988-8_3

he frequently sang his favourite arias in everyday life, often to take the edge off small fights or nervous days.

Now that we are adults, with lives made of work, friends and loved ones, we struggle even more to understand where he found the energy to do everything he did, with inexhaustible enthusiasm. Maybe the best way to share your love is to give time to others, and he was always available and open to spend time with us, both to play and tell us stories when we were children, and to talk and discuss with us when we were adults.

Our father taught us that it's important to have passions, to know things in depth, to think independently, to live independently, to work seriously, to have many interests, to remain open to the world and to never stop learning. But most of all, with few words but rather by example, he taught us that it is crucial to have integrity, and to be coherent with our ideas and values.

His example has helped us find our way, with the deep understanding that however we chose to live our life was going to be fine, as long as we did it with passion and integrity.

Although he declared not to believe in the importance of emotions, he always made us feel the depth of the love that he had for us, and he made us feel appreciated and loved for who we are.

We miss him so much, but everything he was still lives in us.

Iacopo and Cecilia

Part II
Trails in Modern Theoretical and Mathematical Physics

On Some Deformed Canonical Commutation Relations: The Role of Distributions

Fabio Bagarello

Contents

1 Introduction

According to Gianni's view of research, the role of Mathematics is essential in many fields of science, and in Theoretical Physics in particular. The literature is full of claims which look reasonable, but are not based on theorems or rigorous proofs. And this is true in Condensed Matter, Statistical Physics, Quantum Field Theory, Elementary Particles and so on. The content of this chapter, I believe, could have been of some interest for Gianni, even if, as far as I know, he never worked on what I will discuss here. Therefore, with this in mind, let me start my scientific contribution to this volume.

Among the various tools which play a relevant role in quantum mechanics, ladder operators are quite interesting, and useful. We all know the bosonic annihilation and creation operators, mainly because they are found already very early, while studying the spectrum of a simple quantum harmonic oscillator. But these operators are then used in a different, many-body, context: they describe *bosonic modes*, needed, for instance, in the analysis of quantum fields describing interactions. Fermionic ladder operators are also well known: they are used to model easily the Pauli exclusion principle, but they appear also in some quantum fields describing matter.

F. Bagarello (✉)
Dipartimento di Ingegneria, Università di Palermo, Palermo, Italy
Sezione di Catania, Istituto Nazionale di Fisica Nucleare, Catania, Italy
e-mail: fabio.bagarello@unipa.it

A. Cintio, A. Michelangeli (Eds.), *Trails in Modern Theoretical and Mathematical Physics*, https://doi.org/10.1007/978-3-031-44988-8_4

These two classes of ladded operators are usually defined in terms of suitable commutation rules between a fixed (annihilation) operator and its adjoint, the creation operator. For instance, $[c, c^\dagger] = \mathbb{1}_b$ and $\{d, d^\dagger\} = \mathbb{1}_f$, with $d^2 = 0$, are respectively the canonical commutation and the canonical anti-commutation relations. Here $\mathbb{1}_b$ and $\mathbb{1}_f$ are the identity operators in $\mathcal{H}_b$ and $\mathcal{H}_f$ respectively, the *bosonic* and the *fermionic* Hilbert spaces. It is well known that, for a single mode. $\dim(\mathcal{H}_f) = 2$ and $\dim(\mathcal{H}_b) = \infty$. So the two spaces are truly different. Moreover, $c^\dagger$ is the adjoint of c, and $d^\dagger$ that of d. Of course, these adjoints should be computed with respect to the scalar product in the related Hilbert space. This is an easy task for fermions, since the scalar product is just the one in $\mathbb{C}^2$, while it is not entirely trivial in $\mathcal{H}_b$ which is quite often identified with $\mathcal{L}^2(\mathbb{R})$, due to the fact that c is an unbounded operator. This creates a lot of subtle points to consider, of course, since unbounded operators are not as easy as bounded operators: they have, in particular, domain issues that should be considered to avoid mistakes. Hence, writing $[c, c^\dagger] = \mathbb{1}_b$ is just a formal relation which needs to be made more precise, for instance by making explicit the vectors of $\mathcal{H}_b$ on which this formula is well defined. And we can easily imagine that this problem becomes even more complicated when $[c, c^\dagger] = \mathbb{1}_b$ is replaced by $[a, b] = \mathbb{1}_b$, for some pair of operators a and b with $b \neq a^\dagger$. The analysis of this latter situation is indeed the core of this chapter. This is both because the mathematical properties of these operators can be rather interesting, but also because they appear, in a somehow hidden way, in several applications considered in recent years in the physical literature, mainly in connection with manifestly non self-adjoint Hamiltonians. In particular, as we will show later, removing the constraint that the commutation rule is given between an operator (c) and its adjoint ($c^\dagger$), gives us the possibility to extend the functional framework from $\mathcal{L}^2(\mathbb{R})$ to the space of tempered distributions $S'(\mathbb{R})$. Of course, this extension opens the problem of a correct interpretation of the results from a physical side. This is because the usual probabilistic interpretation of quantum mechanics can be lost. Still, some interesting physically relevant operators appear strongly connected to what we will discuss later and, in this perspective, we believe our framework may have some intriguing consequences.

This chapter is organized as follows: we propose our special deformation of the canonical commutation relations and we discuss some of the mathematical consequences of our definition. Section 2.1 is devoted to a brief list of quantum mechanical systems, considered in the literature in recent years, which can be analyzed in terms of our ladder operators, named *pseudo-bosonic*. In particular we show that a and b, together with $a^\dagger$ and $b^\dagger$, behave as ladder operators and allow the construction of two different families of vectors in $\mathcal{L}^2(\mathbb{R})$ which are biorthonormal and are eigenstates of the pseudo-bosonic number operator $N = ba$, and of its adjoint $N^\dagger$. These two sets are not necessarily bases in $\mathcal{L}^2(\mathbb{R})$, but they are usually total sets. In Sect. 3 we take advantage of the fact that in our pseudo-bosonic commutation rule $[a, b] = \mathbb{1}_b$ a and b can be, in general, quite unrelated to propose a generalized version of the Hilbertian settings proposed in Sect. 2. Hence we construct a general settings for what we call *weak pseudo-bosons* (WPBs). Several appearance of these WPBs are described in the second part of Sect. 3. In particular, in Sect. 3.1 we show how the position and the momentum operators $\hat{x}$ and $\hat{p}$ can be seen as weak pseudo-bosonic ladder operators, and we show that an extended scalar product can be introduced to prove the biorhogonality of the generalized eigenstates of the pseudo-bosonic number operators. In Sect. 3.2 we discuss the role of WPBs in the context of the so-called inverted quantum harmonic oscillator (IQHO), while in Sect. 3.3 we propose a rather general family of pseudo-bosons (PBs) which can be defined in or out of $\mathcal{L}^2(\mathbb{R})$. Section 4 contains our conclusions.

2 Pseudo-Bosons

We begin our analysis by recalling few well known facts on bosonic operators. This is important to fix the notation and later to stress the differences between PBs and ordinary bosons.

Let c be an operator on an Hilbert space[1] $\mathcal{H} = \mathcal{L}^2(\mathbb{R})$ satisfying the canonical commutation relation (CCR) $[c, c^\dagger] = \mathbb{1}$, $c^\dagger$ being the adjoint of c and $\mathbb{1}$ the identity operator on $\mathcal{H}$. Notice that, using for c the representation $c = \frac{1}{\sqrt{2}}(\hat{x} + \frac{d}{dx})$, where $\hat{x}$ is the multiplication operator and $\frac{d}{dx}$ is the derivative operator[2], the set of the test functions on $\mathbb{R}$, $S(\mathbb{R})$, i.e. of all those C^∞ functions which go to zero, together with their derivatives, faster than any inverse power, is stable under the action of c and $c^\dagger$: if $f(x) \in S(\mathbb{R})$, then $cf(x), c^\dagger f(x) \in S(\mathbb{R})$. Now we replace $[c, c^\dagger] = \mathbb{1}$ with its *more complete* version, writing

$$[c, c^\dagger] f(x) = f(x), \tag{1}$$

for all $f \in S(\mathbb{R})$. If we consider a vector $e_0(x) \in S(\mathbb{R})$ which is annihilated by c, $c\, e_0(x) = 0$, it is clear that all the vectors $e_n(x) = \frac{1}{\sqrt{n!}} {c^\dagger}^n e_0(x)$, $n \geq 0$, belong to $S(\mathbb{R})$. The set $\mathcal{F}_e = \{e_n(x),\, n \geq 0\}$ is an orthonormal basis for $\mathcal{H}$:

$$\langle e_n, e_m \rangle = \delta_{n,m}, \quad \text{and} \quad f(x) = \sum_{n=0}^{\infty} \langle e_n, f \rangle\, e_n(x),$$

$\forall f(x) \in \mathcal{L}^2(\mathbb{R})$, so that $\mathcal{L}_e = l.s.\{e_n(x)\}$, the linear span of the $e_n(x)$'s, is dense in $\mathcal{L}^2(\mathbb{R})$. The following Parseval equality holds:

$$\langle f, g \rangle = \sum_{n=0}^{\infty} \langle f, e_n \rangle \langle e_n, g \rangle, \tag{2}$$

$\forall f(x), g(x) \in \mathcal{H}$. The explicit form of $e_n(x)$ is well known:

$$e_n(x) = \frac{1}{\sqrt{2^n n! \sqrt{\pi}}} H_n(x)\, e^{-\frac{x^2}{2}}, \tag{3}$$

[1] From now on we will simply use $\mathcal{H}$ rather than $\mathcal{H}_b$, since $\mathcal{H}_f$ will have no role in the rest of this chapter.

[2] We recall that this is proportional to the momentum operator $\hat{p}$, $\hat{p} = -i\frac{d}{dx}$.

where $H_n(x)$ is the n-th Hermite polynomial. It is evident now that $e_n(x) \in S(\mathbb{R})$, for all $n \geq 0$, so that the (strict) inclusion $\mathcal{L}_e \subset S(\mathbb{R})$ holds.

The set $\mathcal{F}_e$ has interesting features, when considered in connection with c and $c^\dagger$. Indeed we have the following

$$c\, e_n = \sqrt{n}\, e_{n-1}, \quad c^\dagger e_n = \sqrt{n+1}\, e_{n+1}, \tag{4}$$

with the agreement that $e_{-1} = 0$. An immediate consequence of these ladder equations is the following eigenvalue equation

$$N_0\, e_n = n\, e_n, \tag{5}$$

$n \geq 0$, where $N_0 = c^\dagger c$ is called the *number* operator. Because of (4), c is a *lowering* or an *annihilation* operator, while $c^\dagger$ is a *raising* or a *creation* operator. Together they are called *ladder operators*. N_0, c and $c^\dagger$ are all unbounded. In particular, N_0 is symmetric, since $\langle N_0 f, g\rangle = \langle f, N_0 g\rangle$ $\forall f, g \in D(N_0)$, and is positive: $\langle N_0 f, f\rangle = \langle cf, cf\rangle = \|cf\|^2 \geq 0$, $\forall f \in D(N_0)$. Hence, N_0 admits a Friedrichs extension, which we still denote with N_0, which is self-adjoint.

Summarizing, if c satisfies the CCR (1), then we can build up an interesting functional settings: a family of vectors, the $e_n(x)$'s, which are eigenvectors of the self-adjoint operator N_0 with eigenvalues $n \in \mathbb{N}_0 = \mathbb{N} \cup \{0\}$, see (5), which obey some relevant ladder conditions, see (4), and which, all together, produce a set of functions $\mathcal{F}_e$ which is an orthonormal basis for $\mathcal{H}$.

During the past few decades, many physicists realized that some non self-adjoint operators can play a significant role in the analysis of various physical systems, [1–5], since there exist quantum mechanical situations in which the dynamics is better described by Hamiltonians (and other *observables*) which are not self-adjoint. This evidence has produced a huge interest in an extended version of quantum mechanics, where self-adjointness of the observables is not a key aspect. This suggested to consider ladder operators of different kind, not necessarily linked by the usual adjoint operation, and their connected number-like operators. We refer to [6] for some preliminary results and to [7] for a more recent monograph on these topics. This chapter is intended to be a review of some recent results on these generalized ladder operators, and to their weak[3] version in particular. For readers' convenience, we begin our analysis by proposing first our definitions and their consequences in a purely Hilbertian settings, postponing their distributional counterparts to Sect. 3.

Let a and b be two operators on $\mathcal{H}$, with domains $\mathcal{D}(a)$ and $\mathcal{D}(b)$ respectively, $a^\dagger$ and $b^\dagger$ their adjoint, and let $\mathcal{D}$ be a dense subspace of $\mathcal{H}$ such that $a^\sharp\mathcal{D} \subseteq \mathcal{D}$ and $b^\sharp\mathcal{D} \subseteq \mathcal{D}$, where $x^\sharp$ is either x or $x^\dagger$: $\mathcal{D}$ is assumed to be stable under the action of a, b, $a^\dagger$ and $b^\dagger$. Notice that we are not requiring here that $\mathcal{D}$ coincides with, e.g. $\mathcal{D}(a)$ or $\mathcal{D}(b)$. However due to the fact that $a^\sharp f$ is well defined, and belongs to $\mathcal{D}$ for all $f \in \mathcal{D}$, it is clear that $\mathcal{D} \subseteq \mathcal{D}(a^\sharp)$. Analogously, we can also conclude that $\mathcal{D} \subseteq \mathcal{D}(b^\sharp)$. The stability of $\mathcal{D}$ implies that both $a(bf)$ and $b(af)$ are well defined, $\forall f \in \mathcal{D}$.

[3] In the sense of distributions!

Definition 1 The operators (a, b) are $\mathcal{D}$-pseudo-bosonic ($\mathcal{D}$-pb) if, for all $f \in \mathcal{D}$, we have

$$a\, b\, f - b\, a\, f = f. \tag{6}$$

Sometimes, to simplify the notation, rather than (6) we will simply write $[a, b] = \mathbb{1}$. Of course, when $b = a^\dagger$ we go back to CCR, and $a, b \notin \mathcal{B}(\mathcal{H})$, the set of bounded operators on $\mathcal{H}$. a and b are unbounded also when $a \neq b^\dagger$, and this is the reason why the role of $\mathcal{D}$ is so relevant.

Our working assumptions, based on several existing systems in quantum mechanics, are the following:

Assumption $\mathcal{D}$-pb 1 there exists a non-zero $\varphi_0 \in \mathcal{D}$ such that $a\, \varphi_0 = 0$.

Assumption $\mathcal{D}$-pb 2 there exists a non-zero $\Psi_0 \in \mathcal{D}$ such that $b^\dagger\, \Psi_0 = 0$.

It is clear that, if $b = a^\dagger$, these two assumptions collapse into a single one and (6) becomes the ordinary CCR, for which the existence of a vacuum which belongs to an invariant set ($\mathcal{S}(\mathbb{R})$, for instance) is guaranteed. On the other hand, if a and b are uncorrelated, it might easily happen that Assumptions $\mathcal{D}$-pb 1 or $\mathcal{D}$-pb 2, or one of the two, are not satisfied. One important example of this situation will be discussed at length in Sect. 3.1.

The stability of $\mathcal{D}$ under the action of b and $a^\dagger$ implies, in particular, that $\varphi_0 \in D^\infty(b) := \cap_{k\geq 0} D(b^k)$ and that $\Psi_0 \in D^\infty(a^\dagger)$. Here $D^\infty(X)$ is the domain of all the powers of the operator X. Hence

$$\varphi_n := \frac{1}{\sqrt{n!}}\, b^n \varphi_0, \quad \Psi_n := \frac{1}{\sqrt{n!}}\, {a^\dagger}^n\, \Psi_0, \tag{7}$$

$n \geq 0$, are well defined vectors in $\mathcal{D}$ and, therefore, they belong to the domains of $a^\sharp$, $b^\sharp$ and $N^\sharp$, where $N = ba$ and $N^\dagger$ is the adjoint of N. We introduce the sets $\mathcal{F}_\Psi = \{\Psi_n,\, n \geq 0\}$ and $\mathcal{F}_\varphi = \{\varphi_n,\, n \geq 0\}$.

It is now simple to deduce the following lowering and raising relations:

$$\begin{cases} b\, \varphi_n = \sqrt{n+1}\, \varphi_{n+1}, & n \geq 0, \\ a\, \varphi_0 = 0, \quad a\varphi_n = \sqrt{n}\, \varphi_{n-1}, & n \geq 1, \\ a^\dagger \Psi_n = \sqrt{n+1}\, \Psi_{n+1}, & n \geq 0, \\ b^\dagger \Psi_0 = 0, \quad b^\dagger \Psi_n = \sqrt{n}\, \Psi_{n-1}, & n \geq 1, \end{cases} \tag{8}$$

as well as the following eigenvalue equations: $N\varphi_n = n\varphi_n$ and $N^\dagger \Psi_n = n\Psi_n$, $n \geq 0$. Hence, despite of the fact that N and $N^\dagger$ are manifestly non self-adjoint, in general, their eigenvalues are real and, actually, coincide with those of the operator $N_0 = c^\dagger c$. We call $\mathcal{D}$-pseudo-bosons ($\mathcal{D}$-PBs) the *excitations* described by φ_n and Ψ_n, in the same way we call bosons those described by the vectors e_n in (3).

As a consequence of these equations, choosing the normalization of φ_0 and Ψ_0 in such a way $\langle\varphi_0, \Psi_0\rangle = 1$, it can be shown that

$$\langle\varphi_n, \Psi_m\rangle = \delta_{n,m}, \tag{9}$$

for all $n, m \geq 0$. The conclusion is, therefore, that $\mathcal{F}_\varphi$ and $\mathcal{F}_\Psi$ are biorthonormal sets of eigenstates of N and $N^\dagger$, respectively. The properties we have deduced for $\mathcal{F}_\varphi$ and $\mathcal{F}_\Psi$ does not allow us to conclude anything about the fact that they are also (Riesz) bases for $\mathcal{H}$. In fact, it is well known that, in some relevant concrete examples, this is not the case, while in other situations this is true. We will return on this aspect in Sect. 2.1, where several counterexamples will be given. With this in mind, we introduce the following (not always satisfied, in view of what just observed) assumption:

Assumption $\mathcal{D}$-pb 3 $\mathcal{F}_\varphi$ is a basis for $\mathcal{H}$.

This is equivalent to assume that $\mathcal{F}_\Psi$ is a basis as well, [8]. Since this assumption is not always true, it is more reasonable to replace Assumption $\mathcal{D}$-pb 3 with a weaker version, which thought being weaker, still produces several interesting results and, maybe more relevant, is satisfied even when Assumption $\mathcal{D}$-pb 3 does not hold. We ask the following:

Assumption $\mathcal{D}$-pbw 3 $\mathcal{F}_\varphi$ and $\mathcal{F}_\Psi$ are $\mathcal{G}$-quasi bases, for some subspace $\mathcal{G}$ dense[4] in $\mathcal{H}$.

This means that, $\forall f, g \in \mathcal{G}$, the following identities hold

$$\langle f, g\rangle = \sum_{n\geq 0}\langle f, \varphi_n\rangle\langle\Psi_n, g\rangle = \sum_{n\geq 0}\langle f, \Psi_n\rangle\langle\varphi_n, g\rangle, \tag{10}$$

which, as it is clear, extend the standard closure relation in $\mathcal{H}$, also known as *Parseval identity*.

While Assumption $\mathcal{D}$-pb 3 implies (10), the reverse is false. However, if $\mathcal{F}_\varphi$ and $\mathcal{F}_\Psi$ satisfy (10), we still have some (weak) form of resolution of the identity, and we can deduce several useful consequences. For instance, just to state a simple result, if $f \in \mathcal{G}$ is orthogonal to all the Ψ_n's (or to all the φ_n's), then f is necessarily zero: $\mathcal{F}_\Psi$ and $\mathcal{F}_\varphi$ are total in $\mathcal{G}$.

For completeness we briefly discuss the role of two intertwining operators which are intrinsically related to our $\mathcal{D}$-PBs. More details can be found in [6].

Assumption $\mathcal{D}$-pb 3 means that

$$f = \sum_{n=0}^{\infty}\langle\varphi_n, f\rangle\,\Psi_n = \sum_{n=0}^{\infty}\langle\Psi_n, f\rangle\,\varphi_n, \tag{11}$$

[4] $\mathcal{G}$ does not need to coincide with $\mathcal{D}$.

$\forall f \in \mathcal{H}$. Then, it is natural to ask if sums like $S_\varphi f = \sum_{n=0}^{\infty} \langle \varphi_n, f \rangle \varphi_n$ or $S_\Psi f = \sum_{n=0}^{\infty} \langle \Psi_n, f \rangle \Psi_n$ make some sense, or for which vectors they do converge, if any. It is clear that, if $b = a^\dagger$, then $\mathcal{F}_\varphi = \mathcal{F}_\Psi$ and $S_\varphi = S_\Psi = \mathbb{1}$: in this case not only the series for S_φ and S_Ψ converge, but they converge to the identity operator.

If, in particular, $\mathcal{F}_\varphi$ is a Riesz basis, [8], then $\mathcal{F}_\Psi$ is a Riesz basis too, and we know that an orthonormal basis $\mathcal{F}_c = \{c_n\}$ exists, together with a bounded operator R with bounded inverse, such that $\varphi_n = Rc_n$ and $\Psi_n = (R^{-1})^\dagger c_n$, $\forall n$. It is clear that, if $R = \mathbb{1}$, the sums for $S_\varphi f$ and $S_\Psi f$ collapse and converge to f. But, what if $R \neq \mathbb{1}$? In this case, let us take $f \in D(S_\varphi)$, which for the moment we do not assume to be coincident with $\mathcal{H}$. Then

$$S_\varphi f := \sum_n \langle \varphi_n, f \rangle \varphi_n = \sum_n \langle Rc_n, f \rangle Rc_n = R\left(\sum_n \langle c_n, R^\dagger f \rangle c_n\right) = RR^\dagger f,$$

where we have used the facts that $\mathcal{F}_c$ is an orthonormal basis and that R is bounded and, therefore, continuous. Of course $RR^\dagger$ is bounded as well and the above equality can be extended to all of $\mathcal{H}$. Therefore we conclude that $S_\varphi = RR^\dagger$. In a similar way we can deduce that $S_\Psi = (R^\dagger)^{-1} R^{-1} = S_\varphi^{-1}$, which is also bounded. In fact, using the C*-property for $B(\mathcal{H})$, we deduce that $\|S_\varphi\| = \|R\|^2$ and $\|S_\Psi\| = \|R^{-1}\|^2$. In this situation, our $\mathcal{D}$-PBs are called *regular*.

Similar results can also be deduced without introducing the operator R, but simply using the biorthonormality of $\mathcal{F}_\varphi$ and $\mathcal{F}_\Psi$:

$$S_\varphi \Psi_n = \varphi_n, \quad S_\Psi \varphi_n = \Psi_n, \tag{12}$$

for all $n \geq 0$. These equalities together imply that $\Psi_n = (S_\Psi S_\varphi)\Psi_n$ and $\varphi_n = (S_\varphi S_\Psi)\varphi_n$, for all $n \geq 0$. Now, since $S_\varphi, S_\Psi \in B(\mathcal{H})$, we can extend these identities to all of $\mathcal{H}$, and we conclude that

$$S_\Psi S_\varphi = S_\varphi S_\Psi = \mathbb{1} \quad \Rightarrow \quad S_\Psi = S_\varphi^{-1}. \tag{13}$$

In other words, both S_Ψ and S_φ are invertible and one is the inverse of the other. It is also clear that S_φ and S_Ψ are positive operators, and it is interesting to check that they obey the following intertwining relations:

$$S_\Psi N \varphi_n = N^\dagger S_\Psi \varphi_n, \quad N S_\varphi \Psi_n = S_\varphi N^\dagger \Psi_n, \tag{14}$$

Indeed we have, recalling that $N\varphi_n = n\varphi_n$ and $N^\dagger \Psi_n = n\Psi_n$, $S_\Psi N \varphi_n = n(S_\Psi \varphi_n) = n\Psi_n$, as well as $N^\dagger S_\Psi \varphi_n = N^\dagger \Psi_n = n\Psi_n$. The second equality in (14) follows from the first one, simply by left-multiplying $S_\Psi N \varphi_n = N^\dagger S_\Psi \varphi_n$ with S_φ, and using (12). These relations are not surprising, since intertwining relations can be often deduced between operators sharing the same eigenvalues.

The situation is mathematically much more complicated, in particular, for $\mathcal{D}$-PBs which are not regular. This is connected to the fact that S_φ and S_Ψ are not bounded, so that the series $\sum_{n=0}^{\infty} \langle \varphi_n, f \rangle \varphi_n$ and $\sum_{n=0}^{\infty} \langle \Psi_n, f \rangle \Psi_n$ do not converge uniformly on $\mathcal{H}$. This case, together with many other details on PBs, can be found in [6] and in references therein.

2.1 Few Appearances of PBs

During the last few decades a lot of physical systems have been considered, mostly in connection with PT-Quantum Mechanics, [1–5], driven by manifestly non self-adjoint Hamiltonians which can be rewritten in terms of PBs. We briefly list here some of these Hamiltonians, and we refer to [6, 7] for many more mathematical details and physical applications. It is useful to remark that, in what follows, we will be extremely concise, since *ordinary* PBs are not the main object of our review here, but are only needed to provide a better setup for WPBs.

The Extended Quantum Harmonic Oscillator

We begin our list of models with the following Hamiltonian, proposed in [9]

$$H_\nu = \frac{\nu}{2}(\hat{p}^2 + \hat{x}^2) + i\sqrt{2}\,\hat{p},$$

where ν is a strictly positive parameter and $[\hat{x}, \hat{p}] = i\mathbb{1}$. H_ν is manifestly non self-adjoint. However, with some algebra, it can be easily diagonalized in terms of PBs.

For that, we start introducing the (standard) bosonic operators $c = \frac{1}{\sqrt{2}}(\hat{x} + \frac{d}{dx})$, $c^\dagger = \frac{1}{\sqrt{2}}(\hat{x} - \frac{d}{dx})$, $[c, c^\dagger] = \mathbb{1}$, and the related operators $A_\nu = c - \frac{1}{\nu}$, and $B_\nu = c^\dagger + \frac{1}{\nu}$. Then we can rewrite $H_\nu = \nu(B_\nu A_\nu + \gamma_\nu \mathbb{1})$, where $\gamma_\nu = \frac{2+\nu^2}{2\nu^2}$. It is clear that, for all $\nu > 0$, $A_\nu^\dagger \neq B_\nu$ and that $[A_\nu, B_\nu] = \mathbb{1}$. Hence we are dealing, at least formally, with pseudo-bosonic operators. Indeed, we can check that Assumptions $\mathcal{D}$-pb1, $\mathcal{D}$-pb2 and $\mathcal{D}$-pbw3, are satisfied, while Assumption $\mathcal{D}$-pb3 is not, see [6, 10].

The Swanson Model

The starting point is here the non self-adjoint Hamiltonian,

$$H_\theta = \frac{1}{2}(\hat{p}^2 + \hat{x}^2) - \frac{i}{2}\tan(2\theta)(\hat{p}^2 - \hat{x}^2),$$

where θ is a real parameter taking value in $\left(-\frac{\pi}{4}, \frac{\pi}{4}\right) \setminus \{0\} =: I$, [9, 11]. As before, $[\hat{x}, \hat{p}] = i\mathbb{1}$. Of course, $\theta = 0$ is excluded from I just to avoid going back to the standard, self-adjoint, harmonic oscillator, which is not so interesting for us. Notice also that H_θ can be rewritten as

$$H_\theta = \frac{1}{2\cos(2\theta)}\left(\hat{p}^2 e^{-2i\theta} + \hat{x}^2 e^{2i\theta}\right) = \frac{e^{-2i\theta}}{2\cos(2\theta)}\left(\hat{p}^2 + \hat{x}^2 e^{4i\theta}\right),$$

which has, except for an unessential overall complex constant, the same form considered in [12], $H = -\frac{d^2}{dx^2} + z^4\hat{x}^2$, $z \in \mathbb{C}$, taking $z = e^{i\theta}$.

Introducing now the bosonic annihilation and creation operators c, $c^\dagger$, and their linear combinations

$$\begin{cases} A_\theta = \cos(\theta)\, c + i \sin(\theta)\, c^\dagger = \frac{1}{\sqrt{2}}\big(e^{i\theta}\hat{x} + e^{-i\theta}\frac{d}{dx}\big), \\ B_\theta = \cos(\theta)\, c^\dagger + i \sin(\theta)\, c = \frac{1}{\sqrt{2}}\big(e^{i\theta}\hat{x} - e^{-i\theta}\frac{d}{dx}\big), \end{cases}$$

we can write $H_\theta = \omega_\theta\big(B_\theta A_\theta + \frac{1}{2}\mathbb{1}\big)$, where $\omega_\theta = \frac{1}{\cos(2\theta)}$ is well defined because $\cos(2\theta) \neq 0$ for all $\theta \in I$. It is clear that, for θ in this set, $A_\theta^\dagger \neq B_\theta$ and that $[A_\theta, B_\theta] = \mathbb{1}$. Again, we have rewritten the Hamiltonian in terms of PBs, and again, we can check that Assumptions $\mathcal{D}$-pb1, $\mathcal{D}$-pb2 and $\mathcal{D}$-pbw3, are satisfied, while Assumption $\mathcal{D}$-pb3 is not, see [6, 10].

Two Coupled Oscillators

The next example we want to briefly mention was originally introduced by Carl Bender and Hugh Jones in [13] and then considered further in [14]. The starting point is the following, manifestly non self-adjoint, Hamiltonian:

$$H = (\hat{p}_1^2 + \hat{x}_1^2) + (\hat{p}_2^2 + \hat{x}_2^2 + 2i\,\hat{x}_2) + 2\epsilon\hat{x}_1\hat{x}_2, \tag{15}$$

where ϵ is a real constant, with $\epsilon \in]-1, 1[$. Here the following commutation rules are assumed: $[\hat{x}_j, \hat{p}_k] = i\delta_{j,k}\mathbb{1}$, $\mathbb{1}$ being the identity operator on $\mathcal{L}^2(\mathbb{R}^2)$. All the other commutators are zero.

In order to rewrite H in a more convenient form it is possible to perform some changes of variables, [14], starting by introducing the operators P_j, X_j, $j = 1, 2$, via

$$P_1 := \frac{1}{2a}(\hat{p}_1 + \xi\,\hat{p}_2), \quad P_2 := \frac{1}{2b}(\hat{p}_1 - \xi\,\hat{p}_2),$$
$$X_1 := a(\hat{x}_1 + \xi\hat{x}_2), \quad X_2 := b(\hat{x}_1 - \xi\hat{x}_2),$$

where ξ can be ± 1, while a and b are real, non zero, arbitrary constants. These operators satisfy the same canonical commutation rules as the original ones: $[X_j, P_k] = i\delta_{j,k}\mathbb{1}$. Next we put

$$\Pi_1 = P_1, \quad \Pi_2 = P_2, \quad q_1 = X_1 + i\frac{a\xi}{1+\epsilon\,\xi}, \quad q_2 = X_2 - i\frac{b\xi}{1-\epsilon\,\xi},$$

and it is clear that $q_j^\dagger \neq q_j$, $j = 1, 2$. However, the commutation rules are preserved: $[q_j, \Pi_k] = i\delta_{j,k}\mathbb{1}$. Finally, we introduce the operators:

$$\begin{cases} a_1 = \frac{a}{\sqrt[4]{1+\epsilon\,\xi}}\left(i\Pi_1 + \frac{\sqrt{1+\epsilon\,\xi}}{2a^2}\, q_1\right), \\ a_2 = \frac{a}{\sqrt[4]{1-\epsilon\,\xi}}\left(i\Pi_2 + \frac{\sqrt{1-\epsilon\,\xi}}{2b^2}\, q_2\right), \end{cases} \tag{16}$$

and

$$\begin{cases} b_1 = \frac{a}{\sqrt[4]{1+\epsilon\,\xi}}\left(-i\Pi_1 + \frac{\sqrt{1+\epsilon\,\xi}}{2a^2}\,q_1\right), \\ b_2 = \frac{a}{\sqrt[4]{1-\epsilon\,\xi}}\left(-i\Pi_2 + \frac{\sqrt{1-\epsilon\,\xi}}{2b^2}\,q_2\right). \end{cases} \tag{17}$$

It may be worth remarking that $b_j \neq a_j^\dagger$, since the q_j's are not self-adjoint. These operators satisfy, at least formally, the pseudo-bosonic commutation rules

$$[a_j, b_k] = \delta_{j,k}\mathbb{1}, \tag{18}$$

the other commutators being zero.

Going back to H, and introducing the operators $N_j := b_j a_j$, we can write

$$H = H_1 + H_2 + \frac{1}{1-\epsilon^2}\,\mathbb{1}, \quad H_1 = \sqrt{1+\epsilon\,\xi}(2N_1 + \mathbb{1}), \quad H_2 = \sqrt{1-\epsilon\,\xi}(2N_2 + \mathbb{1}). \tag{19}$$

In [6, 15] it has been proved that these operators provide a two-dimensional version of the general framework described in Sect. 2: we are dealing with PBs, but in 2D.

Another 2D Example

The last quantum mechanical model of this short (and very minimal!) list was originally introduced, in our knowledge, in [16]. The starting point is the following manifestly non self-adjoint Hamiltonian,

$$H = \frac{1}{2}(\hat{p}_1^2 + \hat{x}_1^2) + \frac{1}{2}(\hat{p}_2^2 + \hat{x}_2^2) + i[A(\hat{x}_1 + \hat{x}_2) + B(\hat{p}_1 + \hat{p}_2)], \tag{20}$$

where A and B are real constants, while $\hat{x}_j$ and $\hat{p}_j$ are the self-adjoint position and momentum operators, satisfying $[\hat{x}_j, \hat{p}_k] = i\delta_{j,k}\mathbb{1}$. Notice that in [6] a noncommutative version of this system has also been considered.

Let us introduce the shifted operators

$$P_1 = \hat{p}_1 + iB, \quad P_2 = \hat{p}_2 + iB, \quad X_1 = \hat{x}_1 + iA, \quad X_2 = \hat{x}_2 + iA,$$

and then

$$a_j = \frac{1}{\sqrt{2}}(X_j + iP_j), \quad b_j = \frac{1}{\sqrt{2}}(X_j - iP_j), \tag{21}$$

$j = 1, 2$. It is easy to check that $[X_j, P_k] = i\delta_{j,k}\mathbb{1}$, $[a_j, b_k] = \delta_{j,k}\mathbb{1}$, and that, since (if $A \neq 0$ or $B \neq 0$) $X_j^\dagger \neq X_j$ and $P_j^\dagger \neq P_j$, $b_j \neq a_j^\dagger$. Introducing further $N_j = b_j a_j$ we can rewrite H as follows: $H = N_1 + N_2 + (A^2 + B^2 + 1)\mathbb{1}$. Also in this case, we can rewrite H in diagonal form in terms of pseudo-bosonic number operators. We refer to [6] to see the details of our computations, and for the mathematical subtleties connected with the Hamiltonian in (20).

We conclude here this list of concrete appearances of PBs in some quantum mechanical models already existing in the literature, before the analysis given in Sect. 2 was undertaken. It is useful to add that PBs have shown to be useful in the analysis of many other models, and in connection with other interesting situations. We refer to [7], in particular, for some applications of PBs to path integrals.

3 Weak PBs

From now on we will concentrate on a specific class of PBs, the so-called weak PBs, WPBs. These are ladder operators acting on distributions, rather than on square-integrable functions. They have, as we will see, similar properties as those of ordinary PBs, but are maybe more intriguing for their mathematical properties.

We start introducing here, as before, two operators a and b which, together with their adjoints $a^\dagger$ and $b^\dagger$, map a certain dense subset of $\mathcal{H}$, $\mathcal{D}$, into itself. Then we assume that a and b can be extended to a larger set, $\mathcal{E} \supset \mathcal{H}$, which is again stable under their action, and under the action of their adjoints. The existence of such a set $\mathcal{E}$ is, of course, very much model-dependent. Some explicit example will be discussed later in Sect. 3. With this in mind, we propose the following

Definition 2 The operators a and b are *weak* $\mathcal{E}$-pseudo bosonic if

$$[a,b]\, F = F, \tag{22}$$

for all $F \in \mathcal{E}$. When the role of $\mathcal{E}$ is clear we will simply call a and b *weak* pseudo bosonic operators.

As in Sect. 2, the commutator in (22) is just the starting point to construct an interesting mathematical framework. This is exactly what we will do here. In particular, the following two assumptions reflect Assumptions $\mathcal{D}$-pb 1 and $\mathcal{D}$-pb 2:

Assumption $\mathcal{E}$-wpb 1 there exists a non-zero $\varphi_0 \in \mathcal{E}$ such that $a\,\varphi_0 = 0$.

Assumption $\mathcal{E}$-wpb 2 there exists a non-zero $\Psi_0 \in \mathcal{E}$ such that $b^\dagger\,\Psi_0 = 0$.

As before, the invariance of $\mathcal{E}$ under the action of the operators a, b, $a^\dagger$ and $b^\dagger$ implies that $\varphi_0 \in D^\infty(b) := \cap_{k\geq 0} D(b^k)$ and $\Psi_0 \in D^\infty(a^\dagger)$, in the sense of generalized domains, so that the vectors

$$\varphi_n := \frac{1}{\sqrt{n!}}\, b^n \varphi_0, \quad \Psi_n := \frac{1}{\sqrt{n!}}\, a^{\dagger^n} \Psi_0, \tag{23}$$

$n \geq 0$, can be defined and they all belong to $\mathcal{E}$. Defining now the sets $\mathcal{F}_\psi = \{\psi_n,\, n \geq 0\}$ and $\mathcal{F}_\varphi = \{\varphi_n,\, n \geq 0\}$, from (22) and from the definition in (23) we easily deduce the same raising and lowering relations as in (8), together with the eigenvalue equations $N\varphi_n = n\varphi_n$ and $N^\dagger\Psi_n = n\Psi_n$, $n \geq 0$. In the attempt to generalize what we have proved for PBs, it is now natural to assume that, with a suitable choice of the normalization of φ_0 and Ψ_0 which implies that $\langle\varphi_0, \Psi_0\rangle = 1$, then

$$\langle\varphi_n, \Psi_m\rangle = \delta_{n,m}, \tag{24}$$

for all $n, m \geq 0$. This means that $\mathcal{F}_\psi$ and $\mathcal{F}_\varphi$ are requested to be biorthonormal, with respect to a bilinear form $\langle., .\rangle$ which extends the ordinary scalar product to $\mathcal{E}$, and which needs to be identified in concrete situations.

Of course, since the vectors of $\mathcal{F}_\psi$ and $\mathcal{F}_\varphi$ are not, in general, in $\mathcal{H}$, it makes not much sense to require any strong version of the basis property for $\mathcal{F}_\psi$ or $\mathcal{F}_\varphi$. On the other hand, what seems natural to require is that a set $C \subseteq \mathcal{E}$ exists, *consisting of "sufficiently many" functions*, such that

$$\langle F, G\rangle = \sum_{n=0}^{\infty}\langle F, \psi_n\rangle\langle\varphi_n, G\rangle = \sum_{n=0}^{\infty}\langle F, \varphi_n\rangle\langle\psi_n, G\rangle, \tag{25}$$

for all $F, G \in C$. A pragmatic view on C is that it should contains all those (generalized) functions which are *interesting* for us, for some specific physical or mathematical reason.

As in Sect. 2, we can use $\mathcal{F}_\varphi$ and $\mathcal{F}_\psi$ to introduce two operators, S_φ and S_ψ, as follows: let

$$D(S_\varphi) = \{F \in \mathcal{E}\colon S_\varphi F \in \mathcal{E}\}, \quad D(S_\psi) = \{F \in \mathcal{E}\colon S_\psi F \in \mathcal{E}\}.$$

These are to be understood as generalized domains of S_φ and S_ψ, respectively. Most of the properties found for ordinary PBs are recovered. Calling $\mathcal{L}_\varphi$ and $\mathcal{L}_\psi$ respectively the linear spans of the vectors φ_n and ψ_n, we see that $\mathcal{L}_\varphi \subseteq D(S_\psi)$, $\mathcal{L}_\psi \subseteq D(S_\varphi)$, $S_\varphi\colon \mathcal{L}_\psi \to \mathcal{L}_\varphi$, and $S_\psi\colon \mathcal{L}_\varphi \to \mathcal{L}_\psi$. In particular we have

$$S_\varphi\left(\sum_{k=0}^{N} c_k\psi_k\right) = \sum_{k=0}^{N} c_k\varphi_k, \quad S_\psi\left(\sum_{k=0}^{N} c_k\varphi_k\right) = \sum_{k=0}^{N} c_k\psi_k, \tag{26}$$

as well as

$$S_\varphi S_\psi F = F, \quad S_\psi S_\varphi G = G, \tag{27}$$

and

$$N S_\varphi G = S_\varphi N^\dagger G, \quad N^\dagger S_\psi F = S_\psi N F. \tag{28}$$

Moreover

$$a\,F = S_\varphi b^\dagger S_\psi F, \quad b\,F = S_\varphi a^\dagger S_\psi F, \quad a^\dagger G = S_\psi b S_\varphi G, \quad b^\dagger G = S_\psi a S_\varphi G, \tag{29}$$

for all $F \in \mathcal{L}_\varphi$ and $G \in \mathcal{L}_\psi$. Using the same notation proposed in [6], the operators a and $b^\dagger$ could be called S_ψ-conjugate. Conjugate operators are sometimes considered in a Hilbertian context, and produce several interesting results. For this reason, a deeper investigation of these similarities conditions in the distributional sense could be interesting, and this analysis is in progress.

We can then conclude that there is no particular obstacle, in principle, in extending the main ideas and results deduced for PBs to WPBs. Of course, the rather abstract construction proposed so far in the section becomes more interesting if it can be really used, that is if there are (physical) systems which can be analysed in terms of WPBs. This is indeed the case, as we will show in the rest of this section.

3.1 Weak PBs for $\hat{x}$ and $\hat{p}$

Let us consider the following operators defined on $\mathcal{H} = \mathcal{L}^2(\mathbb{R})$: $\hat{x} f(x) = x f(x)$, $(\hat{D} g)(x) = g'(x)$, the derivative of $g(x)$, for all $f(x) \in D(\hat{x}) = \{h(x) \in \mathcal{L}^2(\mathbb{R})$: $xh(x) \in \mathcal{L}^2(\mathbb{R}\}$ and $g(x) \in D(\hat{D}) = \{h(x) \in \mathcal{L}^2(\mathbb{R}: h'(x) \in \mathcal{L}^2(\mathbb{R})\}$. Of course, the set of test functions $\mathcal{S}(\mathbb{R})$ is a subset of both sets above: $\mathcal{S}(\mathbb{R}) \subset D(\hat{x})$ and $\mathcal{S}(\mathbb{R}) \subset D(\hat{D})$. The adjoints of $\hat{x}$ and $\hat{D}$ in $\mathcal{H}$ are $\hat{x}^\dagger = \hat{x}$, $\hat{D}^\dagger = -\hat{D}$. We have $[\hat{D}, x] f(x) = f(x)$, for all $f(x) \in \mathcal{S}(\mathbb{R})$. This suggests that $\hat{x}$ and $\hat{D}$ could be thought as $\mathcal{S}(\mathbb{R})$-pseudo bosons, since they satisfy Definition 1 and since $\mathcal{S}(\mathbb{R})$ is stable under their action, and the action of their adjoints. However, if we look for the vacua of $a = \hat{D}$ and $b = \hat{x}$, we easily find that $\varphi_0(x) = 1$ and $\psi_0(x) = \delta(x)$, with a suitable choice of the *normalizations*[5]. It is clear, therefore, that neither $\varphi_0(x)$ nor $\psi_0(x)$ belong to $\mathcal{S}(\mathbb{R})$. Also, they not even belong to $\mathcal{L}^2(\mathbb{R})$. Nonetheless, it is interesting to see what can be recovered of the framework proposed in Sect. 2, or if it can be extended, and how. In fact, we will show how the general settings proposed in the first part of Sect. 3 work for our operators $\hat{x}$ and $\hat{p}$.

First of all, let us check if (7) still makes some sense. We have

$$\varphi_n(x) = \frac{b^n}{\sqrt{n!}} \varphi_0(x) = \frac{x^n}{\sqrt{n!}}, \quad \psi_n(x) = \frac{(a^\dagger)^n}{\sqrt{n!}} \psi_0(x) = \frac{(-1)^n}{\sqrt{n!}} \delta^{(n)}(x), \tag{30}$$

for all $n = 0, 1, 2, 3, \ldots$ Here $\delta^{(n)}(x)$ is the n-th weak derivative of the Dirac delta function. We can check that $\varphi_n(x), \psi_n(x) \in \mathcal{S}'(\mathbb{R})$, the set of the tempered distributions, [17], that is the continuous linear functionals on $\mathcal{S}(\mathbb{R})$. This suggests to consider $a^\dagger$ and b as linear operators acting on $\mathcal{S}'(\mathbb{R})$. This is possible since the action of $\hat{x}$ and $\hat{D}$ can be extended outside $\mathcal{L}^2(\mathbb{R})$, to $\mathcal{S}'(\mathbb{R})$, which is stable under the action of these operators. In other words: $a, b, a^\dagger$ and $b^\dagger$ all map $\mathcal{S}'(\mathbb{R})$ into itself. This is exactly what required before Definition 2, with $\mathcal{S}'(\mathbb{R})$ playing the role of $\mathcal{E}$. Then we can extend the pseudo-bosonic commutation relation, originally defined

[5] In fact, to talk of normalization we should have a scalar product but, for the moment, it is not clear what such a scalar product could be in the present context. This will be clarified later in this section.

as $[D, x]f(x) = f(x)$, for all $f(x) \in S(\mathbb{R})$, to the space of tempered distributions:

$$[a, b]\varphi(x) = \varphi(x), \tag{31}$$

for all $\varphi(x) \in S'(\mathbb{R})$.

From (30) it follows that b and $a^\dagger$ act as raising operators, respectively on the sets $\mathcal{F}_\varphi = \{\varphi_n(x)\}$ and $\mathcal{F}_\psi = \{\psi_n(x)\}$:

$$b\varphi_k(x) = \sqrt{k+1}\varphi_{k+1}(x), \quad a^\dagger\psi_k(x) = \sqrt{k+1}\psi_{k+1}(x), \tag{32}$$

$k = 0, 1, 2, 3, \ldots$ Moreover, from (31), we deduce that $b^\dagger$ and a act as lowering operators on these sets:

$$a\varphi_k(x) = \sqrt{k}\varphi_{k-1}(x), \quad b^\dagger\psi_k(x) = \sqrt{k}\psi_{k-1}(x), \tag{33}$$

$k = 0, 1, 2, 3, \ldots$, with the understanding that $a\varphi_0(x) = b^\dagger\psi_0(x) = 0$. It is now clear that, calling $N = ba = \hat{x}\hat{D}$, $N\varphi_k(x) = k\varphi_k(x)$, for all $k = 0, 1, 2, 3, \ldots$ This is because $N\varphi_k(x) = b(a\varphi_k(x)) = \sqrt{k}\, b\varphi_{k-1}(x) = k\varphi_k(x)$. But the same result can also be found in a different, more explicit, way:

$$N\varphi_k(x) = \hat{x}\,\hat{D}\,\frac{x^k}{\sqrt{k!}} = \hat{x}\,\frac{kx^{k-1}}{\sqrt{k!}} = \frac{kx^k}{\sqrt{k!}} = k\varphi_k(x).$$

The distributions $\psi_k(x)$ are also (generalized) eigenstates of a number-like operator. In fact, calling $N^\dagger = a^\dagger b^\dagger$, and using formulas (32) and (33), one proves that $N^\dagger\psi_k(x) = k\psi_k(x)$. Again, this can be checked explicitly by computing

$$N^\dagger\psi_k(x) = -\hat{D}\hat{x}\left(\frac{(-1)^k}{\sqrt{k!}}\,\delta^{(k)}(x)\right) = \frac{(-1)^{k+1}}{\sqrt{k!}}\,(x\delta^{(k)}(x))' = k\psi_k(x),$$

since the weak derivative of $x\delta^{(k)}(x)$ can be easily computed and we have $(x\delta^{(k)}(x))' = -k\delta^{(k)}(x)$, for all $k = 0, 1, 2, 3, \ldots$ Summarizing, we have that

$$N\varphi_k(x) = k\varphi_k(x), \quad N^\dagger\psi_k(x) = k\psi_k(x), \tag{34}$$

for all $k = 0, 1, 2, 3, \ldots$ This formula, together with (32) and (33), are analogous to those deduced in Sect. 2. Hence, this suggests that a framework close to that of PBs can be extended, for $\hat{x}$ and $\hat{D}$, from the Hilbert space $\mathcal{L}^2(\mathbb{R})$ to the set of tempered distributions.

The next step consists in checking, if possible, the biorthogonality of the sets $\mathcal{F}_\varphi$ and $\mathcal{F}_\psi$, and their basis properties, if any. In other words, we are interested in understanding whether equations (24) and (25), or some similar expressions, can be deduced for our families of tempered distributions.

Of course, to talk of biorthogonality, we should first define some sort of scalar product. But this is impossible for distributions, in general. However, there are pairs of distributions for which such an operation can be defined, as we will discuss now. We should also stress that this *extended scalar product* is not unique: other choices are possible, and a different choice was recently proposed in [18].

First we observe that the scalar product between two *good* functions, for instance $f(x), g(x) \in S(\mathbb{R})$, can be written in terms of a convolution between $\overline{f(x)}$ and the function $\tilde{g}(x) = g(-x)$. Indeed we have $\langle f, g\rangle = (\overline{f} * \tilde{g})(0)$. In the same way we define the scalar product between two elements $F(x), G(x) \in S'(\mathbb{R})$ as the following convolution:

$$\langle F, G\rangle = (\overline{F} * \tilde{G})(0), \tag{35}$$

whenever this convolution exists. This existence issue is discussed, for instance, in [19]. As we will see, this will not be a problem for us. In order to compute $\langle F, G\rangle$, it is necessary to compute $(\overline{F} * \tilde{G})[f]$, $f(x) \in S(\mathbb{R})$, and this can be computed by using the equality[6] $(\overline{F} * \tilde{G})[f] = \langle F, G * f\rangle$.

In our situation we have $F(x) = x^n$ and $G(x) = \delta^{(m)}(x)$, $n, m = 0, 1, 2, 3, \ldots$ Hence $(G * f)(x) = \int_{\mathbb{R}} \delta^{(m)}(y) f(x-y)dy = f^{(m)}(x)$, where $f^{(m)}(x)$ is the ordinary m-th derivative of the test function $f(x)$. Then we have

$$(\overline{F} * \tilde{G})[f] = \langle F, G * f\rangle = \int_{\mathbb{R}} \overline{F(x)} f^{(m)}(x)\, dx = \int_{\mathbb{R}} x^n \frac{d^m f(x)}{dx^m}\, dx$$
$$= (-1)^m \int_{\mathbb{R}} \frac{d^m x^n}{dx^m} f(x)\, dx.$$

But

$$\frac{d^m x^n}{dx^m} = \begin{cases} 0 & \text{if } m > n \\ n! & \text{if } m = n \\ \frac{n!}{(n-m)!} x^{n-m} & \text{if } m < n, \end{cases}$$

and therefore

$$(\overline{F} * \tilde{G})[f] = \begin{cases} 0 & \text{if } m > n \\ (-1)^n n! \int_{\mathbb{R}} f(x)\, dx & \text{if } m = n \\ (-1)^m \frac{n!}{(n-m)!} \int_{\mathbb{R}} x^{n-m} f(x)\, dx & \text{if } m < n. \end{cases}$$

Hence

$$(\overline{F} * \tilde{G})(x) = \begin{cases} 0 & \text{if } m > n \\ (-1)^n n! & \text{if } m = n \\ (-1)^m \frac{n!}{(n-m)!} x^{n-m} & \text{if } m < n, \end{cases}$$

and therefore that $(\overline{F} * \tilde{G})(0) = (-1)^n n! \delta_{n,m}$. Putting all these results together, we conclude that not only $\langle \varphi_n, \psi_m\rangle$ exists, but also that

$$\langle \varphi_n, \psi_m\rangle = \delta_{n,m}, \tag{36}$$

[6] We stress once more that $(\overline{F} * \tilde{G})[f]$ is not always defined, but there exist useful situations when it is. This is the case when $\langle F, G * f\rangle$ exists. It is maybe useful to stress that $(\overline{F} * \tilde{G})[f]$ represents the action of $(\overline{F} * \tilde{G})(x)$ on the function $f(x)$.

as claimed before. Notice that our original choice of normalization for $\varphi_0(x)$ and $\psi_0(x)$ guarantees the biorthonormality (and not only the biorthogonality) of the families $\mathcal{F}_\varphi$ and $\mathcal{F}_\psi$.

Remark It is clear that $\langle .,. \rangle$ cannot satisfy all the properties of an *ordinary* scalar product. In particular, it could be impossible to check that $\langle F, F \rangle \geq 0$ for all tempered distributions F, and that $\langle F, F \rangle = 0$ if, and only if, $F = 0$. The reason is simple: there is no guarantee that $\langle F, F \rangle$ does even exist, indeed. However, $\langle .,. \rangle$ has all the properties of an ordinary scalar product when restricted, for instance, to $S(\mathbb{R})$ since, in this case, $\langle .,. \rangle$ coincides with the ordinary scalar product in $\mathcal{L}^2(\mathbb{R})$.

It is clear that it makes no much sense to check if $\mathcal{F}_\varphi$ or $\mathcal{F}_\psi$, or both, are bases in $\mathcal{H}$. This is because none of the $\varphi_n(x)$ and $\psi_n(x)$ even belongs to $\mathcal{L}^2(\mathbb{R})$. However, the pair $(\mathcal{F}_\varphi, \mathcal{F}_\psi)$ can still be used to expand a certain class of functions, those which admit expansion in Taylor series. In fact we have

$$\sum_{n=0}^{\infty} \langle \psi_n, f \rangle \varphi_n(x) = \sum_{n=0}^{\infty} \frac{(-1)^n}{n!} \langle \delta^{(n)}, f \rangle x^n = \sum_{n=0}^{\infty} \frac{1}{n!} f^{(n)}(0)\, x^n = f(x),$$

for all $f(x)$ admitting this kind of expansion. However, if we invert the role of $\mathcal{F}_\psi$ and $\mathcal{F}_\varphi$, the result is more complicated:

$$\sum_{n=0}^{\infty} \langle \varphi_n, f \rangle \psi_n(x) = \sum_{n=0}^{\infty} \frac{(-1)^n}{n!} \langle x^n, f \rangle \delta^{(n)}(x).$$

This is, in principle, an infinite series of derivatives of delta, called *dual Taylor series*, see [20, 21], for instance. It is known that the series does not define in general an element of $D'(\mathbb{R})$, a distribution, (hence it cannot define a tempered distribution) except when the number of non zero moments of $f(x)$, $\langle x^n, f \rangle$, is finite, since, in this case, the series above returns a finite sum, which is indeed a (tempered) distribution.

This preliminary analysis shows that the pair $(\mathcal{F}_\varphi, \mathcal{F}_\psi)$ obeys a sort of *weak basis property*, at least for very special functions or distributions. What we will do next is to check if, and for which objects, a formula like that in (25) can be written. In this perspective, let us introduce the following set of functions:

$$\mathcal{D} = \mathcal{L}^1(\mathbb{R}) \cap \mathcal{L}^\infty(\mathbb{R}) \cap A(\mathbb{R}), \tag{37}$$

where $A(\mathbb{R})$ is the set of entire real analytic functions, which admit expansion in Taylor series, everywhere convergent in $\mathbb{R}$. It might be useful to notice that $\mathcal{D}$ contains many functions of $S(\mathbb{R})$, but not all.

Let now $f(x), g(x) \in \mathcal{D}$, and let us consider the following sequence of functions: $R_N(x) = \overline{f(x)} \sum_{n=0}^{N} \frac{g^{(n)}(0)}{n!} x^n$. It is clear, first of all, that $R_N(x)$ converges to $\overline{f(x)}\, g(x)$ almost everywhere (a.e.) in $\mathbb{R}$. Of course, it also converges with respect to stronger topologies, but this is not relevant for us. The second useful property is that $R_N(x)$ can be estimated as follows:

$$|R_N(x)| \leq R(x) \equiv |f(x)|(M + \|g\|_\infty), \tag{38}$$

for some fixed $M > 0$ and for all N large enough. It is clear that $R(x) \in \mathcal{L}^1(\mathbb{R})$. To prove the estimate in (38) it is enough to observe that, a.e. in x,

$$|R_N(x)| \leq |f(x)| \left(\left| \sum_{n=0}^{N} \frac{g^{(n)}(0)}{n!} x^n - g(x) \right| + |g(x)| \right) \leq |f(x)|(M + \|g\|_\infty),$$

where M surely exists (independently of x) due to the uniform convergence of $\sum_{n=0}^{N} \frac{g^{(n)}(0)}{n!} x^n$ to $g(x)$. Then we can apply the Lebesgue dominated convergence theorem to conclude that

$$\lim_{N,\infty} \int_{\mathbb{R}} R_N(x)dx = \int_{\mathbb{R}} \overline{f(x)}\, g(x)dx = \langle f, g \rangle.$$

Incidentally we observe that, since $f, g \in \mathcal{D}$, $|\langle f, g \rangle| \leq \|f\|_1 \|g\|_\infty$, which ensures that $\langle f, g \rangle$ is well defined. Now,

$$\langle f, g \rangle = \lim_{N,\infty} \int_{\mathbb{R}} R_N(x)dx = \sum_{n=0}^{\infty} \frac{1}{n!} g^{(n)}(0) \langle f, x^n \rangle = \sum_{n=0}^{\infty} \frac{(-1)^n}{n!} \langle f, x^n \rangle \langle \delta^{(n)}, g \rangle =$$
$$= \sum_{n=0}^{\infty} \langle f, \varphi_n \rangle \langle \psi_n, g \rangle.$$

In a similar way we can also check that, for the same $f(x)$ and $g(x)$,

$$\langle f, g \rangle = \sum_{n=0}^{\infty} \langle f, \psi_n \rangle \langle \varphi_n, g \rangle.$$

Hence we conclude that $(\mathcal{F}_\varphi, \mathcal{F}_\psi)$ are $\mathcal{D}$-quasi bases. It should be stressed that it is not clear if $\mathcal{D}$ is dense or not in $\mathcal{H}$, but this is not particularly relevant in the present context, where the role of the Hilbert space is only marginal. Moreover, there are also distributions which satisfy (half of) formula (25). For instance, if $f(x) = \sum_{k=0}^{M} a_k \psi_k(x)$ for some complex a_k and fixed M, the equality $\langle f, g \rangle = \sum_{n=0}^{\infty} \langle f, \varphi_n \rangle \langle \psi_n, g \rangle$ is automatically satisfied, while it is not even clear if $\sum_{n=0}^{\infty} \langle f, \psi_n \rangle \langle \varphi_n, g \rangle$ does converge or not. Similarly, if we take $g(x) = \sum_{k=0}^{L} b_k \varphi_k(x)$ for some complex b_k and fixed L, $\langle f, g \rangle = \sum_{n=0}^{\infty} \langle f, \varphi_n \rangle \langle \psi_n, g \rangle$ is true, while $\sum_{n=0}^{\infty} \langle f, \psi_n \rangle \langle \varphi_n, g \rangle$ could be not even convergent.

In analogy with what we have done in Sect. 2, we can use $\mathcal{F}_\varphi$ and $\mathcal{F}_\psi$ to introduce two operators, S_φ and S_ψ, which we formally write, for the moment,

$$S_\varphi = \sum_n |\varphi_n >< \varphi_n|, \quad S_\psi = \sum_n |\psi_n >< \psi_n|. \tag{39}$$

We have seen that these operators have interesting properties, and it makes sense to understand if they can be extended, and in which sense, to the present distributional context. In particular, it is interesting to check formulas (26)-(29).

First of all, we introduce the following subsets of $S'(\mathbb{R})$:

$$D(S_\varphi) = \{F(x) \in S'(\mathbb{R}) \colon (S_\varphi F)(x) \in S'(\mathbb{R})\}$$

and

$$D(S_\psi) = \{F(x) \in S'(\mathbb{R}) \colon (S_\psi F)(x) \in S'(\mathbb{R})\}.$$

As always, we call these sets the *generalized domains* of S_φ and S_ψ, respectively. It is easy to see that $\mathcal{L}_\varphi \subseteq D(S_\psi)$ and $\mathcal{L}_\psi \subseteq D(S_\varphi)$ and that $S_\varphi \colon \mathcal{L}_\psi \to \mathcal{L}_\varphi$, while $S_\psi \colon \mathcal{L}_\varphi \to \mathcal{L}_\psi$. In particular we have

$$S_\varphi\left(\sum_{k=0}^{N} c_k \psi_k\right) = \sum_{k=0}^{N} c_k \varphi_k, \quad S_\psi\left(\sum_{k=0}^{N} c_k \varphi_k\right) = \sum_{k=0}^{N} c_k \psi_k, \tag{40}$$

as well as

$$S_\varphi S_\psi F = F, \quad S_\psi S_\varphi G = G, \tag{41}$$

and

$$N S_\varphi G = S_\varphi N^\dagger G, \quad N^\dagger S_\psi F = S_\psi N F, \tag{42}$$

for $F(x) \in \mathcal{L}_\varphi$, $G(x) \in \mathcal{L}_\psi$. Furthermore, it is possible to see that $\mathcal{L}_\psi \neq D(S_\varphi)$. In fact, for F to belong to $D(S_\varphi)$, it is sufficient that the series $\sum_{n=0}^{\infty} \langle \varphi_n, F\rangle \varphi_n(x) = \sum_{n=0}^{\infty} \alpha_n x^n$, $\alpha_n = \frac{1}{n!}\langle x^n, F\rangle$, converges. For instance, if $F(x)$ is equal to 1 for $x \in [0, 1]$ and zero otherwise, the series converges for all $x \in \mathbb{R}$, even if $F(x) \notin \mathcal{L}_\psi$.

We refer to [22] for more results on this specific example of WPBs.

3.2 *Weak PBs for the Inverted Quantum Harmonic Oscillator*

This section is devoted to another appearance of WPBs. In this case, this will occur while studying a particular Hamiltonian which looks like a rotated version of the harmonic oscillator. Once again, we will see that distributions are relevant for our system.

We start considering the Hamiltonian

$$H_\theta = \frac{1}{2}\left(\hat{p}^2 + e^{2i\theta}\Omega^2\hat{x}^2\right), \tag{43}$$

for $\theta \in [-\pi, \pi]$, for the moment, and $\Omega > 0$. Here, as usual, $[\hat{x}, \hat{p}] = i\mathbb{1}$, $\hat{x} = \hat{x}^\dagger$ and $\hat{p} = \hat{p}^\dagger$. It is clear that, if $\theta = \pm\frac{\pi}{2}$, H_θ becomes the Hamiltonian of the IQHO, $H_+ = H_- = \frac{1}{2}\left(\hat{p}^2 - \Omega^2\hat{x}^2\right) =: H$, which is what we are really interested in.

Let us introduce the operators

$$A_\theta = \frac{1}{\sqrt{2\Omega}}\left(e^{i\theta/2}\Omega\,\hat{x} + i\,e^{-i\theta/2}\hat{p}\right), \quad B_\theta = \frac{1}{\sqrt{2\Omega}}\left(e^{i\theta/2}\Omega\,\hat{x} - i\,e^{-i\theta/2}\hat{p}\right), \quad (44)$$

for all admissible θ. It is clear that A_θ and B_θ are densely defined in $\mathcal{L}^2(\mathbb{R})$, since in particular any test function $f(x) \in S(\mathbb{R})$ belongs to the domains of both these operators: $S(\mathbb{R}) \subseteq D(A_\theta)$ and $S(\mathbb{R}) \subseteq D(B_\theta)$, for all θ. It is also clear that $A_\theta^\dagger \neq B_\theta$. Indeed we can check that, for instance on $S(\mathbb{R})$,

$$A_\theta^\dagger = \frac{1}{\sqrt{2\Omega}}\left(e^{-i\theta/2}\Omega\,\hat{x} - i\,e^{i\theta/2}\hat{p}\right), \quad B_\theta^\dagger = \frac{1}{\sqrt{2\Omega}}\left(e^{-i\theta/2}\Omega\,\hat{x} + i\,e^{i\theta/2}\hat{p}\right). \quad (45)$$

The set $S(\mathbb{R})$ is stable under the action of all these operators. Formulas (45) show that

$$A_\theta^\dagger = B_{-\theta}, \quad B_\theta^\dagger = A_{-\theta}. \quad (46)$$

Moreover, it is easy to see that these operators obey pseudo-bosonic commutation rules, [6]:

$$[A_\theta, B_\theta]f(x) = f(x) \quad (47)$$

for all $f(x) \in S(\mathbb{R})$, and for all values of $\theta \in [-\pi, \pi]$. This is in agreement with the fact that, if $\theta = 0$, we go back to the ordinary bosonic operators $d = \frac{\Omega\hat{x}+i\hat{p}}{\sqrt{2\Omega}}$ and $d^\dagger = \frac{\Omega\hat{x}-i\hat{p}}{\sqrt{2\Omega}}$, $[d, d^\dagger] = \mathbb{1}$. Indeed we have

$$A_0 = B_0^\dagger = d, \quad B_0 = A_0^\dagger = d^\dagger.$$

In terms of the operators in (44) H_θ can be rewritten as

$$H_\theta = \Omega e^{i\theta}\left(B_\theta A_\theta + \frac{1}{2}\mathbb{1}\right). \quad (48)$$

Then, because of (46), we have that

$$H_\theta^\dagger = \Omega e^{-i\theta}\left(A_\theta^\dagger B_\theta^\dagger + \frac{1}{2}\mathbb{1}\right) = H_{-\theta}, \quad (49)$$

on $S(\mathbb{R})$. Now the eigensystems of H_θ and $H_\theta^\dagger$ can be constructed by using the strategy adopted for PBs, see Sect. 2, and for WPBs, as shown in the first part of Sect. 3: we should first look for the ground state of the two annihilation operators A_θ and $B_\theta^\dagger$. But, since $B_\theta^\dagger = A_{-\theta}$, it is sufficient to solve the differential equation $A_\theta\varphi_0^{(\theta)}(x) = 0$, since the solution of $B_\theta^\dagger\psi_0^{(\theta)}(x) = 0$ is simply $\psi_0^{(\theta)}(x) = \varphi_0^{(-\theta)}(x)$. Hence, recalling that $\hat{p} = -i\frac{d}{dx}$, we find:

$$\varphi_0^{(\theta)}(x) = N^{(\theta)}e^{-\frac{1}{2}\Omega e^{i\theta}x^2}, \quad \psi_0^{(\theta)}(x) = N^{(-\theta)}e^{-\frac{1}{2}\Omega e^{-i\theta}x^2}, \quad (50)$$

where $N^{(\pm\theta)}$ are normalization constants which will be fixed later. From (50) we see that the vacua are in $\mathcal{L}^2(\mathbb{R})$ if $\Re(e^{\pm i\theta}) = \cos(\theta) > 0$. For this reason, from now on, we will restrict to $\theta \in I = \left]-\frac{\pi}{2}, \frac{\pi}{2}\right[$. This constraint reminds very much the similar one for the Swanson model, where it was needed both for ensuring square-integrability of the eigenstates of the Hamiltonian, but also to work with a well defined Hamiltonian, [9, 10].

With this in mind, and using again the usual pseudo-bosonic approach, we can construct two families of functions, $\mathcal{F}_\varphi^{(\theta)} = \{\varphi_n^{(\theta)}(x),\, n = 0, 1, 2, \ldots\}$ and $\mathcal{F}_\psi^{(\theta)} = \{\psi_n^{(\theta)}(x),\, n = 0, 1, 2, \ldots\}$, where

$$\varphi_n^{(\theta)}(x) = \frac{B_\theta^n}{\sqrt{n!}}\,\varphi_0^{(\theta)}(x) = \frac{N^{(\theta)}}{\sqrt{2^n\, n!}}\, H_n\left(e^{i\theta/2}\sqrt{\Omega}\, x\right) e^{-\frac{1}{2}\Omega e^{i\theta} x^2}, \tag{51}$$

$$\psi_n^{(\theta)}(x) = \frac{A_\theta^{\dagger\, n}}{\sqrt{n!}}\,\psi_0^{(\theta)}(x) = \varphi_n^{(-\theta)}(x) = \frac{N^{(-\theta)}}{\sqrt{2^n\, n!}}\, H_n\left(e^{-i\theta/2}\sqrt{\Omega}\, x\right) e^{-\frac{1}{2}\Omega e^{-i\theta} x^2}. \tag{52}$$

Here $H_n(x)$ is the n-th Hermite polynomial. The proof of these formulas is given in [26].

It is clear that, for $\theta \in I$, $\varphi_n^{(\theta)}(x), \psi_n^{(\theta)}(x) \in \mathcal{L}^2(\mathbb{R})$, for all $n \geq 0$. Also, these functions belong to the domain of A_θ, B_θ and of their adjoints, and we have ladder and eigenvalue equations as those in Sect. 2, see (8) in particular:

$$\begin{cases} B_\theta\, \varphi_n^{(\theta)}(x) = \sqrt{n+1}\,\varphi_{n+1}^{(\theta)}(x), & n \geq 0, \\ A_\theta\, \varphi_0^{(\theta)}(x) = 0, \quad A_\theta\, \varphi_n^{(\theta)}(x) = \sqrt{n}\,\varphi_{n-1}^{(\theta)}(x), & n \geq 1, \\ A_\theta^\dagger\, \psi_n^{(\theta)}(x) = \sqrt{n+1}\,\psi_{n+1}^{(\theta)}(x), & n \geq 0, \\ B_\theta^\dagger\, \psi_0^{(\theta)}(x) = 0, \quad B_\theta^\dagger\, \psi_n^{(\theta)}(x) = \sqrt{n}\,\psi_{n-1}^{(\theta)}(x), & n \geq 1, \\ N^{(\theta)}\varphi_n^{(\theta)}(x) = n\,\varphi_n^{(\theta)}(x), & n \geq 0, \\ N^{(\theta)\dagger}\psi_n^{(\theta)}(x) = n\,\psi_n^{(\theta)}(x), & n \geq 0, \end{cases} \tag{53}$$

where $N^{(\theta)} = B_\theta A_\theta$ and $N^{(\theta)\dagger}$ is its adjoint. Then, using (48) and (49), we conclude that

$$H_\theta \varphi_n^{(\theta)}(x) = E_n^{(\theta)} \varphi_n^{(\theta)}(x), \quad H_\theta^\dagger \psi_n^{(\theta)}(x) = E_n^{(-\theta)} \psi_n^{(\theta)}(x), \tag{54}$$

where $E_n^{(\theta)} = \omega e^{i\theta}\left(n + \frac{1}{2}\right)$. Notice that $E_n^{(-\theta)} = \overline{E_n^{(\theta)}}$. Hence the eigenvalues of H_θ and $H_\theta^\dagger$ have, for generic $\theta \in I$, a non zero real and a non zero imaginary part.

Remark If $\theta = 0$ everything collapses to the usual quantum harmonic oscillator, as it is clear from (43). In this case, if we take $N^{(0)} = \left(\frac{\Omega}{\pi}\right)^{1/4}$,

$$\varphi_n^{(0)}(x) = \psi_n^{(0)}(x) = e_n(x) = \frac{1}{\sqrt{2^n\, n!}}\left(\frac{\Omega}{\pi}\right)^{1/4} H_n\left(\sqrt{\Omega}\, x\right) e^{-\frac{1}{2}\Omega x^2}, \tag{55}$$

which is the well known n-th eigenstate of the quantum harmonic oscillator, as expected.

Another, also expected, feature of the families $\mathcal{F}_\varphi^{(\theta)}$ and $\mathcal{F}_\psi^{(\theta)}$ is that, with a proper choice of normalization, their vectors are mutually biorthonormal. Indeed if we fix

$$N^{(\theta)} = \left(\frac{\Omega}{\pi}\right)^{1/4} e^{i\theta/4}, \tag{56}$$

we can check that

$$\langle \varphi_n^{(\theta)}, \psi_m^{(\theta)} \rangle = \delta_{n,m}, \tag{57}$$

for all $n, m \geq 0$ and for all $\theta \in I$. Incidentally we observe that (56) gives back the right normalization of $e_n(x)$ when $\theta = 0$.

It is interesting to observe that the functions $\varphi_n^{(\theta)}(x)$ and $\psi_n^{(\theta)}(x)$ are essentially the rotated versions of the eigenstates $e_n(x)$ in (55):

$$\varphi_n^{(\theta)}(x) = e^{i\theta/4} e_n(e^{i\theta/2}x), \quad \psi_n^{(\theta)}(x) = e^{-i\theta/4} e_n(e^{-i\theta/2}x), \tag{58}$$

for all $n \geq 0$. This is in agreement with (57):

$$\begin{aligned}\langle \varphi_n^{(\theta)}, \psi_m^{(\theta)} \rangle &= \int_{\mathbb{R}} \overline{\varphi_n^{(\theta)}(x)}\, \psi_m^{(\theta)}(x) dx = \int_{\Gamma_\theta} e_n(z) e_m(z) dz \\ &= \int_{\mathbb{R}} e_n(x) e_m(x) dx = \langle e_n, e_m \rangle = \delta_{n,m},\end{aligned}$$

using well known results in complex integration, see [26] for the details.

Next we can check that $\mathcal{F}_\varphi^{(\theta)}$ and $\mathcal{F}_\psi^{(\theta)}$ are complete (or, as some authors prefer, total) in $\mathcal{L}^2(\mathbb{R})$. This follows from a standard argument adopted in several papers, see [6] for instance, and originally proposed, in our knowledge, in [27]: if $\rho(x)$ is a Lebesgue-measurable function which is different from zero almost everywhere (a.e.) in $\mathbb{R}$, and if there exist two positive constants δ, C such that $|\rho(x)| \leq C\, e^{-\delta|x|}$ a.e. in $\mathbb{R}$, then the set $\{x^n \rho(x)\}$ is complete in $\mathcal{L}^2(\mathbb{R})$. We refer to [6] for some physical applications of this result. Because of their completeness, the sets $\mathcal{L}_\varphi^{(\theta)} = l.s.\{\varphi_n^{(\theta)}(x)\}$ and $\mathcal{L}_\psi^{(\theta)} = l.s.\{\psi_n^{(\theta)}(x)\}$, i.e. the linear spans of the functions in $\mathcal{F}_\varphi^{(\theta)}$ and in $\mathcal{F}_\psi^{(\theta)}$, are both dense in $\mathcal{L}^2(\mathbb{R})$. Now, (57) implies that

$$\sum_{n=0}^{\infty} \langle f, \varphi_n^{(\theta)} \rangle \langle \psi_n^{(\theta)}, g \rangle = \langle f, g \rangle, \tag{59}$$

$\forall f(x) \in \mathcal{L}_\psi^{(\theta)}$ and $\forall g(x) \in \mathcal{L}_\varphi^{(\theta)}$, which is our usual weak version of the resolution of the identity.

Incidentally we observe that what we have discussed here is, in fact, another concrete example of PBs, not particularly different from the Swanson model briefly described in Sect. 2.1.

We refer to [26] for more result on coherent states and for the analysis of a similarity operator which can be used in the analysis of the Hamiltonian in (43). Here we are more interested in discussing how to connect what we have deduced for H_θ to similar results for the IQHO.

From $\mathcal{L}^2(\mathbb{R})$ to Distributions

The Hamiltonian we want to consider in this section is the following:

$$H = \frac{1}{2}(\hat{p}^2 - \Omega^2 \hat{x}^2), \tag{60}$$

where, as in (43), $\Omega > 0$. This is what, in the literature, is called an inverted harmonic oscillator: we have a quadratic potential that, rather being convex, is concave, see, e.g., [23–25]. Hence it is reasonable to expect that there are no bound, square integrable, eigenstates. This is, indeed, what we are going to deduce here. We have already seen that H can be formally deduced by H_θ fixing θ either to $\frac{\pi}{2}$ or to $-\frac{\pi}{2}$. For this reason it is natural to define, see (58),

$$\varphi_n^{(\pm)}(x) = \varphi_n^{(\pm\frac{\pi}{2})}(x) = \frac{e^{\pm i\pi/8}}{\sqrt{2^n\, n!}} \left(\frac{\Omega}{\pi}\right)^{1/4} H_n\left(e^{\pm i\pi/4}\sqrt{\Omega}\, x\right) e^{\mp\frac{i}{2}\Omega x^2} \tag{61}$$

and

$$\begin{aligned}\psi_n^{(\pm)}(x) &= \psi_n^{(\pm\frac{\pi}{2})}(x) = \varphi_n^{(\mp)}(x)\\ &= \frac{e^{\mp i\pi/8}}{\sqrt{2^n\, n!}} \left(\frac{\Omega}{\pi}\right)^{1/4} H_n\left(e^{\mp i\pi/4}\sqrt{\Omega}\, x\right) e^{\pm\frac{i}{2}\Omega x^2}.\end{aligned} \tag{62}$$

It is clear that

$$\|\varphi_n^{(\pm)}\| = \|\psi_n^{(\pm)}\| = \infty,$$

so that none of these functions is square-integrable. However, even if they are not in $\mathcal{L}^2(\mathbb{R})$, they are connected to the operators $A_\pm$, $B_\pm$ and their adjoints, where

$$\begin{aligned} A_\pm &= A_{\pm\frac{\pi}{2}} = \frac{1}{\sqrt{2\Omega}}\left(e^{\pm i\pi/4}\Omega\, \hat{x} + i\, e^{\mp i\pi/4}\hat{p}\right),\\ B_\pm &= B_{\pm\frac{\pi}{2}} = \frac{1}{\sqrt{2\Omega}}\left(e^{\pm i\pi/4}\Omega\, \hat{x} - i\, e^{\mp i\pi/4}\hat{p}\right), \end{aligned} \tag{63}$$

and

$$B_\pm^\dagger = A_\mp, \quad A_\pm^\dagger = B_\mp. \tag{64}$$

These operators can all be written in terms of the ordinary bosonic operators d and $d^\dagger$ introduced before as follows:

$$A_\pm = \frac{d \pm i d^\dagger}{\sqrt{2}}, \quad B_\pm = \frac{d^\dagger \pm i d}{\sqrt{2}}, \tag{65}$$

with $A_\pm^\dagger$ and $B_\pm^\dagger$ deduced as in (64). All these operators leave $S(\mathbb{R})$ stable. Then we have

$$[A_\pm, B_\pm] f(x) = f(x), \tag{66}$$

for all $f(x) \in S(\mathbb{R})$. Moreover, these operators can also be applied to functions which are outside $S(\mathbb{R})$, and even outside $\mathcal{L}^2(\mathbb{R})$. In fact, these operators can also act on $\varphi_n^{(\pm)}(x)$ and $\psi_n^{(\pm)}(x)$ and satisfy ladder equations of the same kind as those given in (53):

$$\begin{cases} A_\pm \, \varphi_0^{(\pm)}(x) = 0, \quad A_\pm \, \varphi_n^{(\pm)}(x) = \sqrt{n} \, \varphi_{n-1}^{(\pm)}(x), & n \geq 1, \\ B_\pm \, \varphi_n^{(\pm)}(x) = \sqrt{n+1} \, \varphi_{n+1}^{(\pm)}(x), & n \geq 0, \end{cases} \tag{67}$$

and

$$\begin{cases} B_\pm^\dagger \, \psi_0^{(\pm)}(x) = 0, \quad B_\pm^\dagger \, \psi_n^{(\pm)}(x) = \sqrt{n} \, \psi_{n-1}^{(\pm)}(x), & n \geq 1, \\ A_\pm^\dagger \, \psi_n^{(\pm)}(x) = \sqrt{n+1} \, \psi_{n+1}^{(\pm)}(x), & n \geq 0. \end{cases} \tag{68}$$

Hence the set $\mathcal{E}$ in Definition 2 surely contains $S(\mathbb{R})$ and the set of all the finite linear combinations of the functions $\psi_n^{(\pm)}(x)$ and $\varphi_n^{(\pm)}(x)$.

Some easy computations show that H in (60) can be written in terms of these ladder operators. To simplify the notation we give the results in an operatorial form[7]. Specializing H_θ in (43) by taking $\theta = \pm\frac{\pi}{2}$ we put

$$H_\pm = \pm i \Omega \left(B_\pm A_\pm + \frac{1}{2} \mathbb{1} \right). \tag{69}$$

We have, as expected,

$$H = H_+ = H_-. \tag{70}$$

Using (64) we conclude that $H_+ = H_+^\dagger$, at least formally. Furthermore

$$H_\pm \varphi_n^{(\pm)}(x) = \pm i \Omega \left(n + \frac{1}{2} \right) \varphi_n^{(\pm)}(x), \tag{71}$$

$\forall n \geq 0$. Hence the eigenvalues of the IQHO are purely imaginary with both a positive and a negative imaginary part. Of course the functions $\psi_n^{(\pm)}(x)$, which are usually the eigenstates of the adjoint of the *original* Hamiltonian, see (54), are not

[7] All the operators we are considering in this section can be applied to functions of $S(\mathbb{R})$, but not necessarily: they can also act on $\varphi_n^{(\pm)}(x)$ and $\psi_n^{(\pm)}(x)$, and to their linear combinations.

so relevant here since the adjoint of H_+ is H_+ itself. This is not surprising since, see (62), $\psi_n^{(\pm)}(x) = \varphi_n^{(\mp)}(x)$.

To put the eigenfunctions of H in a more interesting mathematical settings we start defining the following quantities:

$$\Phi_n^{(\pm)}[f] = \langle \varphi_n^{(\pm)}, f \rangle, \quad \Psi_n^{(\pm)}[g] = \langle \psi_n^{(\pm)}, g \rangle, \tag{72}$$

$\forall f(x), g(x) \in S(\mathbb{R})$ and $\forall n \geq 0$. Here $\langle .,. \rangle$ is the form with extend the ordinary scalar product to *compatible pairs*, i.e. to pairs of functions which are, when multiplied together, integrable, but separately they are not (or, at least, one is not). Compatible pairs have been considered in several contributions in the literature. We refer to [28] for their appearance in *partial inner product spaces*, and to [7] for some consideration closer (in spirit) to what we are doing here.

It is not hard to prove that $\Phi_n^{(\pm)}[f]$ and $\Psi_n^{(\pm)}[g]$ are well defined, linear, and continuous in the natural topology τ_S in $S(\mathbb{R})$. In few words, they are tempered distributions, $\Phi_n^{(\pm)}, \Psi_n^{(\pm)} \in S'(\mathbb{R})$. We will only prove this claim for $\Phi_n^{(+)}$, since for $\Phi_n^{(-)}$ and for $\Psi_n^{(\pm)}$ not many differences appear.

To check that $\Phi_n^{(+)}[f]$ is well defined, we observe that

$$\left|\Phi_n^{(+)}[f]\right| \leq \frac{(\Omega/\pi)^{1/4}}{\sqrt{2^n\, n!}} \int_{\mathbb{R}} \left| H_n\left(e^{i\pi/4}\sqrt{\Omega}\, x\right) f(x) \right| dx \leq M_n \sup_{x\in\mathbb{R}} (1+|x|)^{n+2} |f(x)|. \tag{73}$$

Here we have defined

$$M_n = \frac{(\Omega/\pi)^{1/4}}{\sqrt{2^n\, n!}} \int_{\mathbb{R}} \frac{|H_n(e^{i\pi/4}\sqrt{\Omega}\, x)|}{(1+|x|)^{n+2}}\, dx.$$

As we see, in this computation we have multiplied and divided the original integrand function $\left| H_n\left(e^{i\pi/4}\sqrt{\Omega}\, x\right) f(x) \right|$ for $(1+|x|)^{n+2}$. In this way, since the ratio $\frac{|H_n(e^{i\pi/4}\sqrt{\Omega}\, x)|}{(1+|x|)^{n+2}}$ has no singularity and decreases to zero for $|x|$ divergent as $|x|^{-2}$, we can conclude that M_n is finite (and positive). Moreover,

$$\sup_{x\in\mathbb{R}} (1+|x|)^{n+2} |f(x)| = \sup_{x\in\mathbb{R}} \sum_{k=0}^{n+2} \binom{n+2}{k} |x|^k |f(x)| = \sum_{k=0}^{n+2} \binom{n+2}{k} p_{k,0}(f),$$

where $p_{k,0}(.)$ is one of the seminorms defining the topology τ_S, see [29] for instance: $p_{k,l}(f) = \sup_{x\in\mathbb{R}} |x|^k |f^{(l)}(x)|$, $k, l = 0, 1, 2, \ldots$ Of course, all these seminorms are finite for all $f(x) \in S(\mathbb{R})$.

Summarizing we have

$$\left|\Phi_n^{(+)}[f]\right| \leq M_n \sum_{k=0}^{n+2} \binom{n+2}{k} p_{k,0}(f),$$

so that $\Phi_n^{(+)}[f]$ is well defined for all $f(x) \in S(\mathbb{R})$, as we had to check.

The linearity of $\Phi_n^{(+)}$ is clear: $\Phi_n^{(+)}[\alpha f + \beta g] = \alpha \Phi_n^{(+)}[f] + \beta \Phi_n^{(+)}[g]$, for all $f(x), g(x) \in S(\mathbb{R})$ and $\alpha, \beta \in \mathbb{C}$.

To conclude that $\Phi_n^{(+)} \in S'(\mathbb{R})$ we still have to prove that $\Phi_n^{(+)}$ is continuous. For that we have to consider a sequence of functions $\{f_k(x) \in S(\mathbb{R})\}$, τ_S-convergent to $f(x) \in S(\mathbb{R})$, and check that $\Phi_n^{(+)}[f_k] \to \Phi_n^{(+)}[f]$ for $k \to \infty$ in $\mathbb{C}$, for all fixed n. The proof of this fact is based on the following lemma, whose proof can be found in [26].

Lemma 1 Given a sequence of functions $\{f_k(x) \in S(\mathbb{R})\}$, τ_S-convergent to $f(x) \in S(\mathbb{R})$, it follows that $|x|^l |f_k(x)|$ converges, in the norm $\|.\|$ of $\mathcal{L}^2(\mathbb{R})$, to $|x|^l |f(x)|$, $\forall l \geq 0$.

Then we have

$$\left|\Phi_n^{(+)}[f_k - f]\right| = \left|\langle \varphi_n^{(+)}, f_k - f\rangle\right| = \left|\left\langle \frac{\varphi_n^{(+)}}{(1+|x|)^{n+1}}, (1+|x|)^{n+1}(f_k - f)\right\rangle\right|,$$

with an obvious manipulation. Now, since both $\frac{\varphi_n^{(+)}(x)}{(1+|x|)^{n+1}}$ and $(1+|x|)^{n+1}(f_k(x) - f(x))$ are in $\mathcal{L}^2(\mathbb{R})$, for all n, k, we can use the Schwarz inequality and we get

$$\left|\Phi_n^{(+)}[f_k - f]\right| \leq \left\| \frac{\varphi_n^{(+)}}{(1+|x|)^{n+1}} \right\| \left\|(1+|x|)^{n+1}(f_k - f)\right\| \to 0$$

when $k \to \infty$, for all fixed $n \geq 0$, because of Lemma 1.

The role of tempered distributions in the context of the IQHO is further clarified by the following result.

Theorem 1 For each fixed $n \geq 0$ the vector $\varphi_n^{(\pm)}(x)$ is a weak limit of $\varphi_n^{(\theta)}(x)$, for $\theta \to \pm\frac{\pi}{2}$:

$$\varphi_n^{(\pm)}(x) = w - \lim_{\theta, \pm\frac{\pi}{2}} \varphi_n^{(\theta)}(x). \tag{74}$$

Analogously,

$$\psi_n^{(\pm)}(x) = w - \lim_{\theta, \pm\frac{\pi}{2}} \psi_n^{(\theta)}(x). \tag{75}$$

Proof It is sufficient to prove that $\varphi_n^{(+)}(x) = w - \lim_{\theta, +\frac{\pi}{2}} \varphi_n^{(\theta)}(x)$, i.e. that

$$\langle \varphi_n^{(+)} - \varphi_n^{(\theta)}, f\rangle \to 0 \tag{76}$$

when $\theta \to \frac{\pi}{2}$, for all fixed $n \geq 0$ and for all $f(x) \in \mathcal{S}(\mathbb{R})$. First of all we observe that,

$$|\varphi_n^{(+)}(x) - \varphi_n^{(\theta)}(x)| \leq \frac{(\Omega/\pi)^{1/4}}{\sqrt{2^n\, n!}}\left(\left|H_n(e^{i\pi/4}\sqrt{\Omega}\,x)\right| + \left|H_n(e^{i\theta/2}\sqrt{\Omega}\,x)\right|\right)$$
$$\leq \frac{(\Omega/\pi)^{1/4}}{\sqrt{2^n\, n!}}\, p_n(x),$$

where $p_n(x)$ is a suitable polynomial in $|x|$ of degree n, independent of θ, whose expression is not particularly relevant[8]. This estimate implies that the function

$$\chi_n^{(\theta)}(x) = \frac{\varphi_n^{(+)}(x) - \varphi_n^{(\theta)}(x)}{(1+|x|)^{n+1}}$$

is square integrable for all fixed n and for all $\theta \in I$. Therefore, since $(1+|x|)^{n+1} f(x) \in \mathcal{L}^2(\mathbb{R})$ as well, due to the fact that $f(x) \in \mathcal{S}(\mathbb{R})$, we have

$$\left|\langle \varphi_n^{(+)} - \varphi_n^{(\theta)}, f\rangle\right| = \left|\left\langle \frac{\varphi_n^{(+)} - \varphi_n^{(\theta)}}{(1+|x|)^{n+1}}, (1+|x|)^{n+1} f \right\rangle\right| \leq \|\chi_n^{(\theta)}\| \, \|(1+|x|)^{n+1} f\|,$$

using the Schwarz inequality. Now, to conclude as in (76), it is sufficient to show that $\|\chi_n^{(\theta)}\| \to 0$ when $\theta \to \frac{\pi}{2}$, i.e. that

$$\lim_{\theta,\frac{\pi}{2}} \int_{\mathbb{R}} |\chi_n^{(\theta)}(x)|^2 \, dx = 0.$$

This is a consequence of the Lebesgue dominated convergence theorem, since it is clear first that $\lim_{\theta,\frac{\pi}{2}} \chi_n^{(\theta)}(x) = 0$ a.e. in x and since $|\chi_n^{(\theta)}(x)|^2$ is bounded by an $\mathcal{L}^1(\mathbb{R})$ function, in view of what we have shown before. Indeed we have

$$|\chi_n^{(\theta)}(x)|^2 = \frac{|\varphi_n^{(+)}(x) - \varphi_n^{(\theta)}(x)|^2}{(1+|x|)^{2n+2}} \leq \frac{(\Omega/\pi)^{1/2}}{2^n\, n!} \frac{p_n^2(x)}{(1+|x|)^{2n+2}},$$

which goes to zero for $|x|$ divergent as $|x|^{-2}$. □

Summarizing the results proved so far we can write that *the eigenstates of the IQHO are not square integrable. They define tempered distributions and can be obtained as weak limits of the eigenstates of the Swanson-like Hamiltonian introduced in (43).*

We refer to [26] for more results also on coherent states associated to the IQHO.

---

[8] To clarify this aspect of the proof, let us consider, for instance $H_3(x) = 8x^3 - 12x$. Hence $|H_3(x)| \leq 8|x|^3 + 12|x|$ and, therefore $|H_3(e^{i\theta/2}\sqrt{\Omega}\,x)| \leq 8(\Omega)^{3/2}|x|^3 + 12\sqrt{\Omega}|x| = p_3(x)$, for instance.

3.3 A General Class of Pseudo-Bosonic Operators

Another interesting class of first order differential operators connected to PBs and to WPBs are of the form

$$a = \alpha_a(x)\frac{d}{dx} + \beta_a(x), \quad b = -\frac{d}{dx}\alpha_b(x) + \beta_b(x), \tag{77}$$

for some suitable functions $\alpha_j(x)$ and $\beta_j(x)$, $j = a, b$, which, for convenience, will be assumed to be C^∞ functions. This is what happens in concrete models: for ordinary bosons, for instance, we have $\alpha_a(x) = \alpha_b(x) = \frac{1}{\sqrt{2}}$, and $\beta_a(x) = \beta_b(x) = \frac{1}{\sqrt{2}}x$. For the shifted harmonic oscillator, see [6] and references therein, we have $a = c + \alpha\mathbb{1}$ and $b = c^\dagger + \beta\mathbb{1}$, for some complex α and β with $\alpha \neq \overline{\beta}$, and therefore $\alpha_a(x) = \alpha_b(x) = \frac{1}{\sqrt{2}}$ as before, while $\beta_a(x) = \frac{1}{\sqrt{2}}x + \alpha$ and $\beta_b(x) = \frac{1}{\sqrt{2}}x + \beta$. For the Swanson model, see again [6] and Sect. 2.1, $\alpha_a(x) = \alpha_b(x) = \frac{e^{-i\theta}}{\sqrt{2}}$, while $\beta_a(x) = \beta_b(x) = \frac{e^{i\theta}x}{\sqrt{2}}$.

More recently, [7, 30], a rather general class of pseudo-bosonic operators A and B have been considered, where $A = \frac{d}{dx} + w_A(x)$ and $B = -\frac{d}{dx} + w_B(x)$. In this case $\alpha_a(x) = \alpha_b(x) = 1$, while $w_A(x)$ and $w_B(x)$ have been called *pseudo-bosonic superpotentials* (PBSs) and they must satisfy $(w_A(x) + w_B(x))' = 1$, where the prime is the first x-derivative. In particular, in this last example, different choices of C^∞ functions $w_A(x)$ and $w_B(x)$ give rise to different families of functions, $\varphi_n(x)$ and $\Psi_n(x)$, constructed as in Sect. 2, which may, or may not, be square-integrable. However, see [30], we have proven the following result:

Proposition 1 If $w_A(x)$ and $w_B(x)$ are C^∞ PBSs, then $\varphi_n(x)\,\overline{\Psi_m(x)} \in \mathcal{L}^1(\mathbb{R})$ and $\langle\Psi_m, \varphi_n\rangle = \delta_{n,m}$, for all $n, m \geq 0$. □

This is another case, see also (72), in which the functions $\varphi_n(x)$ and $\Psi_n(x)$ are called *compatible*, in the sense of PIP-spaces, [28]. In this perspective it is useful to recall that two functions $h_1(x) \in \mathcal{L}^p(\mathbb{R})$ and $h_2(x) \in \mathcal{L}^q(\mathbb{R})$ can be multiplied producing a third function $h(x) = h_1(x)h_2(x)$ which is integrable, $h(x) \in \mathcal{L}^1(\mathbb{R})$, if $\frac{1}{p} + \frac{1}{q} = 1$. Hence, a *compatibility form* between $h_1(x)$ and $h_2(x)$ can be introduced, whose functional expression is the same as a scalar product in $\mathcal{L}^2(\mathbb{R})$, to which it reduces if $p = q = 2$. It is clear that, for those functions which are compatible, a generalized notion of biorthonormality can also be introduced.

In what follows, we are interested in extending all the particular cases listed above using the general forms of the operators in (77). Of course, our results will be strongly connected to the functions $\alpha_j(x)$ and $\beta_j(x)$.

To proceed in this direction we first compute the commutator $[a, b]$ on some sufficiently regular function $f(x)$. In particular, if not explicitly said, we will assume $f(x)$ to be at least C^2, while we will not insist much on $f(x)$ being or not square-integrable. Of course, this requirement could be relaxed if we interpret $\frac{d}{dx}$ as the weak derivative, but this will not be done here. An easy computation shows that, under this mild condition on $f(x)$, $[a,b]f(x)$ does make sense, and $[a,b]f(x)=f(x)$ if $\alpha_j(x)$ and $\beta_j(x)$, $j=a,b$, satisfy the following equalities

$$\begin{cases}\alpha_a(x)\alpha_b'(x)=\alpha_a'(x)\alpha_b(x),\\ \alpha_a(x)\beta_b'(x)+\alpha_b(x)\beta_a'(x)=1+\alpha_a(x)\alpha_b''(x).\end{cases} \tag{78}$$

It is easy to check that all the examples listed at the beginning of this section satisfy indeed these two conditions, in agreement with their nature of pseudo-bosonic operators. In particular the first equation in (78) is clearly satisfied by any constant choice of $\alpha_a(x)$ and $\alpha_b(x)$. Moreover, in this case, the second equation in (78) can be rewritten as $(\alpha_a\beta_b(x)+\alpha_b\beta_a(x))'=1$, which implies that $\alpha_a\beta_b(x)+\alpha_b\beta_a(x)=x+k$, for some constant k. This is essentially the situation described in terms of the PBSs $w_A(x)$ and $w_B(x)$ in [30]. Incidentally it is also clear that, if $\alpha_a(x)=\alpha_a\neq 0$, constant, then (78) implies that $\alpha_a(x)\alpha_b'(x)=\alpha_a\alpha_b'(x)=0$, which means that $\alpha_b(x)$ must also be constant. For this reason, to avoid going back to PBSs, in the rest of this section we will mainly focus on the situation in which both $\alpha_a(x)$ and $\alpha_b(x)$ depend on x in a non trivial way. Moreover, it is convenient for what follows to assume that they are never zero: $\alpha_j(x)\neq 0$, $\forall x\in\mathbb{R}$, $j=a,b$.

Under this assumption it is easy to deduce the vacua of a and of $b^\dagger$, as in Sects. 2 and 3. In what follows the following expressions are used for the adjoint in $\mathcal{H}$ of a and b:

$$a^\dagger=-\frac{d}{dx}\,\overline{\alpha_a(x)}+\overline{\beta_a(x)},\quad b^\dagger=\overline{\alpha_b(x)}\,\frac{d}{dx}+\overline{\beta_b(x)}. \tag{79}$$

The vacua of a and $b^\dagger$ are the solutions of $a\varphi_0(x)=0$ and $b^\dagger\psi_0(x)=0$, which turn out to be:

$$\varphi_0(x)=N_\varphi\exp\left\{-\int\frac{\beta_a(x)}{\alpha_a(x)}\,dx\right\},\quad \psi_0(x)=N_\psi\exp\left\{-\int\frac{\overline{\beta_b(x)}}{\overline{\alpha_b(x)}}\,dx\right\}, \tag{80}$$

and are well defined under our assumptions on $\alpha_j(x)$ and $\beta_j(x)$. Here N_φ and N_ψ are normalization constants which will be fixed later. If we now introduce $\varphi_n(x)$ and $\psi_n(x)$ as in (7),

$$\varphi_n(x)=\frac{1}{\sqrt{n!}}\,b^n\varphi_0(x),\quad \psi_n(x)=\frac{1}{\sqrt{n!}}\,{a^\dagger}^n\psi_0(x), \tag{81}$$

$n\geq 0$, we can prove the following, see [31]:

Proposition 2 Calling $\theta(x)=\alpha_a(x)\beta_b(x)+\alpha_b(x)\beta_a(x)$ we have

$$\varphi_n(x)=\frac{1}{\sqrt{n!}}\,\pi_n(x)\varphi_0(x),\quad \psi_n(x)=\frac{1}{\sqrt{n!}}\,\sigma_n(x)\psi_0(x), \tag{82}$$

$n \geq 0$, where $\pi_n(x)$ and $\sigma_n(x)$ are defined recursively as follows:

$$\pi_0(x) = \sigma_0(x) = 1, \tag{83}$$

and

$$\pi_n(x) = \left(\frac{\theta(x)}{\alpha_a(x)} - \alpha_b'(x)\right)\pi_{n-1}(x) - \alpha_b(x)\pi_{n-1}'(x), \tag{84}$$

$$\sigma_n(x) = \overline{\left(\frac{\theta(x)}{\alpha_b(x)} - \alpha_a'(x)\right)}\,\sigma_{n-1}(x) - \overline{\alpha_a(x)}\,\sigma_{n-1}'(x), \tag{85}$$

$n \geq 1$. □

A Special Case: Constant $\alpha_j(x)$

We have already commented that taking $\alpha_a(x) = \alpha_a$ and $\alpha_b(x) = \alpha_b$ is not new, compared to what was done in [30]. However, it is still an interesting exercise, and for this reason we briefly discuss this case first. In this situation, $\alpha_a(x)$ and $\alpha_b(x)$ are always different from zero, at least if $\alpha_a\alpha_b \neq 0$. Formulas (84) and (85) simplify significantly now since, in particular, as we have already deduced before, $\theta(x) = \alpha_a\beta_b(x) + \alpha_b\beta_a(x) = x + k$. Hence we find

$$\pi_n(x) = \frac{1}{\alpha_a}(x+k)\,\pi_{n-1}(x) - \alpha_b\,\pi_{n-1}'(x),$$
$$\sigma_n(x) = \frac{1}{\overline{\alpha}_a}(x+\overline{k})\,\sigma_{n-1}(x) - \overline{\alpha}_a\,\sigma_{n-1}'(x), \tag{86}$$

The case $\alpha_a = \alpha_b = 1$ has been considered in [30], while $\alpha_a = \alpha_b = \frac{1}{\sqrt{2}}$ is discussed in [7]. If α_a is not necessarily equal to α_b, similar conclusions can still be deduced. In particular from (86) we find that

$$\pi_n(x) = \sqrt{\left(\frac{\alpha_b}{2\alpha_a}\right)^n} H_n\left(\frac{x+k}{\sqrt{2\alpha_a\alpha_b}}\right), \quad \sigma_n(x) = \sqrt{\left(\frac{\overline{\alpha}_b}{2\overline{\alpha}_a}\right)^n} H_n\left(\frac{x+\overline{k}}{\sqrt{2\overline{\alpha}_a\overline{\alpha}_b}}\right). \tag{87}$$

Here $H_n(x)$ is the n-th Hermite polynomial, and the square root of the complex quantities are taken to be their principal determinations.

As for the functions in (80) we get $\varphi_0(x) = N_\varphi \exp\left\{-\frac{1}{\alpha_a}\int \beta_a(x)\,dx\right\}$, and $\psi_0(x) = N_\psi \exp\left\{-\frac{1}{\overline{\alpha_b}}\int \overline{\beta_b(x)}\,dx\right\}$, where $\beta_a(x)$ and $\beta_b(x)$ are only required to satisfy the condition $\alpha_a\beta_b(x) + \alpha_b\beta_a(x) = x + k$. Now, it is easy to show that $\varphi_n(x)\,\overline{\Psi_m(x)} \in \mathcal{L}^1(\mathbb{R})$, for all $n, m \geq 0$, as in Proposition 1 above, if $\alpha_a\alpha_b > 0$. The proof is based on the fact that $\varphi_n(x)\,\overline{\Psi_m(x)}$ is (a part some normalization constants), the product of a polynomial of degree $n+m$ times the following exponential

$$\exp\left\{-\int\left(\frac{\beta_a(x)}{\alpha_a}+\frac{\beta_b(x)}{\alpha_b}\right)dx\right\}=\exp\left\{-\frac{1}{\alpha_a\alpha_b}\int\theta(x)\,dx\right\}$$
$$=\exp\left\{-\frac{1}{\alpha_a\alpha_b}\int(x+k)\,dx\right\}$$
$$=\exp\left\{-\frac{1}{\alpha_a\alpha_b}\left(\frac{x^2}{2}+kx+\tilde{k}\right)\right\},$$

for some integration constant $\tilde{k}$. Notice that this is a gaussian term under our assumption on $\alpha_a\alpha_b$. We refer to [30] for the analysis of the biorthonormality (with a little abuse of language) of $\mathcal{F}_\varphi=\{\varphi_n(x)\}$ and $\mathcal{F}_\psi=\{\psi_n(x)\}$ in this specific case of constant $\alpha_j(x)$.

A General Example

The situation we will now consider is when $\alpha_a(x)=\alpha_b(x)=\alpha(x)$, where $\alpha(x)\neq 0$ for all $x\in\mathbb{R}$. In this case the first equation in (78) is automatically true, independently of the particular form of $\alpha(x)$. The second equation becomes $(\beta_a(x)+\beta_b(x))'=\frac{1}{\alpha(x)}+\alpha''(x)$, so that

$$\beta_a(x)+\beta_b(x)=\int\frac{dx}{\alpha(x)}+\alpha'(x). \tag{88}$$

From now on we will identify $\beta_a(x)$ and $\beta_b(x)$ as follows:

$$\beta_a(x)=\int\frac{dx}{\alpha(x)},\qquad \beta_b(x)=\alpha'(x). \tag{89}$$

Of course, other possible choices exist, like that in which the role of $\beta_a(x)$ and $\beta_b(x)$ are simply exchanged. But we could also consider $\beta_a(x)=\int\frac{dx}{\alpha(x)}+\Phi(x)$ and $\beta_b(x)=\alpha'(x)-\Phi(x)$, for some fixed, sufficiently regular, $\Phi(x)$. This can produce interesting results, depending on how $\Phi(x)$ is fixed. However, to simplify our analysis here, we will take $\Phi(x)=0$ in what follows. Similarly, we will also fix to zero all the integration constants, except when explicitly stated. The function $\theta(x)$ introduced in Proposition 2 becomes $\theta(x)=\alpha(x)(\beta_a(x)+\beta_b(x))$, so that

$$\theta(x)=\alpha(x)\left(\int\frac{dx}{\alpha(x)}+\alpha'(x)\right), \tag{90}$$

which, when replaced in (84), produces the following sequence of functions: $\pi_0(x)=1$ and

$$\pi_n(x)=\left(\int\frac{dx}{\alpha(x)}\right)\pi_{n-1}(x)-\alpha(x)\pi_{n-1}'(x). \tag{91}$$

Calling $\rho(x) = \int \frac{dx}{\alpha(x)}$ we can rewrite (91) in the following alternative way:

$$\pi_n(x) = \rho(x)\pi_{n-1}(x) - \frac{1}{\rho'(x)}\pi'_{n-1}(x), \tag{92}$$

$n \geq 1$, which can be used to deduce the following expression for $\pi_n(x)$:

$$\pi_n(x) = \frac{1}{\sqrt{2^n}} H_n\left(\frac{\rho(x)}{\sqrt{2}}\right), \tag{93}$$

for all $n \geq 0$, [31].

Remarks

(1) Equation (93) returns the first equation in (87) if $\alpha(x) = \alpha$, constant in x, as it should.

(2) If $\alpha(x)$ is real then, using (89), $\beta_b(x)$ is also real. Also, $\beta_a(x)$ is real if the integration constant is chosen to be real, as we will do always here[9]. In these conditions, $\sigma_n(x) = \pi_n(x)$, $\forall n \geq 0$.

(3) We believe (but we don't have a rigorous result for that) that Hermite polynomials of some "complicated" argument always appear in connection with PBs and WPBs because these are connected to deformed CCR, and CCR gives rise to Hermite polynomials. This is indeed what we have observed along the years, in all the models we have analysed so far.

As for the vacua in (80), using the fact that $\alpha_a(x) = \alpha_b(x) = \alpha(x)$, together with formulas (89), we deduce that

$$\varphi_0(x) = N_\varphi \exp\left\{-\frac{1}{2}(\rho(x))^2\right\}, \quad \psi_0(x) = \frac{N_\psi}{\overline{\alpha}(x)}, \tag{94}$$

or simply $\psi_0(x) = \frac{N_\psi}{\alpha(x)}$ if $\alpha(x)$ is real, as we will assume from now on, to simplify the notation. Putting all together we conclude that

$$\varphi_n(x) = \frac{N_\varphi}{\sqrt{2^n n!}} H_n\left(\frac{\rho(x)}{\sqrt{2}}\right) e^{-\left(\frac{\rho(x)}{\sqrt{2}}\right)^2}, \quad \psi_n(x) = \frac{N_\psi}{\sqrt{2^n n!}} H_n\left(\frac{\rho(x)}{\sqrt{2}}\right) \frac{1}{\alpha(x)}. \tag{95}$$

These formulas suggest that, for many possible choices of $\alpha(x)$, it is quite easy that $\psi_n(x) \notin \mathcal{L}^2(\mathbb{R})$, even if maybe not for all the values of n. On the contrary, we could easily imagine that, for the same choice of $\alpha(x)$, $\varphi_n(x) \in \mathcal{L}^2(\mathbb{R})$.

It is now very easy to prove that, under very mild assumption on $\alpha(x)$, the families $\mathcal{F}_\varphi$ and $\mathcal{F}_\psi$ are compatible and biorthonormal (in our slightly extended meaning), even when the functions $\varphi_n(x)$ or $\psi_n(x)$ do not both belong to $\mathcal{L}^2(\mathbb{R})$. To prove this claim, it is useful to assume that $\rho(x)$ is increasing in x and that, calling

[9] Actually, as already stated, we will often fix to zero this integration constant.

$s = \frac{\rho(x)}{\sqrt{2}}$, $s \to \pm\infty$ when $x \to \pm\infty$. It is clear then that ρ can be inverted, and that $x = \rho^{-1}(\sqrt{2}s)$. Since $\rho'(x) = \frac{1}{\alpha(x)}$, it follows that $\rho(x)$ is always increasing if $\alpha(x) > 0$. However, this is not enough to ensure that s diverges with x, and therefore this must also be required.

Now, to prove that $\varphi_n(x)$ and $\psi_m(x)$ are compatible (and biorthonormal), we compute the compatibility form:

$$\langle \psi_m, \varphi_n \rangle = \frac{\overline{N_\psi} N_\varphi}{\sqrt{2^{n+m}\, n!\, m!}} \int_{-\infty}^{\infty} H_m\left(\frac{\rho(x)}{\sqrt{2}}\right) H_n\left(\frac{\rho(x)}{\sqrt{2}}\right) e^{-\left(\frac{\rho(x)}{\sqrt{2}}\right)^2} \frac{dx}{\alpha(x)}.$$

This integral can be easily rewritten in terms of s. In fact, recalling the definition of $\rho(x)$, we first observe that $\frac{ds}{dx} = \frac{1}{\sqrt{2}\,\alpha(x)}$, so that $\frac{dx}{\alpha(x)} = \sqrt{2}\, ds$. Hence we have

$$\langle \psi_m, \varphi_n \rangle = \frac{\overline{N_\psi} N_\varphi}{\sqrt{2^{n+m-1}\, n!\, m!}} \int_{-\infty}^{\infty} H_m(s) H_n(s) e^{-s^2} ds = \sqrt{2\pi}\, \overline{N_\psi} N_\varphi\, \delta_{n,m},$$

which returns

$$\langle \psi_m, \varphi_n \rangle = \delta_{n,m}, \quad \text{if} \quad \overline{N_\psi} N_\varphi = \frac{1}{\sqrt{2\pi}}, \tag{96}$$

as will be assumed in the rest of this section. This is what we had to prove.

An Example Let us fix $\alpha(x) = \frac{1}{1+x^2}$. This function is always strictly positive, and produces, using (89) and the definition of $\rho(x)$, the functions $\beta_a(x) = \rho(x) = x + \frac{x^3}{3}$ and $\beta_b(x) = \frac{-2x}{(1+x^2)^2}$. We see that $\rho(x) \to \pm\infty$ when $x \to \pm\infty$. Also, the inverse of ρ exists and can be computed explicitly looking for the only real solution of the equation $\sqrt{2}s = x + \frac{x^3}{3}$. We get

$$x = \rho^{-1}(\sqrt{2}\, s) = \left(\frac{2}{-3\sqrt{2}s + \sqrt{2}\sqrt{2+9s^2}}\right)^{1/3} - \left(\frac{-3\sqrt{2}s + \sqrt{2}\sqrt{2+9s^2}}{2}\right)^{1/3}.$$

The functions in (94) turn out to be

$$\varphi_0(x) = N_\varphi \exp\left\{-\frac{1}{2}(x + x^3/3)^2\right\}, \quad \psi_0(x) = N_\psi\, (1 + x^2). \tag{97}$$

It is clear that $\varphi_0(x) \in \mathcal{L}^2(\mathbb{R})$, while $\psi_0(x)$ is not square-integrable. Furthermore, see (92), we have

$$\pi_n(x) = \left(x + \frac{x^3}{3}\right)\pi_{n-1}(x) - \frac{1}{(1+x^2)}\, \pi'_{n-1}(x),$$

with $\pi_0(x) = 1$, and a similar expression for $\sigma_n(x)$. More explicitly we get

$$\pi_n(x) = \sigma_n(x) = \frac{1}{\sqrt{2^n}} H_n\left(\frac{x + x^3/3}{\sqrt{2}}\right),$$

and

$$\varphi_n(x) = \frac{N_\varphi}{\sqrt{2^n n!}} H_n\left(\frac{x + x^3/3}{\sqrt{2}}\right) e^{-\frac{1}{2}(x+x^3/3)^2},$$
$$\psi_n(x) = \frac{N_\psi}{\sqrt{2^n n!}} H_n\left(\frac{x + x^3/3}{\sqrt{2}}\right)(1 + x^2), \tag{98}$$

$n \geq 0$. Hence $\varphi_n(x) \in \mathcal{L}^2(\mathbb{R})$, while $\psi_n(x) \notin \mathcal{L}^2(\mathbb{R})$, for all $n = 0, 1, 2, \ldots$

The fact that these functions are compatible follows from the speed of decay of $\varphi_n(x)$, when compared with the speed of divergence of $\psi_m(x)$. In particular, formula (96) shows that these functions are biorthonormal if $N_\psi N_\varphi = \frac{1}{\sqrt{2\pi}}$: $\langle \psi_m, \varphi_n \rangle = \delta_{n,m}$, $\forall n, m \geq 0$.

Notice that, for our particular operators in (77), there is no need to move to $S'(\mathbb{R})$. However, we see that $\mathcal{L}^2(\mathbb{R})$ is not enough, in general, and we have to use *compatible spaces*, with a compatibility form which extends the ordinary scalar product in $\mathcal{L}^2(\mathbb{R})$. This is different from what we have seen in Sects. 3.1 and 3.2, where the role of $S'(\mathbb{R})$ was more relevant, if not essential. In other words, WPBs are not intrinsically connected with distributions; they can appear when $\mathcal{L}^2(\mathbb{R})$ is not sufficient in the analysis of our pseudo-bosonic operators.

Since, as the example above shows, the functions $\varphi_n(x)$ and $\psi_n(x)$ are not necessarily square-integrable, it is clear that there is no reason for $\mathcal{F}_\varphi$ and $\mathcal{F}_\psi$ to be bases for $\mathcal{L}^2(\mathbb{R})$. However, despite of the fact that $\varphi_n(x)$ and $\psi_n(x)$ are not necessarily square-integrable, we will show that a set $\mathcal{W}$, dense in $\mathcal{L}^2(\mathbb{R})$, does indeed exist such that $\mathcal{F}_\varphi$ and $\mathcal{F}_\psi$ are $\mathcal{W}$-quasi bases.

Let us introduce the set

$$\mathcal{W} = \left\{h(s) \in \mathcal{L}^2(\mathbb{R}) : h_-(s) := h(\rho^{-1}(\sqrt{2}s))\, e^{s^2/2} \in \mathcal{L}^2(\mathbb{R})\right\} \tag{99}$$

This set is dense in $\mathcal{L}^2(\mathbb{R})$, since it contains the set $\mathcal{D}(\mathbb{R})$ of all the compactly supported C^∞ functions, [31]. It is useful to observe that, if $h(x) \in \mathcal{W}$, then the function $h_+(s) := h(\rho^{-1}(\sqrt{2}s))\, \alpha(\rho^{-1}(\sqrt{2}s))\, e^{-s^2/2} \in \mathcal{L}^2(\mathbb{R})$ as well, at least under very general conditions on $\alpha(x)$. This is because $|h_+(s)|^2 = |h_-(s)|^2 |g(s)|^2$, where $g(s) = \alpha(\rho^{-1}(\sqrt{2}s))\, e^{-s^2}$. Now, it is sufficient that $g(s) \in \mathcal{L}^\infty(\mathbb{R})$ to conclude that $h_+(s) \in \mathcal{L}^2(\mathbb{R})$. But, because of the presence of e^{-s^2} in $g(s)$, this is true for many choices of $\alpha(x)$, like for instance the one proposed in the previous example, $\alpha(x) = \frac{1}{1+x^2}$. However, even if $\alpha(x)$ diverges very fast, if $h(x) \in \mathcal{D}(\mathbb{R})$ then $h_+(s) \in \mathcal{L}^2(\mathbb{R})$ anyhow, which is what we will use in the following.

Theorem 2 $(\mathcal{F}_\varphi, \mathcal{F}_\psi)$ are $\mathcal{W}$-quasi bases.

Proof Let us take $f(x), g(x) \in \mathcal{W}$. It is possible to check that the following equalities hold:

$$\langle f, \varphi_n\rangle = N_\varphi\, \pi^{1/4}\sqrt{2}\langle f_+, e_n\rangle, \quad \langle \psi_n, g\rangle = \overline{N}_\psi\, \pi^{1/4}\sqrt{2}\langle e_n, g_-\rangle. \tag{100}$$

Here $e_n(s) = \frac{1}{\sqrt{2^n n!\sqrt{\pi}}} H_n(s) e^{-s^2/2}$ is the n-th eigenstate of the quantum harmonic oscillator already considered several times in this chapter, while $f_+(s)$ and $g_-(s)$ should be constructed from $f(s)$ and $g(s)$ as shown before. The equalities in (100) show, in particular, that the pairs $(f(x), \varphi_n(x))$ and $(g(x), \psi_n(x))$ are compatible, $\forall n \geq 0$, since all the functions involved in the right-hand sides of the equalities in (100), $e_n(s)$, $f_+(s)$ and $g_-(s)$, are square integrable. The proof of these identities is based on the change of variable $s = \frac{\rho(x)}{\sqrt{2}}$, which has already been used before, to prove (96). Now, since $\mathcal{F}_e = \{e_n(s),\ n \geq 0\}$ is an orthonormal basis for $\mathcal{L}^2(\mathbb{R})$, we have

$$\sum_{n=0}^{\infty} \langle f, \varphi_n\rangle\langle \psi_n, g\rangle = \overline{N}_\psi\, N_\varphi\, 2\sqrt{\pi}\sum_{n=0}^{\infty}\langle f_+, e_n\rangle\langle e_n, g_-\rangle = \sqrt{2}\langle f_+, g_-\rangle,$$

using (96) and the Parseval identity for $\mathcal{F}_e$. Next we have

$$\begin{aligned}
\langle f_+, g_-\rangle &= \int_{-\infty}^{\infty} \overline{f_+(s)}\, g_-(s)\, ds \\
&= \int_{-\infty}^{\infty} \overline{f(\rho^{-1}(\sqrt{2}s))\,\alpha(\rho^{-1}(\sqrt{2}s)) e^{-s^2/2}}\, g(\rho^{-1}(\sqrt{2}s)) e^{s^2/2}\, ds \\
&= \int_{-\infty}^{\infty} \overline{f(\rho^{-1}(\sqrt{2}s))\,\alpha(\rho^{-1}(\sqrt{2}s))}\, g(\rho^{-1}(\sqrt{2}s))\, ds = \frac{1}{\sqrt{2}}\langle f, g\rangle,
\end{aligned}$$

introducing the new variable $x = \rho^{-1}(\sqrt{2}s)$ in the integral. Summarizing,

$$\sum_{n=0}^{\infty}\langle f, \varphi_n\rangle\langle \psi_n, g\rangle = \langle f, g\rangle,$$

and, with similar computations, $\sum_{n=0}^{\infty}\langle f, \psi_n\rangle\langle \varphi_n, g\rangle = \langle f, g\rangle$. □

The conclusion is therefore that, even if $(\mathcal{F}_\varphi, \mathcal{F}_\psi)$ are not necessarily made of functions in $\mathcal{L}^2(\mathbb{R})$, they can be used, together, to deduce a resolution (better, two resolutions) of the identity on $\mathcal{W}$.

More results and explicit examples of these WPBs, together with some application to bi-coherent states, can be found in [31].

4 Conclusions

In this chapter we have reviewed some general aspects and applications of WPBs, and we have discussed how distribution theory and compatible spaces are relevant in this context, and how ladder operators can be extended outside a purely Hilbertian settings. We have not discussed here several aspects of this general framework. In particular, we have not considered the role of coherent states in connection with lowering operators of pseudo-bosonic type. We refer to [6, 7] for many results on this, but more recent results can also be found, for instance in [22].

As we have noticed, biorthonormality of the eigenstates of our number-like operators refers to some bilinear form which cannot be the ordinary scalar product in $\mathcal{L}^2(\mathbb{R})$. In particular, the one proposed in Sect. 3.1 is only one possibility, among many. In [32] we have proposed a new extension of the scalar product not related to convolutions, and we proved that this class of multiplications can be flexible enough to succeed where the convolution cannot really be useful. More on this new definition of multiplication, and its role in connection with the properties of the adjoint of an operator and with the consequences of its definition, is work in progress.

Our approach, thought being mathematically already interesting by itself (in our opinion, at least!), needs some extra effort in the attempt of connecting it with physics, and in particular with the probabilistic interpretation of the wave function. This is another open aspect of our approach, and surely deserve further investigation. This is also an active line of research.

Acknowledgements This work is dedicated to Gianni, with friendship and sincere gratitude. The author acknowledges partial financial support from Palermo University (via FFR2021 "Bagarello") and from G.N.F.M. of the INdAM.

References

1. Mostafazadeh, A.: Pseudo-hermitian quantum mechanics. Int. J. Geom. Methods Mod. Phys. **7**, 1191–1306 (2010)
2. Bender, C., Fring, A., Günther, U., Jones, H. (eds.): Special issue on quantum physics with non-Hermitian operators. J. Phys. A: Math. and Ther., **45** (2012)
3. Bagarello, F., Gazeau, J.P., Szafraniec, F.H., Znojil, M. (eds.): Non-selfadjoint operators in quantum physics: Mathematical aspects. John Wiley & Sons (2015)
4. Bagarello, F., Passante, R., Trapani, C.: Non-Hermitian Hamiltonians in Quantum Physics. Selected Contributions from the 15th International Conference on Non-Hermitian Hamiltonians in Quantum Physics, Palermo, Italy, 18–23 May 2015. Springer (2016)
5. Bender, C.M.: PT Symmetry in quantum and classical physics. World Scientific (2019)
6. Bagarello, F.: Deformed canonical (anti-)commutation relations and non hermitian Hamiltonians. In: Bagarello, F., Gazeau, J.P., Szafraniec, F.H., Znojil, M. (eds.) Non-selfadjoint operators in quantum physics: Mathematical aspects. John Wiley & Sons (2015)
7. Bagarello, F.: Pseudo-bosons and their coherent states. Springer (2022)
8. Heil, C.: A basis theory primer: expanded edition. Springer, New York (2010)
9. da Providência, J., Bebiano, N., da Providência, J.P.: Non hermitian operators with real spectrum in quantum mechanics. ELA, vol. 21., pp. 98–109 (2010)

10. Bagarello, F.: Examples of Pseudo-bosons in quantum mechanics. Phys. Lett. A. **374**, 3823–3827 (2010)
11. Swanson, M.S.: Transition elements for a non-Hermitian quadratic hamiltonian. J. Math. Phys. **45**, 585–601 (2004)
12. Davies, E.B., Kuijlaars, B.J.: Spectral asymptotics of the non-self-adjoint harmonic oscillator. J. London Math. Soc. **70**, 420–426 (2004)
13. Bender, C.M., Jones, H.F.: Interactions of Hermitian and non-Hermitian Hamiltonians. J. Phys. A **41**, 244006 (2008)
14. Li, J.-Q., Li, Q.Y.-G.M.: Investigation of PT-symmetric Hamiltonian Systems from an Alternative Point of View. Commun. Theor. Phys. **58**, 497 (2012)
15. Bagarello, F., Lattuca, M.: $\mathcal{D}$ pseudo-bosons in quantum models. Phys. Lett. A. **377**, 3199–3204 (2013)
16. Li, J.-Q., Miao, Y.-G.Z.X.: A Possible Method for Non-Hermitian and Non-PT-Symmetric Hamiltonian Systems. PLoS ONE **9**(6), e97107 (2014)
17. Gelfand, I.M., Shilov, G.E.: Generalized Functions, vol. I. Academic Press, New York and London (1964)
18. Bagarello, F.: Multiplication of distributions in a linear gain and loss system. ZAMP **74**, 136 (2023)
19. Vladimirov, V.S.: Le distribuzioni nella fisica matematica. MIR, Moscow (1981)
20. Kanwal, R.P.: Delta series solutions of differential and integral equations. Int. Transf. Spec. Funct. **6**(1–4), 49–62 (1998)
21. Estrada, R., Kanwal, R.P.: A Distributional Approach to Asymptotics Theory and Applications. Birkhäuser, Boston (2002)
22. Bagarello, F., Gargano, F.: Bi-coherent states as generalized eigenstates of the position and the momentum operators. ZAMP **73**, 119 (2022)
23. Barton, G.: Quantum mechanics of the inverted oscillator potential. Ann. Phys. **166**, 322–363 (1986)
24. Krason, P., Milewski, J.: On eigenproblem for inverted harmonic oscillators. Banach Cent. Publ. **124**, 61–73 (2021)
25. Subramanyan, V., Hegde, S.S., Vishveshwara, S., Bradlyn, B.: Physics of the inverted harmonic oscillator: from the lowest Landau level to event horizons. Ann. Phys. **435**, 168470 (2021)
26. Bagarello, F.: A Swanson-like Hamiltonian and the inverted harmonic oscillator. J. Phys. A **55**, 225204 (2022)
27. Kolmogorov, A., Fomine, S.: Eléments de la théorie des fonctions et de l'analyse fonctionelle. Mir (1973)
28. Antoine, J.-P., Trapani, C.: Reproducing pairs of measurable functions and partial inner product spaces. Adv. Op. Th. **2**, 126–146 (2017)
29. Reed, S., Simon, B.: Methods of modern mathematical physics, Vol I: Functional analysis. Academic Press, New York (1972)
30. Bagarello, F.: Pseudo-bosons and bi-coherent states out of $\mathcal{L}^2(\mathbf{R})$. J. Phys.: Conf. Ser. **2038**, 12001 (2021)
31. Bagarello, F.: A class of weak pseudo-bosons and their bi-coherent states. JMAA **516**(2), 126531 (2022)
32. Bagarello, F.: Multiplication of distributions in a linear gain and loss system. ZAMP **74**, 136 (2023)

The Dichotomy of Forms and Operators and the Role of Green's Forms

Bernhelm Booß-Bavnbek, Yihan Ji, Alessandro Portaluri, and Chaofeng Zhu

Contents

Abstract We consider a sesquilinear form on sections of a Hermitian bundle over a (not necessarily compact) smooth Riemannian manifold with boundary. We assume that in local coordinates, the form can be written as the integral of the inner product of differential expressions of the sections. We assign a linear operator to the form and show that it is a uniquely determined differential operator. We prove the existence and uniqueness of a differential operator on the boundary which makes a variant of Green's formula hold.

B. Booß-Bavnbek (✉)
Department of Sciences and Environment, Roskilde University, Roskilde, Denmark
e-mail: booss@ruc.dk

Y. Ji
Chern Institute of Mathematics and Laboratory of Pure Mathematics and Combinatorics (LPMC), Nankai University, Tianjin, China
e-mail: 2120200014@mail.nankai.edu.cn

A. Portaluri
Dipartimento di Scienze Agrarie, Forestali e Alimentari - DISAFA, Università degli studi di Torino, Grugliasco (TO), Italy
e-mail: alessandro.portaluri@unito.it

C. Zhu
Chern Institute of Mathematics and Laboratory of Pure Mathematics and Combinatorics (LPMC), Nankai University, Tianjin, China
e-mail: zhucf@nankai.edu.cn

A. Cintio, A. Michelangeli (Eds.), *Trails in Modern Theoretical and Mathematical Physics*, https://doi.org/10.1007/978-3-031-44988-8_5

1 Outline

Mathematicians admire the ingenuity of physicists in using mathematical concepts and results and their virtuosity in switching between seemingly different approaches: the one day, e.g., they achieve their success by variational calculus, i.e., optimizing functionals like the energy functional. The other day, they choose a different approach, e.g., looking for ground solutions of differential operators.

In this Note, we shall elaborate on some of the intimate *purely mathematical* correspondences and cross connections between these two mindsets of physicists, the variational and the steady state approach, roughly speaking: forms and operators.

Our approach may be considered as a first, purely technical step towards bridging the dichotomy between variational approaches, based on the concept of forms, like the Hessian of the energy integral, and operator approaches, yielding, e.g., spectral invariants, the Maslov index, and Morse indices for critical points (i.e., curves and surfaces) of the forms. Then we obtain left Green's form (see (15) below) and right Green's form (see (14) below). The classical Green's form is just the difference between them (see (16) below).

Clearly, we wish that our calculations will be used for a particular part of Gianni Morchio's legacy expressed at the end of [14] in a common program with his collaborators, to study "the variation of the ζ-determinant under the change of the boundary conditions". At that time, we could not find a suitable frame for going forward with Gianni's program. Now, 25 years later, it seems to us that it may be helpful to consider our old and unfinished program just as another variational problem for functionals with the ζ-regularized determinants as criticl points of the Hessian of energy integrals on the Grassmannian of pseudodifferential projections over the boundary of a smooth compact Riemannian manifold.

Actually, C. Zhu offered in 2006 in [15] a 1-dimensional pilot study of such a program (dealing only with curves). A. Portaluri and N. Waterstraat provided in 2015 in [12] a related higher-dimensional case study. For a wider encouragement of our endeavour see [4]: V.I. Girsanov's lecture notes are devoted to a like-minded approach to classical extremal problems of optimisation and control, the discovery "that, notwithstanding the great diversity of these problems, they can be attacked by a unified functional-analytic approach."

In this Note, our path of assumptions and arguments wiggles forward between the open plains of abstract generalization and the rugged cliffs of concrete specialization. In Sect. 2, we summarize various *classical representation theorems* for sesquilinear forms, whose role will become clear in the sequel. Mostly we follow Kato [8, VI.2.6, pp. 331 ff] and Lions and Magenes [9]. In Sect. 2.1, we shall recall and explain the bijective, isometric—and elementary—correspondence between bounded operators between Hilbert spaces X, Y and bounded, sesquilinear forms on $X \times Y$ (our Lemma 1, taken from Pedersen [11, Sect. 3.2]). With concrete applications in mind, we generalize that simple result to the more advanced not necessarily bounded case and, at the same time, specialize it by demanding additional and rather restrictive conditions to be satisfied. In that way we achieve (reproduce) the corresponding result for unbounded forms (called by T. Kato *The First Representation Theorem*, our Lemma 2).

In Sect. 2.2, we shall recall and explain a much sharper result obtainable in a concrete analytic setting, here with forms and operators over open subsets of $\mathbb{R}^n$. Interestingly, by that specialization and concretisation we get rid of the costs of generalization, i.e., the restrictive assumptions of Lemma 2

In Sect. 2.3, we shall turn to the main object of our interest, Green's formula for differential operators over smooth Riemannian manifolds with boundary, though in this section in the classical setting only, i.e., with a given differential operator and not yet beginning with a form.

In Sect. 3 we address the representation problem of forms and operators for the special case of these objects acting over a C^k *Riemannian manifold with boundary* and present our new results regarding the related generalized Green's form. Here the point is the transition of results, achieved in local coordinates in the style of Sect. 2.2, to global results. Exploiting the geometric data admits the wanted sharpening of the results obtained previously in [8, 9].

2 Classical Correspondences Between Sesquilinear Forms and Operators

Main theorems of functional analysis like *The Second Representation Theorem* (see Kato [8, VI.2.6, pp. 331 ff], Lions and Magenes [9, Ch. 2, Sect. 2.2 and Sect. 2.4]) point to various correspondences between sesquilinear forms and operators.

In mathematics, we like representation theorems. We recall: For a mathematical system $\mathcal{A}$, a mapping from $\mathcal{A}$ to a similar (but in general "more concrete") system preserving the structure of $\mathcal{A}$ is called a *representation* of $\mathcal{A}$. In this Note, we consider the representation of linear operators by sesquilinear forms and vice-versa.

2.1 The Purely Functional-Analytic Setting

Throughout this section, $(X, \langle\cdot,\cdot\rangle_X)$ and $(Y, \langle\cdot,\cdot\rangle_Y)$ will denote two complex Hilbert spaces and $\mathcal{B}(X, Y)$ the Banach space of bounded operators between X and Y. We denote by $\langle\cdot,\cdot\rangle := \langle\cdot,\cdot\rangle_X$ if there is no confusion. We denote by $\mathcal{B}(X) := \mathcal{B}(X, X)$.

As usual in mathematics, we call a form $\mathrm{t}\colon X \times Y \to \mathbb{C}$ *sesquilinear*, if it is linear in the first variable and conjugate linear in the second. Recall Pedersen's *definition*, [11, p. 80]: "A mathematical physicist is a mathematician believing that a sesquilinear form is conjugate linear in the first variable and linear in the second." We say that a form t is *bounded* $\Longleftrightarrow$ $\|\mathrm{t}\| := \sup\{|\mathrm{t}[x, y]|;\ \|x\| \leq 1,\ \|y\| \leq 1\} < \infty$.

Much analysis in Hilbert spaces bears on the following simple representation theorem:

Lemma 1 There is a bijective, isometric correspondence between operators in $\mathcal{B}(X, Y)$ and bounded, sesquilinear forms $X \times Y \to \mathbb{C}$ given by $A \mapsto \mathrm{t}_A$, where $\mathrm{t}_A[x, y] := \langle Ax, y\rangle_Y$. □

For our application to differential operators, considered as unbounded operators in L^2 we need a wider version. Many authors generalized Lemma 1 for an unbounded form t, assuming that t is densely defined, sectorial and closed with domain of its components $\mathrm{D}(\mathrm{t}) \subset X$. The operator $A\colon X \supset \mathrm{D}(A) \to X$ that appears will be sectorial as is expected from the sectorial boundedness of t. Actually, A turns out to be m-sectorial (for the basic concepts of the spectral theory of unbounded operators, like *accretive*, *m-accretive*, and *m-sectorial*, see [8, V, Sect. 3.10]), which implies that A is closed and the resolvent set $\mathrm{P}(A)$ covers the exterior of the *numerical range* $\Theta(A) := \{\langle Au, u\rangle;\ u \in \mathrm{D}(A), \|u\| = 1\}$. In particular, A is selfadjoint and bounded from below if t is a densely defined, closed sectorial symmetric form.

We recall: If A is closed, then $\Theta(A)$ is always a connected convex subset of $\mathbb{C}$ with $\sigma(A) \subset \overline{\Theta(A)}$, where $\sigma(A) := \mathbb{C} \setminus \mathrm{P}(A)$ denotes the spectrum of A. We distinguish four special cases for $\overline{\Theta(A)}$: it can be bounded; sectorial; making a kind of halfplane; or a vertical band.

The precise result is given in [8, Ch. 6, Sect. 2.1] by

Lemma 2 (First Representation Theorem) Let $\mathrm{t}\colon (u, v) \mapsto \mathrm{t}[u, v]$ be a densely defined, closed, sectorial sesquilinear form in X. There exists an m-sectorial operator A such that

i) $\mathrm{D}(A) \subset \mathrm{D}(\mathrm{t})$ and $\mathrm{t}[u, v] = \langle Au, v\rangle$ for every $u \in \mathrm{D}(A)$ and $v \in \mathrm{D}(\mathrm{t})$;
ii) $\mathrm{D}(A)$ is a core of t (it is well-defined, e.g., in [8, Ch. 6, Sect. 1.4], when we call a linear submanifold of the domain of a closed sectorial form t a core of that form);
iii) if $u \in \mathrm{D}(\mathrm{t})$, $w \in X$ and $\mathrm{t}[u, v] = \langle w, v\rangle$ holds for every v belonging to a core of t, then $u \in \mathrm{D}(A)$ and $Au = w$. In particular, the m-sectorial operator A is uniquely determined by the condition i). □

2.2 *The Euclidean Case*

Going over from the abstract Hilbert space representation theorems to the more concrete case of functions, forms and operators over open sets in Euclidean $\mathbb{R}^n$ widens the field of investigation and opens up for separating events happening at the boundary from events happening in the interior of the underlying space.

The *strong* side of the abstract purely functional-analytic approach in the preceding subsection is the validity and simplicity of the representation formula

$$\mathrm{t}[u, v] = \langle Au, v\rangle \text{ for every } u \in \mathrm{D}(A) \text{ and } v \in \mathrm{D}(\mathrm{t}), \tag{1}$$

that connects forms and corresponding operators and vice versa. The *weak* side of the abstract approach is that its validity is severely restricted: for Lemma 1 to bounded operators and forms; for Lemma 2 to operators and forms satisfying a sectorial condition, that, e.g., excludes operators with an infinite number of eigenvalues on both sides of the real line like the Dirac operator.

In [9, Ch. 2, Sect. 2.2 and 2.4], J.L.Lions and E. Magenes succeeded and gave a generalized representation formula for forms and operators on functions over $\mathbb{R}^n$ of wide validity, without sectorial assumptions, by adding an error term to (1) and achieving that operators and associated forms act on identical domains, while the abstract Lemma 2 ii), only obtained that $\mathrm{D}(A)$ is a core of $\mathfrak{t}$. Lions and Magenes paid for their strong results, however, by demanding a concrete Euclidean setting with ellipticity and even order of an operator acting only on functions; and compactness and smoothness of the underlying submanifold of $\mathbb{R}^n$ with boundary.

More precisely, this is the classical Euclidean setting for the correspondence between sesquilinear forms and operators, taken from [9, Ch. 2, Eq. 2.19]: Let Ω be an open set in $\mathbb{R}^n$. We assume Ω to be bounded with boundary Γ, a $(n-1)$-dimensional manifold, Ω being locally on one side of Γ, i.e. we consider $\overline{\Omega}$ to be a compact submanifold of codimension 0 with boundary. Here we follow [9, Ch. 2, Sect. 2.3] and impose very strong regularity conditions on Ω and on the coefficients of A, namely smoothness.

Theorem 1 (Euclidean Representation Theorem) Let A be a linear elliptic differential operator of order $2m$, $m \geq 1$ over a compact smooth submanifold $\overline{\Omega} \subset \mathbb{R}^n$ of codimension 0 with boundary Γ. We assume that A is given in divergence form as

$$Au = \sum_{|p|,|q| \leq m} (-1)^{|p|} \partial^p \big(a_{pq}(x) \partial^q u\big),$$

where $a_{pq} \in C^\infty(\overline{\Omega})$. Then

1. We assign to A the sesquilinear form $a[\cdot,\cdot]$ defined as

$$a[u, v] = \int_\Omega \sum_{|p|,|q| \leq m} a_{pq}(x) \partial^p u \overline{\partial^q v} dx \text{ for all } u, v \in C^\infty(\overline{\Omega}).$$

   By definition, A and a have the same domain $C^\infty(\overline{\Omega})$.
2. We can define a Dirichlet system of order m $\{F_j\}_{j=0}^{m-1}$ with smooth coefficients on Γ, and a system $\{\Phi_j\}_{j=0}^{m-1}$ which is normal on Γ and with smooth coefficients and $\mathrm{order}(F_j) + \mathrm{order}(\Phi_j) = 2m + 1$, such that

$$a[u, v] = \int_\Omega (Au)\overline{v} dx - \sum_{j=0}^{m-1} \int_\Gamma \Phi_j u \overline{F_j v} d\sigma \quad \text{for all } u, v \in C^\infty(\overline{\Omega}). \tag{2}$$

Remark 1 Roughly speaking, the technical achievement of this Note is that we prove a much wider validity of the preceding theorem, namely for not-necessarily smooth neither compact Riemannian manifolds, not-necessarily elliptic operators, not-necessarily with smooth coefficients, and that act on sections of vector bundles and not-necessarily solely on functions, see our Sect. 3 below.

2.3 *Green's Formula*

Green's formulae and Green's functions appear at many different places in the treatment of linear partial differential equations. Particularly, for hyperbolic and parabolic equations, they are a mean to establish uniqueness of solutions and to provide solutions, at least in integral form. The idea is always to provide a desuspension formula, i.e., to relate (typically differential) expressions on the whole of a manifold to (typically Riemannian symplectic and integral-)forms on the boundary, i.e., descending one dimension.

The classical result is easily explained in the following set-up. Let (M, g) be a complete Riemannian manifold, Γ is the smooth boundary of a compact submanifold $\Omega \subset M$ of codimension 0. Let $\pi_1 \colon E \to M$, $\pi_2 \colon F \to M$ be Hermitian vector bundles. We set $E' := E|_\Gamma$, $F' := F|_\Gamma$. Then for all $j \geq 0$, the j-th jet trace map $\gamma_j \colon C^\infty(M; E) \to C^\infty(\Gamma; E')$ is defined as $\gamma_j u := (\nabla_\nu^E)^j u|_\Gamma$, where ν denotes the inward unit normal vector, yielding the trace map of order $d - 1 \geq 0$ by $\rho^d := (\gamma_0, \cdots, \gamma_{d-1})$.

It is well known that *Green's Formula* is an important theorem about differential operators, see [3, Proposition 1.1.2]:

Theorem 2 For each differential operator $A : C^\infty(M; E) \to C^\infty(M; F)$ of order d, there exists a unique differential operator $J \colon C^\infty(\Gamma; E'^d) \to C^\infty(\Gamma; F'^d)$ such that for any $u \in C^\infty(\Omega; E), v \in C^\infty(\Omega; F)$ we have

$$\langle Au, v\rangle_{L^2(\Omega,F)} - \langle u, A^t v\rangle_{L^2(\Omega,E)} = \langle J\rho^d u, \rho^d v\rangle_{L^2(\Gamma,F'^d)}$$

There are many proofs of the preceding Green's Formula, typically reducing it to the classical Euclidean claim that is proved, e.g., in [9, Ch. 2, Sect. 2.2]. For a short suggestive proof for Dirac type operators we refer to [1, Proposition 3.4]. There is a rich literature of calculating Green's forms in special and even very delicate situations, see for example [2, 5, 6, 10, 13]. All these calculations take their point of departure with an operator, as it is usual with PDEs, while our start point is a given form, as it is usual in variational calculus.

3 A Variant of Green's Formula

We consider sesquilinear forms on sections of a Hermitian bundle over a (not necessarily compact) C^k Riemannian manifold with boundary. We assume that in local coordinates, the forms can be written as the integral of the inner product of differential expressions of the sections with compact supports. We call them *differential sesquilinear forms* (see Definition 1 below).

We assign a linear operator to each differential sesquilinear form. Then we show that it is a uniquely determined differential operator. We prove the existence and uniqueness of a differential operator on the boundary which makes that a variant of Green's formula holds.

3.1 Differential Sesquilinear Forms

Let (M, g) be an n-dimensional C^k Riemannian manifold, where n and k are positive integers. For a C^k local chart (φ, U) of $p_0 \in M$ and $p \in U$, we denote by $x = (x_1, \ldots, x_n) = \varphi(p)$, $g_{ij} := g(\partial x_i, \partial x_j)$, $|g| := \det(g_{ij})$, and the density $d\,\mathrm{vol} := \sqrt{|g|}|dx|$ respectively. We denote by $L(X, Y)$ the space of complex linear maps $X \to Y$ for two complex linear spaces X and Y and by $L(X) = L(X, Y)$.

Let $r, s \in \mathbb{Z}$ be two non-negative integers with $\max\{r+s, s+1\} \leq k$. Let $\pi_1 \colon E \to M$ be an m-dimensional Hermitian vector bundle of class C^{r+s} and $\pi_2 \colon F \to M$ be an m-dimensional Hermitian vector bundle of class C^s. For a C^k local chart (φ, U) of $p_0 \in M$ and $p \in U$ such that $E|U$ is trival with trivalization ψ_1 and $F|U$ is trival with trivalization ψ_2, we denote by $\varphi_{*,\psi_1} : E_p \to \mathbb{C}^m$, $\varphi_{*,\psi_2} : F_p \to \mathbb{C}^m$ the linear isomorphism maps induced by φ, and

$$(\varphi_{*,\psi_1} u_1, \varphi_{*,\psi_1} u_2)_{E_x} := (u_1, u_2)_{E_p} \quad \text{for } u_1, u_2 \in E_p,$$
$$(\varphi_{*,\psi_2} v_1, \varphi_{*,\psi_2} v_2)_{F_x} := (v_1, v_2)_{F_p} \quad \text{for } v_1, v_2 \in F_p.$$

For integers $h_1 \in [0, r]$ and $h_2 \in [0, s]$, we denote by $C_c^{h_1}(M; E)$ the h_1 times continuously differentiable sections of E with compact support and by $C^{h_2}(M; F)$ the h_2 times continuously differentiable sections of F respectively. For each section u we denote by $\mathrm{supp}(u)$ the closure of the set $\{x \in M;\ u(x) \neq 0\}$.

Definition 1 A sesquilinear form $\mathrm{t}[\cdot, \cdot] \colon C_c^r(M; E) \times C^s(M; F) \to \mathbb{C}$ is called a *differential sesquilinear form of type* (r, s) *on* (E, F), if there exists an atlas $\{U_\alpha, \varphi_\alpha\}_{\alpha \in \Lambda}$ of M, C^r trivilizations $\psi_{1\alpha} \colon \pi_1^{-1}(U_\alpha) \to U_\alpha \times \mathbb{C}^m$, C^s trivilizations $\psi_{2\alpha} \colon \pi_2^{-1}(U_\alpha) \to U_\alpha \times \mathbb{C}^m$ and a C^k partition of unity $\{\chi_\alpha\}_{\alpha \in \Lambda}$ subordinate to $\{U_\alpha, \varphi_\alpha\}_{\alpha \in \Lambda}$ such that

$$\mathrm{t}[u, v] = \sum_\alpha \mathrm{t}_\alpha[\chi_\alpha u, v]. \tag{3}$$

Here the local form t_α has the form

$$\mathrm{t}_\alpha[u,v] := \int_{U_\alpha} \sum_{|i|\le r,|j|\le s} \left(a^\alpha_{ji}(x)\partial^i u_\alpha(x), \partial^j v_\alpha(x)\right)_{F_x} d\,\mathrm{vol}, \quad (4)$$

where $a^\alpha_{ji} \in C^{|j|}(U_\alpha, L(\mathbb{C}^m))$ for multiple indices i, j with $|i| \le r, |j| \le s$ if $u \in C^r_c(M;E)$, $\mathrm{supp}(u) \subset U_\alpha$, $v \in C^s(M;F)$, and u_α, v_α denote the expression of u and v in local charts $(U_\alpha, \varphi_\alpha)$ respectively. We denote by $\mathrm{DSM}_{r,s}(M;E,F)$ the linear space of all differential sesquilinear forms of type (r,s) on (E,F) and $\mathrm{DSM}_{r,s}(M;E) := \mathrm{DSM}_{r,s}(M;E,E)$.

Remark 2 (a) Note that with some possibly different coeffients, we can always assume that

$$\mathrm{t}[u,v] = \mathrm{t}_\alpha[u,v] \quad (5)$$

if $\mathrm{supp}(u) \subset U_\alpha$, $v \in C^s(M;F)$. We assume (5) in the rest of the Note.

(b) Definition 1 is consistent if we make different choices of local charts and local trivilizations of Hermitian bundles.

The following lemma shows that the right hand side of (3) is a finite sum.

Lemma 3 Let X be a topological space. Let $\{U_\alpha\}_\alpha$ be a locally finite family of subsets of X. Let K be a compact subset of X. Then the set $\{\alpha \in \Lambda;\ U_\alpha \cap K \ne \emptyset\}$ is finite.

Proof Since $\{U_\alpha\}_\alpha$ is a locally finite family of subsets of X, for each $x \in X$ there exists a neighborhood V_x of x such that the set $\Lambda_x := \{\alpha \in \Lambda;\ U_\alpha \cap V_x \ne \emptyset\}$ is finite. Since K is a compact subset of X, there exists a finite subset B of X such that $K \subset \cup_{x\in B} U_x$.

Note that

$$U_\alpha \cap K \subset U_\alpha \cap (\cup_{x\in B} V_x) = \cup_{x\in B}(U_\alpha \cap V_x).$$

Then the set $\{\alpha \in \Lambda;\ U_\alpha \cap K \ne \emptyset\}$ is a subset of $\cup_{x\in B}\Lambda_x$. So it is finite. □

The differential sesquilinear forms satisfy the following local property.

Lemma 4 Let $u \in C^r_c(M;E)$, $v \in C^s(M;F)$ be two sections. Assume that $\mathrm{supp}(u) \cap \mathrm{supp}(v)$ has measure 0. Then we have

$$\mathrm{t}[u,v] = 0. \quad (6)$$

Proof For each $\alpha \in \Lambda$ we have $\text{supp}(\chi_\alpha u_\alpha) \subset \text{supp}(u)$ and $\text{supp}(v_\alpha) \subset \text{supp}(v)$. Since $\text{supp}(u) \cap \text{supp}(v)$ has measure 0, $\text{supp}(\chi_\alpha u_\alpha) \cap \text{supp}(v_\alpha)$ also has measure 0. By (4) we have $\mathfrak{t}_\alpha[\chi_\alpha u, v] = 0$. By (3) we have $\mathfrak{t}_\alpha[u, v] = 0$. □

3.2 The Existence and Uniqueness of the Assigned Differential Operator

For the form defined by Definition 1, we will assign a linear operator to each differential sesquilinear form and show that it is a uniquely determined differential operator. As usual we denote by $H^h(M; E)$ the Soblev space of sections of E with exponent h and by $L^2(M; E)$ the L^2 space of sections of E respectively. Then $H_0^h(M; E)$ denotes the H^h closure of $C_0^k(M; E)$ with $k \geq h$, $H_{\text{loc}}^h(M; E)$ denotes the space of sections of E with finite H^h norms on each compact subset of M. We denote by $\text{DO}_h(M; E, F)$ the linear space of differential operators of order h from E to F.

Theorem 3 Let (M, g) be an n-dimensional C^k Riemannian manifold with C^{k-1} Riemannian structure g. Let $r, s \in \mathbb{Z}$ be non-negative integers such that $k \geq \max\{r + s, s+1\}$. Let $\pi_1 \colon E \to M$ be a m-dimensional Hermitian vector bundle of class C^{r+s} and $\pi_2 \colon F \to M$ be a m-dimensional Hermitian vector bundle of class C^s. Then we have the following.

(a) There is a linear injective map which assigns a differential operator $A_\mathfrak{t} \in \text{DO}_{r+s}(M; E, F)$ to each $\mathfrak{t} \in \text{DSM}_{r,s}(M; E, F)$ such that

$$\mathfrak{t}[u, v] = \int_K (A_\mathfrak{t} u, v)_{F_x} d\text{vol}, \quad \forall u \in H_0^{r+s}(K; E|_K), v \in H_{\text{loc}}^s(M; F) \tag{7}$$

holds for each compact subset $K \subset M^o$, where M^o denotes the interior of M.

(b) Assume further that the form $\mathfrak{t}$ satisfies (5). If $\text{supp}(u) \subset U_\alpha$ holds, we have

$$A_\mathfrak{t} u = \sqrt{|g|^{-1}} \sum_{|i| \leq r, |j| \leq s} (-1)^{|j|} \partial^j \left(\sqrt{|g|} a_{ji}^\alpha(x) \partial^i(u)\right). \tag{8}$$

Note The space $\text{DSM}_{r,s}(M; E, F)$ and the mapping $\mathfrak{t} \mapsto A_\mathfrak{t}$ depend on the metric structures on M, E and F.

Proof We divide the proof into five steps.

Step 1. Denote by $A_{\mathfrak{t},\alpha} u$ the right hand side of (8). If $\text{supp}(u) \subset U_\alpha \cap K$ holds, the operator $A_{\mathfrak{t},\alpha} u$ makes (7) holds if we replace $A_\mathfrak{t}$ by $A_{\mathfrak{t},\alpha}$.

Since $\operatorname{supp}(u) \subset U_\alpha \cap K$, by (4) and (5) we have for $v \in C^s(M, F)$,

$$\begin{aligned}
\mathfrak{t}[u, v] &= \int\limits_{U_\alpha} \sum_{|i|\le r, |j|\le s} \Big(a^\alpha_{ji}(x)\partial^i(u_\alpha(x)), \partial^j(v_\alpha(x))\Big)_{F_x} d\,\mathrm{vol} \\
&= \int\limits_{U_\alpha} \sqrt{|g|^{-1}} \sum_{|i|\le r, |j|\le s} (-1)^{|j|}\Big(\partial^j\big(\sqrt{|g|}a^\alpha_{ji}(x)\partial^i(u_\alpha(x))\big), v_\alpha(x)\Big)_{F_x} d\,\mathrm{vol} \\
&= \int\limits_K (A_{\mathfrak{t},\alpha}u, v)_{F_x} d\,\mathrm{vol}.
\end{aligned}$$

Then our claim follows.

Step 2. Given a compact subset $K \subset M^o$, there exists a unique bounded linear operator $A_{\mathfrak{t},K} : H_0^{r+s}(K; E|_K) \to L^2(K; F|_K)$ such that

$$\mathfrak{t}[u, v] = \int\limits_K (A_{\mathfrak{t},K}u, v)_{F_x} d\,\mathrm{vol}.$$

By Lemma 3, the set $D_K := \{\alpha \in \Lambda;\ \operatorname{supp}(\chi_\alpha) \cap K \neq \emptyset\}$ is a finite subset of Λ. Note that $A^\alpha_{ji} \in C^{|j|}(U_\alpha, L(\mathbb{C}^m))$ and $|g| \in C^s(U_\alpha, \mathbb{R})$. By (3), Step 1 and Cauchy–Buniakowsky–Schwarz inequality we have

$$\begin{aligned}
\big|\mathfrak{t}[u, v]\big| &\le \sum_{\alpha\in D_K} C_\alpha \|\chi_\alpha u\|_{H_0^{r+s}(K;E|_K)} \|v\|_{L^2(K;F|_K)} \\
&\le C\,\|u\|_{H_0^{r+s}(K;E|_K)} \|v\|_{L^2(K;F|_K)}
\end{aligned}$$

with some finite set $\{C_\alpha\}_{\alpha\in D_K}$ of positive constants and a single positive constant C. Note that $C^s(M; F)$ is dense in $H^s_{\mathrm{loc}}(M; F)$. By Riesz's representation theorem [8, p. 253], for each $u \in H_0^{r+s}(K; E|_K)$ there exists a unique $A_{\mathfrak{t},K}u \in L^2(K; F|_K)$ satisfying $\mathfrak{t}[u, v] = \int_K (A_{\mathfrak{t},K}u, v)_{F_x} d\,\mathrm{vol}$. Moreover, $A_{\mathfrak{t},K}$ is a bounded linear operator (see Lemma 1) with $\|A_{\mathfrak{t},K}\| \le C$.

Step 3. Existence, uniqueness and locality of $A_{\mathfrak{t}}$.

Let K_1 and K_2 be two compact subset of M. Then $K_1 \cup K_2$ is compact. Let $u \in H_0^{r+s}(K_1; E|_{K_1}) \cap H_0^{r+s}(K_2; E|_{K_2})$. By Step 2 we have

$$A_{\mathfrak{t},K_1}u = A_{\mathfrak{t},K_1\cup K_2}u = A_{\mathfrak{t},K_2}u.$$

Then we define $A_{\mathfrak{t}}u := A_{\mathfrak{t},K}u$ if $u \in H_0^{r+s}(K; E|_K)$ for some compact subset K of M. Thus $A_{\mathfrak{t}}$ is a well-defined operator and satisfies (7).

For each $p \in M$, let U_p be a neighborhood of p in M with $U_p \subset K_3^o \subset K_3 \subset K_4^o \subset U_\alpha$ for some compact subsets K_3 and K_4 of p and $\alpha \in \Lambda$. Then there exists two C^k function χ_h on M, where $h = 1, 2$ such that

$$0 \le \chi_h \le 1, \quad \chi_1|_{U_x} = 1, \quad \chi_1|_{M\setminus K_3} = 0, \quad \chi_2|_{K_4} = 1, \quad \chi_2|_{M\setminus K_4} = 0.$$

Then we have $\chi_1(1-\chi_2) = 0$. Therefore we obtain

$$\langle A_{\mathrm{t},K_4}(1-\chi_2)u, \chi_1 v\rangle_{L^2(K_4;F)} = \mathrm{t}[(1-\chi_2)u, \chi_1 v] = 0$$

for all $v \in H^s_{\mathrm{loc}}(M; F)$. Thus we have $\chi_1 A_{\mathrm{t},K}((1-\chi_2)u) = 0$, $\chi_1 A_{\mathrm{t}}((1-\chi_2)u) = 0$ and $(A_{\mathrm{t}}u)|_{U_x} = (A_{\mathrm{t}}\chi_2 u)|_{U_x}$. By Step 1, A_{t} is a differential operator of order $r+s$ such that (8) holds.

Step 4. By Lemma 4 and properties of $A_{\mathrm{t},\alpha}$, we obtain the required properties of A_{t}. Then the assignment $\mathrm{t} \mapsto A_{\mathrm{t}}$ is a well-defined map.

Step 5. Since (7) is linear, the assignment $\mathrm{t} \mapsto A_{\mathrm{t}}$ is a linear map. If $A_{\mathrm{t}} = 0$, by (7) we have $\mathrm{t} = 0$. So the assignment $\mathrm{t} \mapsto A_{\mathrm{t}}$ is a linear injective map. □

Remark 3 We denote by $\mathrm{i} := \sqrt{-1}$ and $\xi^i := \xi_1^{i_1} \dots \xi_n^{i_n}$. For $p \in U_\alpha$ with $\alpha \in \Lambda$, $\xi \in T_p^* M$ and $x = \varphi_\alpha(p)$, we have

$$(\varphi_\alpha)_{*,\psi_{1\alpha}} \circ \sigma_{r+s}(A_{\mathrm{t}})(p,\xi) \circ ((\varphi_\alpha)_{*,\psi_{2\alpha}})^{-1} = (-1)^s \mathrm{i}^{r+s} \sum_{|i|=r,|j|=s} \xi^{i+j} a^\alpha_{ji}(x).$$

Denote by 0_M the zero section of T^*M. We recall that if the principal symbol $\sigma_{r+s}(A_{\mathrm{t}})(x,\xi)$ of A_{t} is nondegenerate for each $(x,\xi) \in T^*M \setminus 0_M$, the operator A_{t} is an elliptic differential operator of order $r+s$.

Remark 4 Assume that $k \geq \max\{r+s, r+1, s+1\}$ and $\pi_1 \colon E \to M$, $\pi_2 \colon F \to M$ are two m-dimensional Hermitian vector bundle of C^{r+s}. The formal adjoint of A_{t} is the operator with local expression

$$A^t_{\mathrm{t}} = \sqrt{|g|^{-1}} \sum_{|i|\leq r, |j|\leq s} (-1)^{|i|} \partial^i \left(\sqrt{|g|} a^\alpha_{ji}(x)^t \partial^j\right). \tag{9}$$

Indeed, we have

$$\begin{aligned}
&\int_K (A_{\mathrm{t}}u, v)_{F_x} d\,\mathrm{vol} \\
&= \sum_\beta \int_{U_\beta} \sum_{|i|\leq r, |j|\leq s} \left(a^\beta_{ji}(x)\partial^i(\chi_\beta u(x)), \partial^j v(x)\right)_{F_x} d\,\mathrm{vol} \\
&= \sum_\beta \int_{U_\beta} \sum_{|i|\leq r, |j|\leq s} \left(\chi_\beta u(x), (-1)^{|i|}\sqrt{|g|^{-1}}\partial^i\left(\sqrt{|g|}a^\alpha_{ji}(x)^t \partial^j(v(x))\right)\right)_{E_x} d\,\mathrm{vol}
\end{aligned}$$

$$= \sum_{\beta} \int_{U_\alpha} \sum_{|i|\le r, |j|\le s} \left(\chi_\beta u(x), (-1)^{|i|}\sqrt{|g|}^{-1}\partial^i\left(\sqrt{|g|}a^\alpha_{ji}(x)^t\partial^j(v(x))\right)\right)_{E_x} d\,\mathrm{vol}$$

$$= \int_K (u, A^t_{\mathrm{t}}v)_{E_x} d\,\mathrm{vol}$$

for all $u \in H^{r+s}_0(K; E)$, $v \in H^{r+s}(K; F)$ such that $\mathrm{supp}(v) \subset U_\alpha$. Then we obtain

$$\int_K (A_{\mathrm{t}}u, v)_{F_x} d\,\mathrm{vol} = \int_K (u, A^t_{\mathrm{t}}v)_{E_x} d\,\mathrm{vol} \tag{10}$$

for all $u \in H^{r+s}_0(K; E|_K)$, $v \in H^{r+s}_{\mathrm{loc}}(M; F)$.

3.3 A Variant of Green's Formula

Let (M, g) be an n-dimensional C^k Riemannian manifold with C^{k-1} boundary Γ and Riemannian metric g. By Theorem 3, for each differential sesquiliinear form t we assign a uniquely-determined differential operator A_{t} to it such that (7) holds. In this section, we want to study the error terms of (7) for those sections with support not necessary in the interior M^o of M, namely the Green's form. Then the Green's forms yields the correction term between differential sesquilinear forms and differential operators.

Let $r, s \in \mathbb{Z}$ be non-negative integers and $k \ge \max\{r + s, s + 1\}$. Let $\pi_1\colon E \to M$ be a m-dimensional Hermitian vector bundle of C^{r+s} and $\pi_2\colon F \to M$ be a m-dimensional Hermitian vector bundle of C^s.

We denote by $E' := E|_\Gamma$, $F' := F|_\Gamma$. The trace map $\gamma_l\colon C^h(M; E) \to C^{h-l}(\Gamma; E')$ is defined by $\gamma_l u := (\nabla^E_\nu)^l u|_\Gamma$, where ν is the unit inner vector, $l, h \in \mathbb{Z}$ and $l \le h \le k$. For vector bundle $\pi_2\colon F \to M$, we can define the corresponding map γ_l. We set $\rho^d := (\gamma_0, \cdots, \gamma_{d-1})$ for a positive integer $d \le k$.

We denote by $H^h_c(M; E)$ the set of sections $u \in H^h_c(M; E)$ with compact support. Set

$$X_h := \bigoplus_{j=0}^{h-1} H^{h-1-j}_c(\Gamma; E'), \quad Y_h := \bigoplus_{i=0}^{h-1} H^{h-1-i}_c(\Gamma; F'). \tag{11}$$

Theorem 4 (a) There is a linear map which assigns a linear operator $J^L_{\mathrm{t}} : X_{r+s} \to Y_s$ to each $\mathrm{t} \in \mathrm{DSM}_{r,s}(M; E, F)$ such that, for all $u \in H^{r+s}_c(M; E)$, $v \in H^s_{\mathrm{loc}}(M; F)$ we have

$$\langle A_{\mathrm{t}}u, v\rangle_{L^2(M;F)} - \mathrm{t}[u, v] = \langle J^L_{\mathrm{t}}\rho^{r+s}u, \rho^s v\rangle_{L^2(\Gamma;F'^s)} \tag{12}$$

(b) The operator $J^L_{\mathrm{t}} = (J^L_{\mathrm{t},q_1,p_1})_{0\le p_1\le r+s-1, 0\le q_1\le s-1}$ is a matrix of differential operators of order $r + s - 1 - p_1 - q_1$. If $p_1 + q_1 > r + s - 1$, we have $J^L_{\mathrm{t},q_1,p_1} = 0$.

Assume further that the form $\mathfrak{t}$ satisfies (5). If $q_1 = r + s - 1 - p_1$ and $x' \in U_\alpha \cap \Gamma'$ for some $\alpha \in \Lambda$, we have

$$J^L_{\mathfrak{t},r+s-1-p_1,p_1}(x') = (-1)^{r-p_1} a^\alpha_{sr}(x').$$

Note Note the pecularity of our Green's formula (12): While the conventional Green's form yields an error between the action of the operators $A_\mathfrak{t}$ and $A^t_\mathfrak{t}$, as wanted in work with PDEs, searching for solutions, our Green's form (12) yields an error between the operator action and the action of the underlying differential sesquilinear form, as wanted in variational analysis, searching for critical points of functionals.

Proof If $s = 0$, the left hand side of (12) is 0 and the unique operator $J^L$is 0. For the case $s > 0$, we divide the proof into five steps.

Step 1. Local charts near the boundary. Let $\alpha \in \Lambda$ be such that $\mathrm{U}_\alpha \cap \Gamma$ is not an empty set. By the proof of [3, Proposition 1.1.2] (see also[7, Theorem 4.6.1]), we may assume the following. Denote by $\mathbb{H}^n := \{(x_1, \ldots, x_n) \in \mathbb{R}^n;\ x_1 \geq 0\}$ the half space. We have a local chart $\varphi_\alpha : U_\alpha \to \mathbb{H}^n$. For $p \in U_\alpha)$ we denote by $x_i = \varphi^i_\alpha(p)$. We assume that $x_1(p) = d(p, \Gamma)$ and $\partial_1 \perp \partial_j\,(j \geq 2)$. Denote by $x = (x_1, x')$ where $x' = (x_2, \cdots, x_n)$. Denote by $\partial^a_l = \partial^{a_2}_2 \cdots \partial^{a_n}_n$, where $a = (a_2, \cdots, a_n) \in \mathbb{N}^{n-1}$. For each $z \in \mathbb{N}$, we denote by $z = (z, 0, \cdots, 0) \in \mathbb{N}^n$.

We denote by $g' := g(x_1, \cdot)$. We have $|g| = |g'|$ if $x_1 = 0$. We denote by $d\,\mathrm{vol}(x) := \sqrt{|g|}|dx|$ the density on M and $d\,\mathrm{vol}(x') := \sqrt{|g'|}\,|dx'|$ the density on $\{x_1\} \times \Gamma$ respectively.

Step 2. Let $\alpha \in \Lambda$ be such that $\mathrm{U}_\alpha \cap \Gamma$ is not an empty set. We calculate the left hand side of (12) when $\mathrm{supp}(u) \subset U_\alpha$ is compact.

By (3), (5) and (8), we integrate by parts and get

$$\begin{aligned}
&\langle A_\mathfrak{t} u, v\rangle_{L^2(M,F)} - \mathfrak{t}[u, v] \\
&= \int_{\mathbb{H}^n} \sum_{|i|\leq r, |j|\leq s} \Big((-1)^{|j|}\sqrt{|g|^{-1}}\partial^j\big(\sqrt{|g|}A^\alpha_{ji}(x)\partial^i u(x)\big), v(x)\Big) d\,\mathrm{vol}(x) \\
&\quad - \int_{\mathbb{H}^n} \sum_{|i|\leq r, |j|\leq s} \Big(A^\alpha_{ji}(x)\partial^i u(x), \partial^j v(x)\Big) d\,\mathrm{vol}(x) \\
&= \int_{\mathbb{H}^n} \sum_{|i|\leq r, |j|\leq s} \Big((-1)^{|j|}\sqrt{|g|^{-1}}\partial^j\big(\sqrt{|g|}a^\alpha_{ji}(x)\partial^i u(x)\big), v(x)\Big)_{F_x} d\,\mathrm{vol}(x) \\
&\quad - \int_{\mathbb{H}^n} \sum_{|i|\leq r, j'\leq |j|\leq s} \sqrt{|g'|^{-1}}\Big((-1)^{|j|-j'}\partial^{j-j'}_l\big(\sqrt{|g'|}a^\alpha_{ji}(x)\partial^i u(x)\big), \partial^{j'}_1 v(x)\Big)_{F_x} d\,\mathrm{vol}(x)
\end{aligned}$$

$$= \int_{\partial\mathbb{H}^n} \sum_{\substack{|i|\le r\\ j'\le|j|\le s}} \sum_{0\le z\le j'-1} (-1)^{|j|-j'+z}$$
$$\times \left(\gamma_0\partial_1^z\big(\sqrt{|g'|}^{-1}\partial_l^{j-j'}\big(\sqrt{|g'|}a_{ji}^{\alpha}(x)\partial^i u(x)\big)\big), \gamma_0\partial_1^{j'-z-1}v(x)\right)_{F_x} d\,\mathrm{vol}(x').$$

We arrange the right hand side and get

$$\langle A_{\mathfrak{t}}u, v\rangle_{L^2(M,F)} - \mathfrak{t}[u,v]$$
$$= \int_{\partial\mathbb{H}^n} \sum_{\substack{i'\le|i|\le r\\ j'\le|j|\le s}} \sum_{0\le z\le j'-1} (-1)^{|j|-j'+z}$$
$$\times \left(\gamma_0\partial_1^z\big(\sqrt{|g'|}^{-1}\partial_l^{j-j'}\big(\sqrt{|g'|}a_{ji}^{\alpha}(x)\partial_l^{i-i'}\partial_1^{i'}u(x)\big)\big), \gamma_0\partial_1^{j'-z-1}v(x)\right)_{F_x} d\,\mathrm{vol}(x')$$
$$= \int_{\partial\mathbb{H}^n} \sum_{\substack{i'\le|i|\le r\\ j'\le|j|\le s}} \sum_{\substack{0\le z\le j'-1\\ |j''|\le|j-j'|}} (-1)^{|j|-j'+z}$$
$$\times \Big(\gamma_0\partial_1^z\big(\sqrt{|g'|}^{-1}(\partial_l^{j-j'-j''}(\sqrt{|g'|}a_{ji}^{\alpha}(x)))(\partial_l^{i-i'+j''}\partial_1^{i'}u(x))\big),$$
$$\gamma_0\partial_1^{j'-z-1}v(x)\Big)_{F_x} d\,\mathrm{vol}(x')$$
$$= \int_{\partial\mathbb{H}^n} \sum_{\substack{i'\le|i|\le r\\ j'\le|j|\le s}} \sum_{\substack{0\le z'\le z\le j'-1\\ |j''|\le|j-j'|}} (-1)^{|j|-j'+z}$$
$$\times \Big(\gamma_0\big((\partial_1^{z-z'}(\sqrt{|g'|}^{-1}(\partial_l^{j-j'-j''}(\sqrt{|g'|}a_{ji}^{\alpha}(x)))))(\partial_l^{i-i'+j''}\partial_1^{z'+i'}u(x))\big),$$
$$\gamma_0\partial_1^{j'-z-1}v(x)\Big) d\,\mathrm{vol}(x').$$

Finally we obtain

$$\langle A_{\mathfrak{t}}u, v\rangle_{L^2(M,F)} - \mathfrak{t}[u,v]$$
$$= \int_{\partial\mathbb{H}^n} \sum_{\substack{i'\le|i|\le r\\ j'\le|j|\le s}} \sum_{\substack{0\le z'\le z\le j'-1\\ |j''|\le|j-j'|}} (-1)^{|j|-j'+z}$$
$$\times \Big(\big(\gamma_{z-z'}(\sqrt{|g'|}^{-1}(\partial_l^{j-j'-j''}(\sqrt{|g'|}A_{ji}^{\alpha}(x))))\big)\big(\partial_l^{i-i'+j''}\gamma_{z'+i'}u(x')\big),$$
$$\gamma_{j'-z-1}v(x')\Big)_{F_x} d\,\mathrm{vol}(x').$$

Step 3. The formula (12) holds in the case of Step 2 if we replace $J_{\mathfrak{t}}^L$ by $J_{\mathfrak{t}}^{L,\alpha}$.

Set $p_1 := z' + i', q_1 := j' - z - 1$, where $i' \le |i| \le r$, $j' \le |j| \le s$, and $0 \le z' \le z \le j' - 1$. Then we have $0 \le q_1 \le s - 1)$ and $0 \le p_1 \le r + s - 1 - q_1 \le r + s - 1$.

For $\cdot p_1 + q_1 \leq r + s - 1$, we define the operator

$$J^{L,\alpha}_{\mathrm{t},q_1,p_1} := \sum_{\substack{i' \leq |i| \leq r \\ j' \leq |j| \leq s}} \sum_{\substack{p_1 \geq i' \\ q_1 \geq 0 \\ |j''| \leq |j - j'| \\ p_1 + q_1 \leq i' + j' - 1}} (-1)^{|j| - q_1 - 1}$$
$$\times \left(\gamma_{i'+j'-1-p_1-q_1}\left(\sqrt{|g'|}^{-1}(\partial_l^{j-j'-j''}(\sqrt{|g'|}a^{\alpha}_{ji}(x)))\right)\partial_l^{i-i'+j''}\right). \quad (13)$$

Then the operator $J^{L,\alpha}_{\mathrm{t},q_1,p_1}$ is a differential operator of order $r + s - p_1 - q_1 - 1$. If $q_1 = r + s - 1 - p_1$, we have $J^{L,\alpha}_{\mathrm{t},r+s-1-p_1,p_1} = (-1)^{r-p_1} a^{\alpha}_{sr}(x')$. If $p_1 + q_1 > r + s - 1$, we set $J^{L,\alpha}_{\mathrm{t},q_1,p_1} := 0$. We define $J^{L,\alpha}_{\mathrm{t}} : = (J^{L,\alpha}_{\mathrm{t},q_1,p_1})_{s \times (r+s)}$. By Step 2 we have

$$\langle A_{\mathrm{t}} u, v\rangle_{L^2(M,F)} - \mathrm{t}[u, v] = \int_{\partial \mathbb{H}^n} (J^{L,\alpha}_{\mathrm{t}} \rho^{r+s} u, \rho^s v)_{F^s_x} d\,\mathrm{vol}(x').$$

Step 4. Existence, uniqueness and locality of J^L_{t}.

Denote by $b^L_{\mathrm{t}}(u, v)$ the left hand side of (12). Let K be a subset of M. By Lemma 3, the set $D_K := \{\alpha \in \Lambda;\ \mathrm{supp}(\chi_\alpha) \cap K \neq \emptyset\}$ is a finite subset of Λ. Note that $A^{\alpha}_{ji} \in C^{|j|}(U_\alpha, L(\mathbb{C}^m))$ and $|g| \in C^s(U_\alpha, \mathbb{R})$. Denote by

$$X_{r+s,K} : = \bigoplus_{j=0}^{r+s-1} H^{r+s-1-j}\left(\Gamma \cap K; E'|_{\Gamma \cap K}\right),$$

$$Y_{s,K} : = \bigoplus_{i=0}^{s-1} H^{s-1-i}\left(\Gamma \cap K; F'|_{\Gamma \cap K}\right).$$

For sections $u \in H^{r+s}(K; E|_K)$ and $v \in H^s(K; F|_K)$, by (3), Step 3 and Cauchy–Buniakowsky–Schwarz inequality we have

$$\begin{aligned} |b^L_{\mathrm{t}}(u, v)| &= \sum_{\alpha \in D_K} |b^L_{\mathrm{t}}(\chi_\alpha u, v)| \\ &\leq \sum_{\alpha \in D_K} C_\alpha \|\chi_\alpha \rho^{r+s} u\|_{X_{r+s,K}} \|\rho^s u\|_{Y_{s,K}} \\ &\leq C \|\rho^{r+s} u\|_{X_{r+s,K}} \|\rho^s u\|_{Y_{s,K}} \end{aligned}$$

with some finite set $\{C_\alpha\}_{\alpha \in D_K}$ of positive constants and a single positive constant C. From the proof of [3, Theorem 1.1.4] we obtain that the maps

$$\rho^{r+s} : H^{r+s}(K; E|_K) \to X_K, \quad \rho^s : H^s(K; F|_K) \to Y_K$$

are surjective with continuous right inverse $\eta^{r+s} : X_K \to H^{r+s}(K; E|_K)$ and $\eta^s : Y_K \to H^s(K; F|_K)$ respectively. By Lemma 1, there exists a unique bounded linear operator $J^L_{\mathrm{t},K}: X_{r+s,K} \to Y_{s,K}$ such that

$$b^L_{\mathrm{t}}(u, v) = \langle J^L_{\mathrm{t},K} \rho^{r+s} u, \rho^s u \rangle_{L^2(\Gamma; F'^s)}.$$

Similar to Step 3 of the proof of Theorem 3, we define $J^L_{\mathrm{t}} u = J^L_{\mathrm{t},K} u$ for each compact subset K with $\mathrm{supp}(u) \subset K$ and obtain the existence, uniqueness and locality of J^L_{t}.

Step 5. By the locality of J^L_{t} and properties of $J^{L,\alpha}_{\mathrm{t}}$, we obtian the required properties of J^L_{t}. □

We have the following analog of Theorem 4.

Corollary 1 Assume that $k \geq \max\{r+s, r+1, s+1\}$ and $\pi_1 \colon E \to M$, $\pi_2 \colon F \to M$ are two m-dimensional Hermitian vector bundle of C^{r+s}. Then the following hold.

(a) There is a linear map which assigns a linear operator $J^R_{\mathrm{t}}: Y_{r+s} \to X_r$ to each $\mathrm{t} \in \mathrm{DSM}_{r,s}(M; E, F)$ such that, for all $u \in H^r_c(M; E)$, $v \in H^{r+s}_{\mathrm{loc}}(M; F)$ we have

$$\langle u, A^t_{\mathrm{t}} v \rangle_{L^2(M;E)} - \mathrm{t}[u, v] = \langle \rho^r u, J^R \rho^{r+s} v \rangle_{L^2(\Gamma; E'^{r+s})} \tag{14}$$

(b) The operator $J^R_{\mathrm{t}} = \left(J^R_{\mathrm{t},p_2,q_2} \right)_{0 \leq p_2 \leq r-1, 0 \leq q_2 \leq r+s-1}$ is a matrix of differential operators $J^R_{\mathrm{t}p_2,q_2}$ of order $r+s-1-p_2-q_2$. If $p_2+q_2 > r+s-1$, we have $J^R_{\mathrm{t},p_2,q_2} = 0$. Assume further that the form t satisfies (5). If $q_2 = r+s-1-p_2$ and $x' \in U_\alpha \cap \Gamma$ for some $\alpha \in \Lambda$, we have

$$J^{R,\alpha}_{\mathrm{t},r+s-1-p_2,p_2} = (-1)^{r-p_2-1} a^\alpha_{sr}(x')^t.$$

Proof Apply Theorem 4 to the form $\overline{\mathrm{t}[u, v]}$. and interchange the role of E and F. □

Corollary 2 Assume that $k \geq \max\{r+s, r+1, s+1\}$ and $\pi_1 \colon E \to M$, $\pi_2 \colon F \to M$ are two m-dimensional Hermitian vector bundle of C^{r+s}. Then the following hold.

(a) There is a linear map which assigns a linear operator $J_{\mathrm{t}} \colon X_{r+s} \to Y_{r+s}$ to each $\mathrm{t} \in \mathrm{DSM}_{r,s}(M; E, F)$ such that, for all $u \in H^{r+s}_c(M; E)$, $v \in H^{r+s}_{\mathrm{loc}}(M; F)$ we have

$$\langle A_{\mathrm{t}} u, v \rangle_{L^2(M;F)} - \langle u, A^t_{\mathrm{t}} v \rangle_{L^2(M;E)} = \langle J_{\mathrm{t}} A_{\mathrm{t}} \rho^{r+s} u, \rho^{r+s} v \rangle_{L^2(\Gamma; F'^{r+s})} \tag{15}$$

where

$$J_{\mathrm{t}} = \begin{pmatrix} J^L_{\mathrm{t}} \\ 0_{r \times (r+s)} \end{pmatrix} - \begin{pmatrix} (J^R_{\mathrm{t}})^t & 0_{(r+s) \times s} \end{pmatrix} \tag{16}$$

(b) Assume further that the form $\mathfrak{t}$ satisfies (5). If $x' \in U_\alpha \cap \Gamma$ for some $\alpha \in \Lambda$, we have

$$J^{\alpha}_{\mathfrak{t},r+s-1-p,p} = (-1)^{r-p} a^{\alpha}_{sr}(x').$$

Proof By (12) and (14) we obtain (15) for J defined by (16). The properties of $J_\mathfrak{t}$ is obtained from those of $J^L_\mathfrak{t}$ and $J^R_\mathfrak{t}$. The proof of the uniqueness of $J_\mathfrak{t}$ is the same as Step 4 of the proof of Theorem 4. □

Remark 1 Under the condition of Theorem 2, we have $r = d, s = 0$. Then we have

$$J^{R,\alpha}_{\mathfrak{t},p_2,q_2} = \sum_{i' \le |i| \le d} \sum_{\substack{0 \le p_2 \le i'-1 \\ 0 \le q_2 \le i'-1 \\ p_2+q_2 \le i'-1}} (-1)^{|i|-p_2-1} \sqrt{|g'|^{-1}} \partial_l^{i-i'} (\sqrt{|g'|} a^{\alpha}_{ji}(x)^t), \tag{17}$$

and $J^L_\mathfrak{t} = 0$. So there exists a unique linear operator

$$J_\mathfrak{t} = -(J^R_\mathfrak{t})^t : C^\infty(\Gamma; E'^d) \to C^\infty(\Gamma; F'^d)$$

makes Green formula holds. Then Theorem 2 follows.

Corollary 3 Assume that $k \ge 2d$, $r = s = d$, $E = F$ are two m-dimensional Hermitian vector bundle of class C^{2d}. Assume that the form $\mathfrak{t} \in \mathrm{DSM}_{d,d}(M; E)$ is symmetric. Then the following hold.

(a) The operator $A_\mathfrak{t}$ is formally self-adjoint and $J^L_\mathfrak{t} = J^R_\mathfrak{t}$. Moreover, the matrix $J_\mathfrak{t}$ is an upper skew-triangular matrix

$$J_\mathfrak{t} = \begin{pmatrix} J^L_\mathfrak{t} \\ 0_{d\times 2d} \end{pmatrix} - \begin{pmatrix} (J^R_\mathfrak{t})^t & 0_{2d\times d} \end{pmatrix} \tag{18}$$

of order $2d$ and $J_\mathfrak{t} = -J^t_\mathfrak{t}$.

(b) Assume further that the form $\mathfrak{t}$ satisfies (5). If $x' \in U_\alpha \cap \Gamma$ for some $\alpha \in \Lambda$, we have

$$\begin{aligned} J^{\alpha}_{\mathfrak{t},2d-1-p,p} &= \begin{cases} J^{L,\alpha}_{\mathfrak{t},2d-1-p,p}, & 0 \le p \le d-1 \\ -(J^{L,\alpha}_{\mathfrak{t},2d-1-p,p})^t, & d \le p \le 2d-1 \end{cases} \\ &= (-1)^{d-p} a^{\alpha}_{dd}(x'). \end{aligned} \tag{19}$$

Proof (a) Since the form $\mathfrak{t}$ is symmetric, by Theorem 3 we have $A_\mathfrak{t} = A^t_\mathfrak{t}$. By Theorem 4 and Corollary 1 we have $J^L_\mathfrak{t} = J^R_\mathfrak{t}$. By Corollary 2 we have $J_\mathfrak{t} = -J^t_\mathfrak{t}$.

(b) If $x' \in U_\alpha \cap \Gamma$ for some $\alpha \in \Lambda$, by (16) we obtain (19). □

Acknowledgements Alessandro Portaluri is partially supported by: "Stability in Hamiltonian dynamics and beyond" N. 2022FPZEES. Chaofeng Zhu is supported by National Key R&D Program of China (2020YFA0713300), NSFC, China Grants (11971245, 11771331), Nankai Zhide Foundation and Nankai University.

References

1. Booß-Bavnbek, B., Wojciechowski, K.P.: Elliptic boundary problems for Dirac operators. Mathematics: Theory & Applications. Birkhäuser Boston, Inc, Boston, MA (1993). https://doi.org/10.1007/978-1-4612-0337-7
2. Casas, E., Fernández, L.A.: A Green's formula for quasilinear elliptic operators. J Math Anal Appl **142**(1), 62–73 (1989)
3. Frey, C.: On Non-local Boundary Value Problems for Elliptic Operators (2005). Inaugural-Dissertation zur Erlangung des Doktorgrades der Mathematisch-Naturwissenschaftlichen Fakultät der Universität zu Köln (http://d-nb.info/1037490215/34)
4. Girsanov, I.V.: Lectures on Mathematical Theory of Extremum Problems. Lecture Notes in Economics and Mathematical Systems 67, Springer, Berlin, Heidelberg, New York (1972)
5. Grubb, G.: Green's formula and a Dirichlet-to-Neumann operator for fractional-order pseudodifferential operators. Comm. Partial. Differ. Equations **43**(5), 750–789 (2018)
6. Grubb, G.: Exact Green's formula for the fractional Laplacian and perturbations. Math. Scand. **126**(3), 568–592 (2020)
7. Hirsch, M.W.: Differential Topology, vol. 33. Springer Science & Business Media (1990)
8. Kato, T.: Perturbation Theory for Linear Operators. Springer, Heidelberg, New York (1966). second, 1976 ed.
9. Lions, J.L., Magenes, E.: Non-homogeneous Boundary Value Problems and Applications I. Springer, Berlin (1972)
10. Nazarov, S.A., Plamenevskiĭ, B.A.: A generalized Green's formula for elliptic problems in domains with edges. Differential and pseudodifferential operators., vol. 73., pp. 674–700 (1995)
11. Pedersen, G.K.: Analysis now. Graduate Texts in Mathematics, vol. 118. Springer, New York (1989)
12. Portaluri, A., Waterstraat, N.: A Morse–Smale index theorem for indefinite elliptic systems and bifurcation. J. Differ. Equations **258**, 1715–1748 (2015)
13. Witt, I.: Green's formulas for cone differential operators. Trans. Amer. Math. Soc. **359**(12), 5669–5696 (2007)
14. Wojciechowski, K.P., Scott, S.G., Morchio, G., Booß-Bavnbek, B.: Determinants, manifolds with boundary and Dirac operators. In: D., V. (ed.) Clifford Algebras and Their Apllication in Mathematical Physics, pp. 423–432. Kluwer Academic Publishers (1998)
15. Zhu, C.: A generalized Morse index theorem. In: Analysis, Geometry and Topology of Elliptic Operators, pp. 493–540. World Sci. Publ., Hackensack, NJ (2006). arXiv:math/0504126 [math.DG]

Gauss's Law, the Manifestations of Gauge Fields, and Their Impact on Local Observables

Detlev Buchholz, Fabio Ciolli, Giuseppe Ruzzi, and Ezio Vasselli

Contents

Abstract Within the framework of the universal algebra of the electromagnetic field, the impact of globally neutral configurations of external charges on the field is analyzed. External charges are not affected by the field, but they induce localized automorphisms of the universal algebra. Gauss's law implies that these automorphisms cannot be implemented by unitary operators involving only the electromagnetic field, they are outer automorphisms. The missing degrees of freedom can be incorporated in an enlargement of the universal algebra, which can concretely be represented by exponential functions of gauge fields and an abelian algebra describing the external charges. In this manner, gauge fields manifest themselves in the framework of gauge invariant observables. The action of the automorphisms on the vacuum state gives rise to representations of the electromagnetic field with vanishing global charge, which are locally disjoint from the vacuum representation.

D. Buchholz (✉)
Mathematisches Institut, Universität Göttingen, Göttingen, Germany
e-mail: buchholz@theorie.physik.uni-goettingen.de

F. Ciolli
Dipartimento di Matematica e Informatica, Università della Calabria, Rende, Italy
e-mail: fabio.ciolli@unical.it

G. Ruzzi · E. Vasselli
Dipartimento di Matematica, Università di Roma Tor Vergata, Roma, Italy
e-mail: ruzzi@mat.uniroma2.it

E. Vasselli
e-mail: ezio.vasselli@gmail.com

A. Cintio, A. Michelangeli (Eds.), *Trails in Modern Theoretical and Mathematical Physics*, https://doi.org/10.1007/978-3-031-44988-8_6

This feature disappears in the enlarged universal algebra of the electromagnetic field. The energy content of the states is well defined in both cases and bounded from below. The passage from these globally neutral states to charged states and the determination of their energy content are also being discussed.

1 Introduction

Gauss's law, relating the electric charge to the surrounding flux of the electromagnetic field, is the most distinctive feature of quantum electrodynamics. Its numerous implications for the structure of the theory have been widely discussed in the literature, most thoroughly by G. Morchio and his longtime scientific companion F. Strocchi, cf. [13] and references quoted there. One of the fundamental insights gained in these studies is the observation that field operators, creating electrically charged states as well as their Coulomb fields from a vacuum state, cannot be localized in compact spacetime regions. Charged states are limits of neutral states, involving pairs of opposite charges, where one of the charges is shifted to spacelike infinity. In contrast to massive theories, there remains in this limit a Coulomb field as an inevitable memory effect that requires a description by non-local operators.

These features are commonly attributed to the fact that quantum electrodynamics is a gauge theory. Yet gauge fields do not have a direct physical significance. So one had to clarify on one hand how these fields can be eliminated in computations of the physical observables of the theory. On the other hand, it raised the question whether the usage of gauge fields is really necessary for the formulation of the theory, or whether it is just a convenient calculational tool without further physical significance, cf. for example the novel approach to quantum electrodynamics in [15].

Morchio and Strocchi commented on this issue by a thoughtful closing remark in Ref. [14]. They wrote: "The validity of local Gauss' laws appears to have a more direct physical meaning than the gauge symmetry, which is non-trivial only on non-observable fields. It is therefore tempting to regard the validity of local Gauss' laws as the basic characteristic feature of gauge field theories, and to consider gauge invariance merely as a useful recipe for writing down Lagrangian functions which automatically lead to the validity of local Gauss' laws."

It is the aim of the present article to shed further light on this issue. In order to understand whether gauge fields are an indispensable part of the theory, one must proceed from observables, while avoiding *a priori* assumptions about non-observable structures. In the case at hand, the basic observable ingredient is the electromagnetic field, satisfying the homogeneous Maxwell equation. It is related to the electric current by the divergence of its Hodge dual. Being an observable, it has also to comply with the condition of locality (Einstein causality). It has been shown in [3] that these features can consistently be incorporated in a universal C^*-algebra of the electromagnetic field that underlies every theory of electromagnetism. This general framework provides the basis for our analysis of the problem at hand.

To analyze the effects of Gauss's law on the structure of the theory, one must first understand whether matter which carries an electric charge can be sufficiently well localized, as opposed to the infinitely extended Coulomb-like field that emanates from it. In a long-term project, Morchio, along with four close colleagues, has been working on this problem. They showed that a sharp localization in compact spacetime regions is possible in the classical Maxwell–Dirac theory and also in some non-interacting quantized models [5, 6]. It turned out, however, that realistic quantized and electrically charged matter cannot be localized in such a manner [6]. This led to a new, but rather heavy general framework covering electrically charged systems [7].

These complications are not necessary, however, to clarify the role of gauge fields. We can rely here on the idealization that the matter, carrying an electric charge, is of an external nature (*i.e.* it is not affected by interactions with the electromagnetic field). As has been shown in [4], one can then describe the impact of the electric charges on the electromagnetic field by automorphisms of the corresponding universal algebra. In case of pairs of opposite charges located in compact spacetime regions, the corresponding automorphisms are well-localized. Even though their global charge is zero, they cannot be unitarily implemented, however, by operators involving only the electromagnetic field; they are outer automorphisms. This shows that the dipole field, connecting the charges, comprises some additional degrees of freedom, indicating the presence of gauge fields. In fact, the automorphisms can be implemented by exponential functions of gauge fields in the Gupta–Bleuler framework, which thus generate the dipole field. The accompanying charged matter can be described by an abelian algebra of local fields which commute with the electromagnetic field, but have non-trivial commutation relations with the gauge charges. The resulting algebra of gauge and matter fields is an improved version of a preliminary proposal made in [4].

It is the aim of the present article to elaborate on these observations in detail. In particular, we will analyze the properties of the states which are obtained by composing the basic (charge-free) vacuum state on the universal algebra with the localized charge-containing automorphisms. The resulting states give rise to representations which are disjoint (even locally) from the basic vacuum representation, in spite of the fact that they have a vanishing global charge. This apparent conflict with the Doplicher–Haag–Roberts approach to sector analysis [8] resolves if one recalls that the local subalgebras of the universal algebra are not weakly closed, *i.e.* they are not factors of type III, as is frequently taken for granted in algebraic quantum field theory. In fact, these algebras contain a primitive two-sided ideal. It is generated by the current, which vanishes in the basic vacuum representation but becomes non-trivial in the presence of genuine charge distributions.

To remedy this undesirable feature, we enlarge the algebra by adding operators which are complementary to the current. It leads again to a version of the universal algebra. The charge-containing states on this algebra are then locally normal relative to each other. Despite these differences, the states have in both cases a well-defined energy content that is bounded from below. We will also discuss how states with non-vanishing global charge arise as limits of these globally neutral states and determine their energy content.

Models of the latter type have been previously discussed in the literature. There gauge and matter fields are generally taken as input, cf. for example [11, 12]; we do not add much to that issue. The principal subject of our investigation is the longstanding question of whether the presence of gauge fields and their physical significance can be uncovered from the local observables [10]. An affirmative answer would corroborate the view that the physically relevant information of a theory is encoded in the observables and can be extracted from them. Such an analysis has been exceedingly successful in case of fields carrying a global gauge charge in massive theories [8, 9]. However, similar convincing results have not been obtained up to now in case of local gauge fields. The present results are a modest step, pointing into the direction of an affirmative answer. Yet they are far from giving a complete solution.

Our article is organized as follows. In the subsequent Sect. 2 we recall the definition of the universal algebra of the electromagnetic field, presented in [3], and improve on the construction of local, charge-containing automorphisms and their unitary implementers, proposed in [4]. In Sect. 3, being the central part of our article, we illustrate the abstract framework by various concrete examples, based on the Gupta–Bleuler formalism. Within this setting, we discuss the properties and the relations between the resulting charge-containing representations of the universal algebra. We establish the existence of meaningful dynamics and determine the energetic properties of the charge-containing states. In Sect. 4 we discuss how charged states are obtained as limits of these globally neutral states and we summarize our findings in the conclusions.

2 Universal Algebra and External Charges

For the convenience of the reader, we recall in this section the definition and basic properties of the universal C*-algebra $\mathfrak{V}$ of the electromagnetic field, introduced in [3]. We outline how local charge measurements are described in this setting and improve on the construction of localized outer automorphisms, considered in [4]. These automorphisms describe the impact of external charges on the electromagnetic field and are shown to be unitarily implemented by the action of local gauge and matter fields.

In heuristic terms, the universal algebra is generated by exponential functions of the electromagnetic field F, which can conveniently be represented by an intrinsic (gauge invariant) vector potential A_I, *viz.* $e^{iF(f)} = e^{iA_I(g)}$. Here $f \in \mathcal{D}_2(\mathbf{R}^4)$ is any real, skew-tensor-valued test function with compact support. The corresponding function g is a solution of the equation $g = \delta f$, where $\delta = \star d \star$ is the exterior co-derivative, $\star$ denotes the Hodge operator, and d is the exterior derivative. The space of the real, vector-valued test functions g, satisfying $\delta g = 0$, is denoted by

$C_1(\mathbf{R}^4)$. Since F satisfies the homogeneous Maxwell equation, the potential $A_I(g)$ is unambiguously defined for any given $F(f)$ by Poincaré's Lemma.

The exponentials of the intrinsic vector potential are described by unitary operators $V(ag)$, where $a \in \mathbf{R}$ and $g \in C_1(\mathbf{R}^4)$; they generate a *-algebra $\mathfrak{V}_0$. These operators are subject to relations, expressing basic algebraic and locality properties of the potential, where $g. \in C_1(\mathbf{R}^4)$, $f. \in \mathcal{D}_2(\mathbf{R}^4)$,

$$V(a_1 g)V(a_2 g) = V((a_1 + a_2)g)\,, \; V(g)^* = V(-g)\,, \; V(0) = \mathbf{1}\,, \tag{1}$$

$$V(\delta f_1)V(\delta f_2) = V(\delta f_1 + \delta f_2) \;\text{ if }\; \operatorname{supp} f_1 \perp \operatorname{supp} f_2\,, \tag{2}$$

$$\lfloor V(g_1), V(g_2) \rfloor \in \mathfrak{V}_0 \cap \mathfrak{V}_0' \;\text{ if }\; \operatorname{supp} g_1 \perp \operatorname{supp} g_2\,. \tag{3}$$

Here the symbol $\perp$ between two regions indicates that they are spacelike separated, $\mathfrak{V}_0 \cap \mathfrak{V}_0'$ is the center of $\mathfrak{V}_0$, and $\lfloor X_1, X_2 \rfloor := X_1 X_2 X_1^* X_2^*$ denotes the group theoretic commutator of X_1, X_2. We refer to [3, 4] for a discussion of the physical significance of these relations.

On the algebra $\mathfrak{V}_0$ there exists a particular faithful state which has all unitaries in its kernel, apart from $\mathbf{1}$ (counting measure). The corresponding GNS-representation determines a C*-norm on this algebra. Proceeding to the completion of $\mathfrak{V}_0$ with regard to its maximal C*-norm, resulting from the set of all of its states, one arrives at the C*-algebra $\mathfrak{V}$, the universal algebra of the electromagnetic field [3]. This algebra admits an automorphic action of the proper orthochronous Poincaré group which is fixed by the relations

$$\alpha_P(V(g)) := V(g_P)\,, \quad P \in \mathcal{L}_+^\uparrow \ltimes \mathbf{R}^4\,, \; g \in C_1(\mathbf{R}^4)\,, \tag{4}$$

where $x \mapsto {g_P}^\mu(x) := L^\mu{}_\nu \, g^\nu(L^{-1}(x - y)) \in C_1(\mathbf{R}^4)$ for $P = (L, y)$.

Whereas the algebra $\mathfrak{V}$ does not include elements which create electric charges, it contains all ingredients for their analysis. To recall this fact, we resort to the heuristic picture (being meaningful in regular states) that the underlying unitaries are exponential functions of the electromagnetic field F. It determines the electric current J by the inhomogeneous Maxwell equation,

$$J(h) := F(dh) = A_I(\delta dh)\,, \quad h \in \mathcal{D}_1(\mathbf{R}^4)\,. \tag{5}$$

Here $\mathcal{D}_1(\mathbf{R}^4)$ is the space of real, vector-valued test functions with compact support. The zero component of the current determines local charge operators for suitable choices of the test functions h. Its expectation values can be changed by linear maps of the intrinsic vector potential of the form

$$A_I(g) \mapsto A_I(g) + \varphi(g)\mathbf{1}\,, \quad g \in C_1(\mathbf{R}^4)\,, \tag{6}$$

where $\varphi\colon C_1(\mathbf{R}^4) \to \mathbf{R}$ is any real linear functional. Applying these maps to the current, one obtains

$$J(h) \mapsto J(h) + \varphi(\delta dh)\mathbf{1}\,, \quad h \in \mathcal{D}_1(\mathbf{R}^4)\,. \tag{7}$$

The choice of functions $h \in \mathcal{D}_1(\mathbf{R}^4)$, corresponding to charge measurements in a given spacetime region, and of functionals $\varphi: C_1(\mathbf{R}^4) \to \mathbf{R}$, describing configurations of external charges, were discussed in detail in [4], cf. also Sect. 4. Choosing a Lorentz frame, charge measurements are described by test functions $h \in \mathcal{D}_1(\mathbf{R}^4)$ with vanishing spatial components and time components of the form $x \mapsto \tau(x_0)\chi(\boldsymbol{x})$, where τ and χ fix the time and spatial region where charges are to be determined. The resulting test functions for the intrinsic vector potential are given by

$$x \mapsto \delta dh(x) = (\tau(x_0)\boldsymbol{\Delta}\chi(\boldsymbol{x}), \, \dot{\tau}(x_0)\boldsymbol{\nabla}\chi(\boldsymbol{x})) \, . \tag{8}$$

Here $\boldsymbol{\nabla}$ denotes the spatial gradient, $\boldsymbol{\Delta}$ the Laplacian, and the dot $\dot{}$ indicates a time derivative.

The functionals of interest here have the form

$$\varphi_m(g) := -\int dx dy \; m^\mu(x) \, D(x-y) \, g_\mu(y) \, , \quad g \in C_1(\mathbf{R}^4) \, , \tag{9}$$

where $m \in \mathcal{D}_1(\mathbf{R}^4)$ and D denotes the zero mass Pauli–Jordan commutator function. Note that they vanish if m and g have spacelike separated supports. The co-derivative δm can be interpreted as a density of external charges, which are not affected by the electromagnetic field, and φ_m describes their impact on this field. (We restrict ourselves to well-behaved functions m for the sake of simplicity, but signed measures can be admitted.) Since m has compact support, the global charge determined by φ_m is zero. Yet if the support of δm consists of spacelike separated, compact regions, each of them may contain a charge which is different from zero. The external charge content in these subregions can be precisely determined by means of the local charge operators, defined above, and Gauss's law, cf. [4] and Sect. 4.

These observations can be transferred to the algebraic framework. The maps φ_m, defined in relation (9), determine automorphisms β_m of the universal algebra $\mathfrak{V}$ which act on the generating unitaries, *i.e.* the exponential functions of the electromagnetic field, according to

$$\beta_m(V(g)) := e^{i\varphi_m(g)} \, V(g) \, , \quad g \in C_1(\mathbf{R}^4) \, . \tag{10}$$

The composition of these automorphisms with Poincaré transformations P satisfies $\alpha_P \beta_m = \beta_{m_P} \alpha_P$, where $m_P(g) = m(g_{P^{-1}})$ for $g \in C_1(\mathbf{R}^4)$.

Identifying the exponential functions of the local charge operators $J(h)$ with the unitaries $V(\delta dh)$, where δdh is given by (8), it follows that

$$\beta_m(V(\delta dh)) = e^{i\varphi_m(\delta dh)} \, V(\delta dh) \, , \quad h \in \mathcal{D}_1(\mathbf{R}^4) \, . \tag{11}$$

It was shown in [4] that for functionals φ_m, based on non-trivial charge distributions δm, and for suitable test functions h, the phase factors in (11) are different from 1. In these cases the maps β_m define outer automorphisms of the universal algebra $\mathfrak{V}$.

The latter assertion follows from the existence of a vacuum representation of $\mathfrak{V}$, describing the non-interacting electromagnetic field [3]. There all unitaries $V(\delta dh)$ are represented by $\mathbf{1}$, whence all local charge operators are equal to zero. This fact excludes the existence of unitary operators in the universal algebra $\mathfrak{V}$ which implement the action of β_m in (11). In order to implement it, one must extend the algebra $\mathfrak{V}$, which is accomplished as follows.

One considers a family of operators $W(m)$, $m \in \mathcal{D}_1(\mathbf{R}^4)$, satisfying for $a_1, a_2 \in \mathbf{R}$ and $m, m_1, m_2 \in \mathcal{D}_1(\mathbf{R}^4)$ the equalities

$$W(a_1 m)W(a_2 m) = W((a_1 + a_2)m)\,, \ W(m)^* = W(-m)\,, \ W(0) = \mathbf{1}\,, \tag{12}$$

$$W(m_1)W(m_2) = W(m_1 + m_2) \ \text{ if } \ \operatorname{supp} m_1 \perp \operatorname{supp} m_2\,. \tag{13}$$

These relations encode the information that each $W(m)$ is a unitary operator with causal localization properties, determined by the support of m. One can also define Poincaré transformations, putting $\alpha_P(W(m)) = W(m_P)$, where m_P is defined similarly to (4). The assumption that the unitaries $W(m)$ induce the automorphisms β_m is encoded in the equalities

$$W(m)V = \beta_m(V)W(m)\,, \quad V \in \mathfrak{V}\,, \ m \in \mathcal{D}(\mathbf{R}^4)\,. \tag{14}$$

One then proceeds from the unitary groups generated by $V(g)$, $g \in C_1(\mathbf{R}^4)$, respectively $W(m)$, $m \in \mathcal{D}_1(\mathbf{R}^4)$, to their semi-direct product, fixed by (14). By a similar procedure as in case of the universal algebra (existence of a counting measure), one obtains a C*-algebra $\mathfrak{W} \supset \mathfrak{V}$. Note that its elements $W(m)$ only describe the effects of external charges. In addition, there could be dynamical charges present which would manifest themselves by further relations within the universal algebra of the electromagnetic field, depending on details of the dynamics.

It turns out that the unitaries $W(m)$, implementing the automorphisms β_m, are not fixed by (14). There exists an abundance of operators inducing the same action. It is an indication that the charged system bears additional local degrees of freedom. From our present point of view, this fact manifests itself by the existence of local gauge transformations. Picking any scalar distribution s on $\mathbf{R}^4$, these transformations are given by

$$\gamma_s(W(m)) := e^{i \int dx\, (ds)_\mu(x)\, m^\mu(x)}\, W(m) = e^{-i \int dx\, s(x)\, \delta m(x)}\, W(m)\,. \tag{15}$$

So γ_s acts trivially on $W(m)$ if $\delta m = 0$, and this triviality of action holds similarly for all elements of $\mathfrak{V}$.

These in the formalism still missing degrees of freedom are described by charged matter fields, which compensate for the gauge charges carried by the operators $W(m)$. Their combined action then defines a unique (*i.e.* gauge-invariant) local operation, inducing automorphisms creating the charge distributions. In the present simple case of external charges, the charged matter fields can be described by the elements of an abelian algebra. Given any real, scalar test functions $\rho \in \mathcal{D}_0(\mathbf{R}^4)$, this algebra is generated by unitary operators $\psi(\rho)$, the matter fields, which are subject to the relations

$$\psi(\rho_1)\psi(\rho_2) = \psi(\rho_1 + \rho_2), \quad \psi(\rho)^* = \psi(-\rho), \quad \psi(0) = \mathbf{1}. \tag{16}$$

The Poincaré transformations P act on the fields by $\alpha_P(\psi(\rho)) = \psi(\rho_P)$ and, most importantly, they transform under gauge transformations according to

$$\gamma_s(\psi(\rho)) = e^{i\int dx\, s(x)\rho(x)}\psi(\rho). \tag{17}$$

Thus the operators $\psi(\delta m)W(m)$ and $W(m)\psi(\delta m)$ are gauge invariant and we assume that they are equal. Yet, in contrast to the discussion in [4], we do not assume from the outset that the matter fields commute with all elements of $\mathfrak{W}$. We only require that they commute with the electromagnetic field,

$$\psi(\rho)V = V\psi(\rho), \quad \rho \in \mathcal{D}_0(\mathbf{R}^4), \ V \in \mathfrak{V}. \tag{18}$$

The physical picture behind this assumption is the idea that the impact of the external charged matter and its electromagnetic tail on the electromagnetic field is fully described by the automorphisms β_m of the universal algebra $\mathfrak{V}$. There are no other interactions between the external matter and the electromagnetic field.

The relations given above define a consistent extension of the universal algebra $\mathfrak{V}$ by gauge and matter fields. This becomes apparent if one notices that all relations involve unitary operators, *i.e.* they determine a unitary group. One can then proceed to a corresponding C*-algebra by the same token as in case of the universal algebra $\mathfrak{V}$, *i.e.* one makes use again of the existence of a faithful state (counting measure). The resulting algebra has a local net structure, fixed by the supports of the underlying test functions, Poincaré transformations act on it covariantly, and the algebra is stable under the automorphic action of gauge transformations. We refrain from proving these statements here. Instead, we present in the subsequent section a concrete representation of the present abstract algebraic framework.

3 Neutral States with Varying Charge Densities

We construct in this section concrete representations of the abstract universal algebra that describe external charge distributions with vanishing global charge. As in [4], we make use of the Gupta–Bleuler formalism. Since this setting does not fix a gauge, *i.e.* incorporates operators that generate non-trivial gauge-transformations, we must also specify commutation relations between the Gupta–Bleuler fields and the matter fields. This step was missing in [4].

We begin by briefly recalling the Gupta–Bleuler framework, cf. [17, 18]. The exponentials of the Gupta–Bleuler fields are denoted by the symbols

$$e^{iA(f)}, \quad f \in \mathcal{D}_1(\mathbf{R}^4). \tag{19}$$

They satisfy the equations for $a \in \mathbf{R}$ and $f, g \in \mathcal{D}_1(\mathbf{R}^4)$

$$e^{iA(f)} e^{iaA(g)} = e^{-i(a/2)\langle f, Dg\rangle} \, e^{iA(f+ag)}, \ \left(e^{iA(f)}\right)^* = e^{-iA(f)}, \ e^{iA(0)} = \mathbf{1}, \quad (20)$$

where we made use of the shorthand notation

$$\langle f, Dg\rangle := \int dxdy \; f_\mu(x) \, D(x-y) \, g^\mu(y) \, . \quad (21)$$

Note that we do not assume that the Gupta–Bleuler fields are solutions of the wave equation. However, since D is a bi-solution of the wave equation, the operators $e^{iA(\Box f)}$, where $\Box$ is the d'Alembertian, commute with all other operators, *i.e.* they are central elements of the Weyl algebra generated by the Gupta–Bleuler fields.

We represent now the abstract operators, defined in the preceding section, by these fields and deal with them in the remainder of this article. With this understanding we put, by some slight abuse of notation,

$$W(m) := e^{iA(m)} , \quad m \in \mathcal{D}_1(\mathbf{R}^4) \, . \quad (22)$$

Assuming that there are only external charges present, we can identify the exponentials for functions in the subspace $C_1(\mathbf{R}^4) \subset \mathcal{D}_1(\mathbf{R}^4)$ with the generating elements of the universal algebra $\mathfrak{V}$,

$$V(g) := e^{iA(g)} = W(g) \, , \quad g \in C_1(\mathbf{R}^4) \, . \quad (23)$$

In view of the underlying (20), the latter unitaries comply with relations (1) to (3). Note that the exponentials of the current $V(\delta dh)$, cf. (5), are elements of the center of $\mathfrak{V}$. The former unitaries $W(m)$ satisfy (12), (13). They generate the algebra $\mathfrak{W}$ in the present setting and induce the action of the automorphisms β_m, in accordance with (14).

The algebraic properties of the Gupta–Bleuler fields imply that for $s \in \mathcal{D}_0(\mathbf{R}^4)$ and $m \in \mathcal{D}_1(\mathbf{R}^4)$ one has

$$W(ds) W(m) W(ds)^* = e^{-i \int dx \, (Ds)(x) \, (\delta m)(x)} \, W(m) \, . \quad (24)$$

Hence the operators $W(ds)$ induce non-trivial gauge transformations of the unitaries $W(m)$ if $\delta m \neq 0$. They amount to shifts of the underlying potential by dDs. Note that the function $x \mapsto (Ds)(x)$, obtained by convolution of D with s, is a solution of the wave equation, whence not a test function.

The matter field $\psi(\rho)$, $\rho \in \mathcal{D}_0(\mathbf{R}^4)$, introduced in the preceding section, was assumed to commute with the electromagnetic field but not with gauge fields. This requires to postulate in the present setting specific commutation relations between the matter fields and the Gupta–Bleuler fields. There we rely on the fact that the Pauli–Jordan distribution D can uniquely be split into a retarded and an advanced part, $D = D_r - D_a$, which are Riesz distributions, cf. [1, Sect. 1.2]. Their products

$D_r^2 := D_r * D_r$ and $D_a^2 := D_a * D_a$, defined by convolution, also belong to this class. Making use of this fact, we impose for $m \in \mathcal{D}_1(\mathbf{R}^4)$ and $\rho \in \mathcal{D}_0(\mathbf{R}^4)$ the relations

$$W(m)\,\psi(\rho)\,W(m)^* = e^{-i\langle\delta m, D^\flat\rho\rangle}\,\psi(\rho)\,, \tag{25}$$

where

$$\langle\delta m, D^\flat\rho\rangle := \int dxdy\,(\delta m)(x)\,(D_r^2(x-y) - D_a^2(x-y))\,\rho(y)\,. \tag{26}$$

Thus $\psi(\rho)$ commutes with $W(m)$ if $\delta m = 0$. Since $\langle\delta m, D^\flat\rho\rangle = -\langle\rho, D^\flat\delta m\rangle$, it is also clear that $\psi(\delta m)$ and $W(m)$ commute for all $m \in \mathcal{D}_1(\mathbf{R}^4)$. Finally, if $m = ds$ with $s \in \mathcal{D}_0(\mathbf{R}^4)$, whence $\delta m = \Box s$, one has $\langle\delta m, D^\flat\rho\rangle = \langle s, \Box D^\flat\rho\rangle$. The relation $\Box\, D^\flat = D$, cf. [1, Prop. 1.2.4], then implies

$$W(ds)\psi(\rho)W(ds)^* = e^{i\int dx\,(Ds)(x)\rho(x)}\,\psi(\rho)\,. \tag{27}$$

Hence the unitaries $W(ds)$ induce on the matter fields the gauge transformations Ds, in accordance with their action on the potentials in (24). In particular, the operators $\psi(\delta m)W(m) = W(m)\psi(\delta m)$ are gauge invariant for any choice of $m \in \mathcal{D}_1(\mathbf{R}^4)$. So, within the present framework, the matter field has all properties postulated in the preceding section. Moreover, since $D^\flat$ is a causal distribution, the operators $\psi(\delta m)W(m)$ are local, *i.e.* their commutators vanish for test functions m with spacelike separated supports.

We denote by $\mathfrak{Z}$ the algebra that is generated by the gauge invariant operators $\psi(\delta m)W(m)$, $m \in \mathcal{D}_1(\mathbf{R}^4)$. It is a proper extension of the algebra $\mathfrak{V}$, which is generated by the restriction of these operators to the subspace of functions m satisfying $\delta m = 0$. The preceding relations and the algebraic properties of the matter fields imply, $m, m_1, m_2 \in \mathcal{D}_1(\mathbf{R}^4)$,

$$\begin{aligned}\psi(\delta m_1)W(m_1)\,\psi(\delta m_2)W(m_2) &= \eta\,\psi(\delta(m_1+m_2))W(m_1+m_2)\\ (\psi(\delta m)W(m))^* &= W(-m)\psi(-\delta m) = \psi(-\delta m)W(-m)\,,\end{aligned} \tag{28}$$

where η are phase factors. It follows that the linear span of these gauge invariant operators is norm dense in $\mathfrak{Z}$.

We rearrange their sums into groups of operators creating the same charge density: any pair of functions $m_1, m_2 \in \mathcal{D}_1(\mathbf{R}^4)$ satisfying $\delta m_1 = \delta m_2$ differs by some element of $C_1(\mathbf{R}^4)$. Thus every sum of gauge invariant operators $\sum_j c_j\,\psi(\delta m_j)W(m_j)$, where all charge densities coincide, $\delta m_j = \delta m$, can be presented with the help of relation (20) in the standard form

$$S(\delta m) := \psi(\delta m)W(m)\sum_i c_i\eta_i V(g_i)\,,\ g_i \in C_1(\mathbf{R}^4)\,. \tag{29}$$

An arbitrary sum of gauge invariant operators can be presented as a sum $\sum_j S(\delta m_j)$ of operators $S(\delta m_j)$ with different charge densities δm_j.

We are now in the position to exhibit a Poincaré invariant state ω_0 on $\mathfrak{Z}$, which extends the non-interacting vacuum state on $\mathfrak{V}$. It is given by linear extension of the functional

$$\omega_0(\psi(\delta m)W(m)) := \begin{cases} e^{(1/2)\langle m, D_+ m\rangle} & \text{if} \quad \delta m = 0 \\ 0 & \text{if} \quad \delta m \neq 0\,, \end{cases} \tag{30}$$

where the Pauli–Jordan distribution D in (21) has been replaced by its positive frequency part D_+. That ω_0 is a positive functional is a consequence of the fact that $\omega_0(S(\delta m_j)^* S(\delta m_k)) = 0$ if $\delta m_j \neq \delta m_k$. Thus

$$\omega_0\Big(\Big(\sum_j S(\delta m_j)\Big)^* \Big(\sum_k S(\delta m_k)\Big)\Big)$$
$$= \sum_j \omega_0(S(\delta m_j)^* S(\delta m_j)) = \sum_j \omega_0\Big(\Big|\sum_l c_{j,l}\eta_{j,l} V(g_{j,l})\Big|^2\Big) \geq 0\,. \tag{31}$$

This lower bound obtains since ω_0 is a state on $\mathfrak{V}$, proving that its extension to $\mathfrak{Z}$ is a state as well. Moreover, it follows from this relation that the (non-separable) Hilbert space, which arises by the GNS-representation from the functional ω_0 on $\mathfrak{Z}$, decomposes into sectors labeled by δm. These sectors are stable under the action of $\mathfrak{V}$.

It is also clear from relation (30) that ω_0 is invariant under Poincaré transformations. Hence these transformations are unitarily implemented in the GNS-representation of $\mathfrak{Z}$. But the resulting unitaries do not depend continuously on these transformations. This is a consequence of the fact that we have chosen in definition (30) as a state the matter fields the singular counting measure. Since we are primarily interested in the properties of the electromagnetic field in the presence of external charges, we do not need to examine here the question of whether there are more regular extensions of the vacuum state on $\mathfrak{V}$ to the algebra $\mathfrak{Z}$.

We turn now to the analysis of the states on $\mathfrak{V}$ which are obtained by the adjoint action of the operators $\psi(\delta m)W(m)$ on the vacuum state ω_0. Since the matter field commutes with all elements of $\mathfrak{V}$, the states are given by, cf. (14) and (10),

$$\omega_m(V(g)) := \omega_0(\beta_m(V(g)) = e^{i\varphi_m(g)}\omega_0(V(g))\,, \quad g \in C_1(\mathbf{R}^4)\,. \tag{32}$$

Considering exponentials of the current, *i.e.* $g = \delta dh$ with $h \in \mathcal{D}_1(\mathbf{R}^4)$, definition (9) yields

$$\varphi_m(\delta dh) = -\int dxdy\,(\delta m)(x)D(x-y)(\delta h)(y)\,. \tag{33}$$

Thus, unless $\delta m = 0$, the functional does not vanish for suitable test functions h. On the other hand one has $\omega_0(V(\delta dh)) = 1$ for any choice of h. It therefore follows from (32) that the states ω_m and ω_0 on $\mathfrak{V}$ are disjoint on regions of Minkowski space where $D\delta m$ is different from 0. The fact that they are disjoint can also be seen by noticing that both states are pure and that β_m acts non-trivially on the central elements of $\mathfrak{V}$.

Turning to the energetic properties of the states ω_m, $m \in \mathcal{D}_1(\mathbf{R}^4)$, we make use of the fact that the GNS-representation of $\mathfrak{V}$, induced by ω_m, is equivalent to the representation on the vacuum Hilbert given by

$$\beta_m(V(g)) := e^{i\varphi_m(g)} V(g) , \quad g \in C_1(\mathbf{R}^4) . \tag{34}$$

Assuming for a moment that the time translations on $\mathfrak{V}$ act in these representations by unitary operators e^{itH_m}, $t \in \mathbf{R}$, one would have that

$$\begin{aligned} \mathrm{Ad}(e^{itH_m})(\beta_m(V(g)) &= \beta_m(V(g_t)) \\ &= e^{i\varphi_{m_{-t}}(g)} \mathrm{Ad}(e^{itH_0})(V(g)) = \mathrm{Ad}(e^{itH_0})(\beta_{m_{-t}}(V(g)) . \end{aligned} \tag{35}$$

Thus the representations given by β_m and $\beta_{m_{-t}}$ would necessarily be equivalent. Since m and m_{-t} have different supports, this is impossible by the preceding remarks if $\delta m \neq 0$, *i.e.* in the presence of a non-trivial charge distribution. That result is also expected on physical grounds: shifting the external charges by operations involving only the electromagnetic field requires an ill-defined amount of energy. In order to cope with this problem, there are two possible strategies.

(I) One may focus on the energy carried by the transversal part of the electromagnetic field (the photons) and ignore the energy needed to shift the elements of the center of $\mathfrak{V}$, such as the current (5). This is accomplished by relying on the operator determining the energy density in the vacuum representation of $\mathfrak{V}$. It is given by the (normal ordered) operator $(1/2) :\boldsymbol{E}^2 + \boldsymbol{B}^2:$, where $\boldsymbol{E}^j := F^{0j}$ and $\boldsymbol{B}^j := (1/2)\epsilon_{jkl} F^{kl}$ are the components of the electric, respectively magnetic, field, $j, k, l = 1, 2, 3$. This density is a local observable which can be used in any representation of $\mathfrak{V}$ in order to determine the energy content of the transversal part of the electromagnetic field. The density commutes with the elements of the center of $\mathfrak{V}$ and thus is insensitive to their energetic properties.

(II) Alternatively, one can enlarge the universal algebra $\mathfrak{V}$ to a gauge invariant algebra with non-trivial center. This is accomplished by fixing a Lorentz frame and proceeding to the algebra $\mathfrak{W}_0$ that is generated by the exponentials of $(\boldsymbol{A} - \nabla \xi)$. These operators consist of the spatial components of the vector potential, amended by the generator ξ of the matter field, *i.e.* $\psi(\rho) = e^{i\,\xi(\rho)}$. Putting $\boldsymbol{E} := -(\dot{\boldsymbol{A}} - \nabla \dot{\xi})$ and $\boldsymbol{B} := \nabla \times (\boldsymbol{A} - \nabla \xi)$, it is apparent that $\boldsymbol{E}$ and $\boldsymbol{B}$ satisfy the homogeneous Maxwell equations, hence they also generate a concrete version of the universal algebra $\mathfrak{V} \subset \mathfrak{W}_0$. It follows from relation (25) and the commutativity of the matter fields that the three components of $(\boldsymbol{A} - \nabla \xi)$ satisfy canonical commutation relations with their time derivatives. Apart from achieving gauge invariance, the matter field ξ has no further algebraic impact on the electromagnetic field and potential. We can therefore proceed from $(\boldsymbol{A} - \nabla \xi)$ to $\boldsymbol{A}$ again in the following discussion.

Disregarding the action of Lorentz transformations, the fields $\boldsymbol{A}$ and $\dot{\boldsymbol{A}}$ may be regarded as the canonical data of three scalar massless fields, satisfying the wave equation. It is clear then that there exists a vacuum state on $\mathfrak{W}_0$. The resulting time translations on $\mathfrak{W}_0$ preserve the subalgebra $\mathfrak{V} \subset \mathfrak{W}_0$, generated by $\boldsymbol{E} = -\dot{\boldsymbol{A}}$ and $\boldsymbol{B} = \nabla \times \boldsymbol{A}$. It acts covariantly on the current, which is no longer an element of the center of $\mathfrak{V}$, however. As a matter of fact, a covariant action on an abelian algebra would not be compatible with the existence of a positive generator of the time translations according to the Borchers–Arveson theorem [2, Thm. 3.2.46]. In the present case, this generator is obtained by integration of the energy density $(1/2)\, {:}\boldsymbol{E}^2 + \sum_k \nabla A_k\, \nabla A_k{:}$ at fixed time over space. With this input, one can then study the dynamics and energy of both, the electromagnetic field and the external charges, which are created by the automorphisms β_m.

We discuss in the following both strategies. Turning to approach (I), we make use of the normal ordering procedure in Minkowski space, relying on the local fields, *i.e.* we put

$$\begin{aligned} &{:}\boldsymbol{E}^2 + \boldsymbol{B}^2{:}\,(x) := \lim_{\varepsilon} \big(\boldsymbol{E}(x+\varepsilon)\boldsymbol{E}(x-\varepsilon) + \boldsymbol{B}(x+\varepsilon)\boldsymbol{B}(x-\varepsilon) \\ &- \omega_0(\boldsymbol{E}(x+\varepsilon)\boldsymbol{E}(x-\varepsilon) + \boldsymbol{B}(x+\varepsilon)\boldsymbol{B}(x-\varepsilon))\mathbf{1}\,. \end{aligned} \tag{36}$$

Here ε is a suitable sequence of spacelike translations, tending to 0. In any state, where this sequence converges to an operator-valued distribution, it defines its electromagnetic energy density relative to the vacuum representation. Proceeding to the representation β_m of $\mathfrak{V}$ on the vacuum Hilbert space, we obtain for the energy density after a straightforward computation

$$\begin{aligned} &\beta_m\big((1/2)\,{:}\boldsymbol{E}^2 + \boldsymbol{B}^2{:}\,(x)\big) \\ &= (1/2)\,{:}\boldsymbol{E}^2 + \boldsymbol{B}^2{:}\,(x) + \boldsymbol{E}(x)\big(\partial_0 \underline{\boldsymbol{m}} - \nabla \underline{m}_0\big)(x) + \boldsymbol{B}(x)\big(\nabla \times \underline{\boldsymbol{m}}\big)(x) \\ &+ (1/2)\big((\partial_0 \underline{\boldsymbol{m}} - \nabla \underline{m}_0)^2 + (\nabla \times \underline{\boldsymbol{m}})^2\big)(x)\,\mathbf{1}\,. \end{aligned} \tag{37}$$

Here $\underline{m} := D\,m$, thus $\underline{m}$ is a solution of the wave equation, and $\underline{m}_0$ and $\underline{\boldsymbol{m}}$ are its time and spatial components. The spatial integral of this density at any given time t is well defined as a quadratic form. Making use of the fact that on the vacuum Hilbert space one has $\boldsymbol{\delta}\boldsymbol{E} = 0$ and $\nabla \times \boldsymbol{B} = \partial_0 \boldsymbol{E}$, one obtains by partial integration

$$\begin{aligned} H_m(t) &:= (1/2) \int d\boldsymbol{x}\; \beta_m\big({:}\boldsymbol{E}^2 + \boldsymbol{B}^2{:}\,(t,\boldsymbol{x})\big) \\ &= H_0 + \int d\boldsymbol{x}\, \big(\boldsymbol{E}\,\partial_0 \underline{\boldsymbol{m}} - (\partial_0 \boldsymbol{E})\,\underline{\boldsymbol{m}}\big)(t,\boldsymbol{x}) \\ &+ (1/2) \int d\boldsymbol{x}\, \big((\partial_0 \underline{\boldsymbol{m}} - \nabla \underline{m}_0)^2 + (\nabla \times \underline{\boldsymbol{m}})^2\big)(t,\boldsymbol{x})\,\mathbf{1}\,. \end{aligned} \tag{38}$$

Since both, $\boldsymbol{E}$ and $\underline{\boldsymbol{m}}$, are solutions of the wave equation, the term in the second line does not depend on t. Thus the time dependence of the Hamiltonian $H_m(t)$ is completely contained in the c-number contribution. It describes the mean energy of the perturbed vacuum state. Apart from this term, $H_m(t)$ is constant in time, so it induces a time independent (autonomous) dynamics of the transversal components of the electromagnetic field. But it leaves the current invariant (being an element of the trivially represented center of $\beta_m(\mathfrak{V})$), hence the obstructions arising from relation (35) do not come into play.

Dynamics of this type have been thoroughly discussed in the literature by functional analytic methods, cf. for example [16] and the discussion in Sect. 4. In the present algebraic setting their properties follow more easily from the representation of the algebra $\mathfrak{Z}$, considered above. Restricting the algebra $\mathfrak{V}$, acting on the non-separable representation space, to the sector corresponding to $\delta m = 0$, one obtains the standard irreducible vacuum representation of the electromagnetic field. There the time translations are implemented by a continuous unitary group $t \mapsto U_0(t)$ with positive generator. By the adjoint action of the unitary operators $W(m)\psi(\delta m)$ on this group one obtains continuous unitary representations $t \mapsto U_m(t)$ of the time translations on the algebras $\beta_m(\mathfrak{V})$ in the sectors attached to δm, $m \in \mathcal{D}_1(\mathbf{R}^4)$. Their generators coincide with the operators $H_m(t)$ given in (38), apart from the c-number term.

Turning to the second approach (II), the corresponding normal-ordered energy density is defined similarly as in (36), where ω_0 is to be replaced by the extended vacuum functional on $\mathfrak{W}_0$. And, similarly as in the preceding discussion, we consider the representations β_m of $\mathfrak{W}_0$ on the corresponding vacuum Hilbert space. In contrast to approach (I), these representations are now given by the adjoint action of unitary operators, *i.e.* they are equivalent to the vacuum representation of $\mathfrak{W}_0$. Hence, the embedding of $\mathfrak{V}$ into the bigger algebra $\mathfrak{W}_0$, involving gauge fields, cures the subtle dependence of its representations on δm.

This assertion requires a comment if m has a non-trivial zero-component m_0 since the component A_0 of the Gupta–Bleuler field is not represented on the underlying Hilbert space. The corresponding map β_{m_0} does not change $\boldsymbol{A}$, but it induces an action on the electric field $\boldsymbol{E}$ given by

$$\beta_{m_0}(e^{i\boldsymbol{E}(\boldsymbol{f})}) = e^{i\int dx\, \underline{m_0}(x)(\delta f)(x)}\, e^{i\boldsymbol{E}(\boldsymbol{f})}\,, \quad f \in \mathcal{D}_1(\mathbf{R}^4)\,, \tag{39}$$

where $\boldsymbol{f}$ are the spatial components of f. It turns out that in the present representation these maps are induced by exponentials involving only the spatial components of the vector potential. To see this, we make use of the fact that the map $\boldsymbol{g} \mapsto e^{i\boldsymbol{A}(\boldsymbol{g})}$ is continuous in the strong operator topology for $g \in \mathcal{D}_1(\mathbf{R}^4)$ varying continuously with respect to the single particle seminorm

$$\|\boldsymbol{g}\|_0^2 := \int \frac{d\boldsymbol{p}}{2|\boldsymbol{p}|}\, |\widetilde{\boldsymbol{g}}(|\boldsymbol{p}|, \boldsymbol{p})|^2\,, \quad g \in \mathcal{D}_1(\mathbf{R}^4)\,. \tag{40}$$

Note that the kernel of this seminorm consists of functions which vanish on the zero-mass shell. So the fields, being solutions of the wave equation, also vanish on these functions. Whence the unitary operators $e^{i\boldsymbol{A}(\boldsymbol{g})}$ can continuously be extended to all real functions in the single particle space. After a moment of reflection, it follows that the (real, vector-valued) function

$$x \mapsto \boldsymbol{n}(x) := \boldsymbol{\nabla}\partial_0\boldsymbol{\Delta}^{-1} m_0(x) \tag{41}$$

is contained in this space and the adjoint action of the unitaries $e^{i A(\boldsymbol{n})}$ on $\mathfrak{W}_0$ coincides with the action of β_{m_0}. Thus all automorphisms β_m, $m \in \mathcal{D}_1(\mathbf{R}^4)$, are unitarily implemented in the vacuum representation of $\mathfrak{W}_0$, as claimed.

In view of this result it is clear from the outset that the time translations are unitarily implemented in all representations β_m of $\mathfrak{W}_0$ and their generators are bounded from below. Nevertheless, it is of interest to have a look at the corresponding Hamiltonians and to compare them with the results in (I). This is accomplished by noting that the unperturbed energy density, given above, can be recasted, disregarding partial derivatives of local operators which vanish upon integration over all space. One has, up to such negligible terms,

$$(1/2)\,{:}\boldsymbol{E}^2 + \sum_k \nabla A_k\, \nabla A_k{:}\,(x) \;=\; (1/2)\,{:}\boldsymbol{E}^2 + \boldsymbol{B}^2 + (\delta A)^2{:}\,(x)\,. \tag{42}$$

Comparing this with (I), it is apparent that the energy density related to proper time translations of the current is encoded in the last term. The spatial integral of this density yields the Hamiltonian in the vacuum representation of $\mathfrak{W}_0$. The Hamiltonians in the representations β_m are obtained from it by the adjoint action of the unitaries $e^{i A(\boldsymbol{n}-\boldsymbol{m})}$, where $\boldsymbol{m}$ are the spatial components of m and $\boldsymbol{n}$ was defined in (41). In particular, the mean energy of the perturbed vacuum states is in general bigger than the proportion of the transversal electromagnetic field, determined in (I). We refrain from presenting here these straightforward computations.

Let us point out in conclusion that the case of external charges, considered here in the Gupta–Bleuler formalism, can also be studied in other approaches to the treatment of gauge fields, which inevitably accompany the charges. The Gupta–Bleuler formulation has the advantage that the localization properties of charge-containing operators can be described in a simple, transparent manner. It leads to a completely local formulation in case of states with vanishing global charge. In the subsequent section we will show how states carrying a non-vanishing global charge can be approximated by these neutral states, thereby developing the long-range localization, known from physical gauges.

4 Passage to Charged States

Using the framework of the preceding Sect. 3, we consider now sequences of neutral states which describe bi-localized external charge distributions. Thereby, one of these distributions is kept fixed and the other (compensating) charge distribution is moved to spacelike infinity. In this way, the influence of the compensating charges on local observables is suppressed and the resulting limit states are charged. We adopt in this analysis arguments given in [4], where we need to appropriately increase the localization regions occupied by the compensating charges in order to maintain control on the energetic properties of the states. As a result, the charged limit states typically differ from the vacuum state in spacelike cones, where fields with an asymptotic behavior of Coulomb-type appear.

Proceeding to the construction, the essential step consists of the proof of existence of suitable functions $m \in \mathcal{D}_1(\mathbf{R}^4)$ that enter in the automorphisms β_m. We begin with the distributions, where $a, b \in \mathbf{R}^4$,

$$x \mapsto m^\mu(a,b)(x) := (b-a)^\mu \int_0^1 du\, \delta(ub + (1-u)a - x)\,. \tag{43}$$

They satisfy the equation $\partial_\mu m^\mu(a,b)(x) = \delta(x-a) - \delta(x-b)$, where δ is the four-dimensional Dirac measure. Integrating $m^\mu(a,b)$ with test functions depending on a, b and having compact support, one obtains a test function with regard to x whose support is contained in the convex hull of all pairs of points a, b in the support of the chosen functions. We pick now a smooth charge distribution $a \mapsto \vartheta(a)$ with compact support, which is kept fixed, and real test functions $b \mapsto \sigma(b)$ whose supports will be moved to spacelike infinity, while their integrals are kept equal to 1. Putting

$$x \mapsto m(\sigma)^\mu(x) := \int da\,db\, \vartheta(a)\sigma(b-a)\, m^\mu(a,b)(x)\,, \tag{44}$$

equation (43) implies

$$\partial_\mu m(\sigma)^\mu(x) = \vartheta(x) - \vartheta * \sigma(x)\,. \tag{45}$$

In the situation of interest here, these distributions occupy two spacelike separated regions, where both components have fixed spacetime integrals with opposite signs. So they compensate each other.

The charge content in the regions can be determined by the current (5) with test functions of the form $x \mapsto h(x) := \tau(x_0)\chi(\boldsymbol{x})$. Here one chooses smooth characteristic functions χ of the regions in $\mathbf{R}^3$ in which the charge is to be determined and test functions τ with support in a small time interval about the time where this is to happen; the integrals of the latter functions are to be equal to 1.

Let $\mathcal{O}_h$ be the causal completion of the convex hull of $\operatorname{supp} h$. It is the spacetime region where the operation of charge measurement takes place. Due to the unavoidable choice of smoothed characteristic functions χ and Dirac measures τ, the charge measurements become effective only in certain specific subregions of $\mathcal{O}_h$. These subregions come close to the surrounding region $\mathcal{O}_h$ by a suitable choice of χ and τ, cf. [4]. Whenever a region $\mathcal{O}_0 \subset \mathbf{R}^4$ is contained in such a subregion, we write $\mathcal{O}_0 \Subset \mathcal{O}_h$. By arguments given in the proof of [4, Lem. 2.2], one obtains for the functionals (9), depending on the given charge distribution (45),

$$\varphi_{m(\sigma)}(h) = \begin{cases} \int dx\, \vartheta(x) & \text{if } \operatorname{supp}\vartheta \Subset \mathcal{O}_h\,,\ \operatorname{supp}\vartheta * \sigma \perp \mathcal{O}_h \\ 0 & \text{if } \operatorname{supp}\vartheta \cup \operatorname{supp}\vartheta * \sigma \Subset \mathcal{O}_h\,, \\ -\int dx\, \vartheta(x) & \text{if } \operatorname{supp}\vartheta * \sigma \Subset \mathcal{O}_h\,,\ \operatorname{supp}\vartheta \perp \mathcal{O}_h \end{cases} \tag{46}$$

These relations are a distinctive consequence of Gauss's law. Since the regions $\mathcal{O}_h$ are bounded, it is apparent that by moving the support of σ to spacelike infinity, only the charge in supp ϑ remains visible in the limit. Thus, composing the vacuum functional with the automorphisms $\beta_{m(\sigma)}$, the resulting limits describe charged states. That these limit states exist for suitable sequences of σ can be seen as follows.

Picking any test function σ with compact support in the spacelike complement of the origin of $\mathbf{R}^4$ and integral 1, one proceeds to the scale-transformed functions $x \mapsto \sigma_r(x) := (1/r^4)\,\sigma(x/r)$ for $r > 0$. The support of these functions approaches spacelike infinity in the limit of large r and the resulting functionals $\varphi_{m(\sigma_r)}$ converge on $\mathcal{D}_1(\mathbf{R}^4)$ in this limit. For the proof one needs to determine the Fourier transforms of $m(\sigma_r)$. Putting $n := (1/|p|)\,p$, where $|p|$ is the Euclidean length of p, and $b \mapsto \sigma^\mu(b) := \sigma(b)b^\mu$, they are given by

$$p \mapsto \widetilde{m}(\sigma_r)^\mu(p) = \widetilde{\vartheta}(p) \int db\, \sigma^\mu(b)\, \frac{e^{irbp} - 1}{ibp}$$

$$= (2\pi)^2\, \widetilde{\vartheta}(p)\, |p|^{-1} \int_0^{r|p|} du\, \widetilde{\sigma}^\mu(u\, n)\,. \tag{47}$$

Thus one obtains for $p \neq 0$ in the limit of large r the singular function

$$p \mapsto \widetilde{m}(\sigma_\infty)^\mu(p) = (2\pi)^2\, \widetilde{\vartheta}(p)\, |p|^{-1} \varrho^\mu(n)\,, \tag{48}$$

where $n \mapsto \varrho^\mu(n) := \int_0^\infty du\, \widetilde{\sigma}^\mu(un)$ is continuous and bounded. By a routine computation this yields for $f \in \mathcal{D}_1(\mathbf{R}^4)$

$$\varphi_{m(\sigma_\infty)}(f) := -\lim_{r\to\infty} \int dx dy\, m(\sigma_r)^\mu(x)\, D(x-y)\, f_\mu(y)$$

$$= -i\,(2\pi)^3 \int dp\, \epsilon(p_0)\delta(p^2)\, \tilde{\vartheta}(p)|p|^{-1}\varrho^\mu(n)\, \widetilde{f}_\mu(-p)$$

$$= (2\pi)^3\, \mathrm{Im}\Big(\int \frac{d\boldsymbol{p}}{\sqrt{2}\,|\boldsymbol{p}|^2}\, \widetilde{\vartheta}(\underline{p})\varrho^\mu(\underline{n})\widetilde{f}_\mu(-\underline{p}) \Big)\,. \tag{49}$$

Here we have put $\underline{p} := (|\boldsymbol{p}|, \boldsymbol{p})$ and $\underline{n} := (1/\sqrt{2})(1, \boldsymbol{p}/|\boldsymbol{p}|)$. The singularity in the integral on the last line is absolutely integrable, so the limit functionals are well-defined on $\mathcal{D}_1(\mathbf{R}^4)$.

It follows that the automorphisms $\beta_{m(\sigma_r)}$ converge pointwise on $\mathfrak{V}$ in the strong operator topology for asymptotic r on the vacuum Hilbert space. The states appearing in the resulting limit representation carry the global charge $\int dx\, \vartheta(x)$. Yet the underlying unitary operators $W(m(\sigma_r))\psi(\delta m(\sigma_r))$, which implement the automorphisms, do not have a meaningful limit. Thus, in order to determine the energetic properties of the limit states, one cannot proceed as in Sect. 3. Instead, one has to rely on functional analytic methods.

We consider first the electromagnetic energy content of the limit states, as described in approach (I) in Sect. 3. In doing so, we make use of the fact that the middle term in the Hamiltonian (38) remains unchanged if one replaces the spatial components $\boldsymbol{m}$ of m by

$$x \mapsto \boldsymbol{g}_m(x) := (\boldsymbol{m} - \boldsymbol{\nabla}\boldsymbol{\Delta}^{-1}\boldsymbol{\delta m})(x) \,. \tag{50}$$

The gradient term does not contribute to the integral as one sees by partial integration, making use of the fact that $\boldsymbol{\delta E} = 0$ in the vacuum representation of $\mathfrak{V}$. The functions $\boldsymbol{g}_m$ are smooth and satisfy $\boldsymbol{\delta g}_m = 0$. Moreover, their seminorm (40) is finite. So the unitary operators $V(g)$, $g \in C_1(\mathbf{R}^4)$, can continuously be extended to unitary operators $V(\boldsymbol{g}_m)$ in the strong operator topology on the vacuum Hilbert space. By a routine computation, one finds that the adjoint action of $V(\boldsymbol{g}_m)$ on the density (26) yields the Hamiltonian (36) after integration, disregarding again the c-number term.

Given any sequence $m(\sigma_r)$, $r > 0$, these observations put us into the position to use the results in [16] for the determination of the electromagnetic energy content of the limit states. Whereas the functions $\boldsymbol{g}_{m(\sigma_r)}$ do not converge for large r with regard to the seminorm (40), they have limits with regard to its mollified version,

$$\|\boldsymbol{g}_{m(\sigma_r)}\|_1^2 := \int \frac{d\boldsymbol{p}}{2|\boldsymbol{p}|} \left(\frac{\boldsymbol{p}^2}{1+\boldsymbol{p}^2}\right)^{1/2} |\widetilde{\boldsymbol{g}}_{m(\sigma_r)}(|\boldsymbol{p}|, \boldsymbol{p})|^2 \,. \tag{51}$$

(The kernel of this seminorm likewise consists of functions that do not contribute in the fields.) Due to the mollifying factor in the integral (51), the singularity of the limit function $\widetilde{\boldsymbol{g}}_{m(\sigma_\infty)}$ at zero momentum is tamed, cf. (48). In fact, the sequence $\widetilde{\boldsymbol{g}}_{m(\sigma_r)}$, $r > 0$, converges with respect to the mollified seminorm to $\widetilde{\boldsymbol{g}}_{m(\sigma_\infty)}$.

Making use of arguments in [16, Prop. 3], it follows that the energy densities, obtained by applying the automorphism $\beta_{m(\sigma_r)}$ to the vacuum density (36), determine by integration over all space a meaningful limit dynamics. In more detail: let H_0 be the Hamiltonian in the vacuum representation of $\mathfrak{V}$. Because of the convergence properties of the functions $\boldsymbol{g}_{m(\sigma_r)}$ for asymptotic r, the sequence of cocycles

$$\begin{aligned} t \mapsto V(\boldsymbol{g}_{m(\sigma_r)})\, e^{itH_0} V(\boldsymbol{g}_{m(\sigma_r)})^* e^{-itH_0} &= V(\boldsymbol{g}_{m(\sigma_r)}) V(\boldsymbol{g}_{m(\sigma_r),t})^* \\ &= e^{-(i/2)\langle \boldsymbol{g}_{m(\sigma_r)},\, D\,(\boldsymbol{g}_{m(\sigma_r)} - \boldsymbol{g}_{m(\sigma_r),t})\rangle}\, V(\boldsymbol{g}_{m(\sigma_r)} - \boldsymbol{g}_{m(\sigma_r),t}) \end{aligned} \tag{52}$$

converges in this limit in the strong operator topology. Note that the Fourier transforms of the differences $(\boldsymbol{g}_{m(\sigma_r)} - \boldsymbol{g}_{m(\sigma_r),t})$ vanish at zero momentum and converge for large r with regard to the single particle seminorm (40), uniformly on compact subsets of t. These facts imply that the limit dynamics exists in approach (I). It is given as a limit in the strong operator topology

$$\begin{aligned} t \mapsto e^{itH_{(I)}(m(\sigma_\infty))} &:= \lim_{r \to \infty} V(\boldsymbol{g}_{m(\sigma_r)})\, e^{itH_0}\, V(\boldsymbol{g}_{m(\sigma_r)})^* \\ &= \lim_{r \to \infty} V(\boldsymbol{g}_{m(\sigma_r)}) V(\boldsymbol{g}_{m(\sigma_r),t})^*\, e^{itH_0} \,. \end{aligned} \tag{53}$$

The limit dynamics is strongly continuous, has a positive generator, but leaves the center of $\beta_{\boldsymbol{m}(\sigma_\infty)}(\mathfrak{V})$ pointwise fixed. So it does not act properly on the charge operators. But it fully describes the dynamics and energy content of the transversal components of the electromagnetic field in the charged sectors.

In a similar manner one proceeds in approach (II). There the dynamics in the vacuum sector of $\mathfrak{W}_0$ leaves $\mathfrak{V} \subset \mathfrak{W}_0$ invariant and acts covariantly on the current. The representations in the locally charged sectors are obtained by acting on $\mathfrak{W}_0$ in the vacuum sector with the automorphisms β_m. One then chooses sequences of functions $m(\sigma_r)$, describing as above bi-localized charge distributions, and considers the resulting automorphisms $\beta_{m(\sigma_r)}$, $r > 0$. As was explained in Sect. 3, these automorphisms are implemented on the vacuum Hilbert space by unitary operators $e^{iA(\boldsymbol{n}(\sigma_r)-\boldsymbol{m}(\sigma_r))}$, $r > 0$. Here $\boldsymbol{m}(\sigma_r)$ denotes again the spatial components of $m(\sigma_r)$, and $\boldsymbol{n}(\sigma_r)$ is determined by the time component $m_0(\sigma_r)$ according to (41).

One then determines the Fourier transforms of $(\boldsymbol{n}(\sigma_r) - \boldsymbol{m}(\sigma_r))$ and finds that these sequences have similar convergence properties as those of the sequences $g_{\boldsymbol{m}(\sigma_r)}$ in approach (I), $r > 0$. They also develop a singularity at zero momentum in the limit whose restriction to the zero-mass shell has the same form as in (48). Thus, by applying the preceding arguments, one arrives at the conclusion that the limit dynamics exists in approach (II) in all charged sectors. It is given by

$$t \mapsto e^{itH_{(\mathrm{II})}(m(\sigma_\infty))} := \lim_{r\to\infty} e^{iA(\boldsymbol{n}(\sigma_r)-\boldsymbol{m}(\sigma_r))}\, e^{itH_0}\, e^{-iA(\boldsymbol{n}(\sigma_r)-\boldsymbol{m}(\sigma_r))}\,, \tag{54}$$

where H_0 is the Hamiltonian in the vacuum representation of $\mathfrak{W}_0$. The resulting unitary group is continuous and has a positive generator. We omit the proof since it completely parallels the preceding discussion.

Let us finally mention that the globally charged representations, obtained in this manner, are locally normal relative to the vacuum representation. Yet, depending on the choice of the approximating functions $m(\sigma_r)$, $r > 0$, there exists an abundance of limit states that carry the same global charge, but are mutually disjoint, nevertheless. These states differ by the asymptotic configurations of the electromagnetic field or, alluding to the particle picture, by infinite clouds of low energy photons. Since these "infrared problems" are widely known, we do not dwell here on this issue any further.

5 Conclusions

Proceeding from the universal algebra of the electromagnetic field, providing a general framework for the discussion of electromagnetism, we have studied the question of whether the presence of gauge fields can be uncovered from the gauge invariant observables. All basic features of the electromagnetic field are encoded in this algebra, in particular the homogeneous Maxwell equations, the existence of a current operator, and the condition of locality. These features are conveniently expressed in a C*-algebraic framework, from which the underlying unbounded field operators can be recovered in regular representations. In order to uncover in this setting possible traces of gauge fields, we have studied the impact of compactly localized external charge distributions on the algebra; they give rise to a vanishing global charge. These charge distributions can be traced and analyzed by means of the current operator in the algebra, making use of Gauss's law.

It turned out that the external charge distributions induce outer automorphisms of the universal algebra. This fact reveals the presence of local degrees of freedom, which are not incorporated in the algebra. In order to represent the automorphisms by the adjoint action of unitary operators, describing these additional degrees of freedom, the universal algebra had to be extended. Yet this first extension did not unambiguously describe the unitaries: they can be modified by an abundance of local transformations without changing their action on the observables. These transformations can be interpreted as local gauge transformations, whereby the generating elements of the extended algebra correspond to the exponentials of gauge fields.

In order to completely fix the unitary implementers (up to phase factors), one has to introduce further operators, which can be interpreted as charged matter fields. They may be of a bosonic, fermionic, or classical nature. In the present case of external charges, we considered the simple case of classical charged matter. It affects the quantized electromagnetic field only through the accompanying tail of gauge fields. The gauge invariant combinations of gauge and matter fields define a physically meaningful extension of the universal algebra. It is generated by local, Poincaré covariant implementers of the charge-containing automorphisms. The appearance of gauge and matter fields is thus encoded in the structure of the universal algebra and manifests itself already in the simple case of external charges.

We have illustrated this abstract framework by representing the gauge and matter fields concretely in a setting of Gupta–Bleuler type. There the universal algebra is canonically embedded in the Weyl algebra generated by the Gupta–Bleuler fields. Moreover, the framework is enriched by some dynamical input, deriving from the underlying wave equation. In this formalism the current does not vanish, but it is affiliated with the center of the universal algebra as a consequence of the dynamical input. Since the Gupta–Bleuler fields incorporate operators which induce gauge transformations, we had also to impose for the sake of consistency commutation relations between the Gupta–Bleuler and matter fields. To the best of our knowledge, these commutation relations have not been considered before in the literature.

Applying the implementers of the charge-containing automorphisms to the non-interacting vacuum state of the universal algebra, one obtains further irreducible representations of this algebra. They are mutually disjoint, even locally, for different choices of charge distributions. Nevertheless, there exist energy operators in these representations that are bounded from below and induce a dynamics on the transversal degrees of freedom of the electromagnetic field (the photons). Let us mention as an aside that these dynamics, together with their spectral properties, exist in all Lorentz frames. In other words, the spacetime translations are unitarily implemented in the representations and satisfy the relativistic spectrum condition. However, they act trivially on the current operator which, being an element of the center, is represented by multiples of the identity.

It is clear from the outset that one cannot have a dynamics with positive generator that acts non-trivially on an abelian algebra. We have therefore enlarged the concretely given universal algebra to ensure a covariant action of the dynamics on the current. This enlargement was accomplished by adding to the algebra the spatial components of the Gupta–Bleuler field. The resulting algebra complies again with the defining properties (axioms) of the universal algebra. As a result of this step, all representations, obtained by applying the charge-generating unitary operators to the vacuum state of the extended algebra, become unitarily equivalent. Moreover, the resulting dynamics has a positive generator, it leaves the universal algebra invariant, and it acts covariantly on the current. Again, this property obtains in all Lorentz frames.

For completeness, we have also discussed the passage from neutral states to states carrying a global charge. This was accomplished by the well-known method to create pairs of compensating charges and to move one of them to spacelike infinity. Just as for localized charge distributions, the resulting representations of the universal algebra have a dynamics with positive generator. In case of the extended algebra, the charged representations are also locally normal with respect to each other. But, for given value of the global charge, they are in general disjoint, remembering the localization properties of the approximating states. This feature is in accordance with the poor localization properties of charged states, known from physical gauges. It is the origin of the notorious infrared problems in quantum electrodynamics.

In summary, starting from the local gauge invariant observables, we have established the occurrence of gauge fields as an inherent feature of the quantized electromagnetic field. First of all, gauge fields inevitably accompany operations that generate local and global charge distributions. They must be amended by local charged matter fields which compensate the gauge charges carried by the gauge fields. Secondly, the gauge fields also enter in a fundamental way into the description of the dynamics. They are needed to describe the temporal evolution of the current operator, which is essential for the verification of Gauss's law. Thus, whereas the gauge fields cannot be observed directly, their impact on observables can be determined by observations. In this way, they become an integral part of physics.

Acknowledgements DB gratefully acknowledges the support and hospitality of Roberto Longo and the University of Rome Tor Vergata, which made this collaboration possible. He is also grateful to Dorothea Bahns and the Mathematics Institute of the University of Göttingen for their continuing hospitality. FC and GR acknowledge the MIUR Excellence Department Project awarded to the Department of Mathematics, University of Rome Tor Vergata, CUP E83C18000100006, the ERC Advanced Grant 669240 QUEST "Quantum Algebraic Structures and Models" and GNAMPA-INdAM. FC was supported in part by MIUR-FARE R16X5RB55W QUEST-NET "Operator Algebras and (non)-equilibrium Thermodynamics in Quantum Field Theory".

References

1. Bär, C., Ginoux, N., Pfäffle, F.: Wave Equations on Lorentzian Manifolds and Quantization. EMS ESI Lect. Math. Phys., vol. 3 (2007)
2. Bratteli, O., Robinson, D.W.: Operator Algebras and Quantum Statistical Mechanics I. Springer (1979)
3. Buchholz, D., Ciolli, F., Ruzzi, G., Vasselli, E.: The universal C*-algebra of the electromagnetic field. Lett. Math. Phys. **106**, 269–285 (2016)
4. Buchholz, D., Ciolli, F., Ruzzi, G., Vasselli, E.: The universal algebra of the electromagnetic field III. Static charges and emergence of gauge fields. Lett. Math. Phys. **112**, 27 (2022)
5. Buchholz, D., Doplicher, S., Morchio, G., Roberts, J.E., Strocchi, F.: A model for charges of electromagnetic type. In: Doplicher, S. (ed.) Proceedings of the conference on Operator Algebras and Quantum Field Theory Rome, 1996. International Press (1997)
6. Buchholz, D., Doplicher, S., Morchio, G., Roberts, J.E., Strocchi, F.: Quantum delocalization of the electric charge. Ann. Phys. **290**, 53–66 (2001)
7. Buchholz, D., Doplicher, S., Morchio, G., Roberts, J.E., Strocchi, F.: Asymptotic abelianness and braided tensor C*-categories. In: Rigorous Quantum Field Theory. A Festschrift for Jacques Bros, Progr. in Math., vol. 251, pp. 49–64 (2007)
8. Doplicher, S., Haag, R., Roberts, J.E.: Local observables and particle statistics I. Commun. Math. Phys. **23**, 199–230 (1971)
9. Doplicher, S., Roberts, J.E.: Why there is a field algebra with a compact gauge group describing the superselection structure in particle physics. Commun. Math. Phys. **131**, 51–107 (1990)
10. Haag, R.: Local algebras. A look back at the early years and at some achievements and missed opportunities. Eur. Phys. J. H **35**, 255–261 (2010)
11. Herdegen, A.: Semidirect product of CCR and CAR algebras and asymptotic states in quantum electrodynamics. J. Math. Phys. **39**, 1788–1817 (1998)
12. Morchio, G.: Charged fields, Higgs phenomenon and confinement. Lesson from soluble models. Ann. H. Poincaré Sect. A **64**, 461–478 (1996)
13. Morchio, G., Strocchi, F.: A non-perturbative approach to the infrared problem in QED: Construction of charged states. Nucl. Phys. B **211**, 471–508 (1983)
14. Morchio, G., Strocchi, F.: Localization and symmetries. J. Phys. A: Math. Th. **40**, 3173–3187 (2007)
15. Mund, J., Rehren, K.-H., Schroer, B.: Gauss' Law and string-localized quantum field theory. J. High Energy Phys. **2020**, 1 (2020)
16. Roepstorff, G.: Coherent photon states and spectral condition. Commun. Math. Phys. **19**, 301–314 (1970)
17. Steinmann, O.: Perturbative Quantum Electrodynamics and Axiomatic Field Theory. Springer (2000)
18. Strocchi, F.: Selected Topics on the General Properties of Quantum Field Theory. LNP, vol. 51. World Scientific (1993)

Classical Representability for Partial Boolean Structures in Quantum Mechanics

Costantino Budroni

Contents

1 Introduction

The quantum mechanical description of a physical system provides a set of predictions for all possible experiments that can be performed on it. Given an observable A, representing a physical quantity, e.g., energy, its expectation value $\langle A \rangle$, or the probability for the observation of a certain outcome, can be obtained from the knowledge of the quantum state through the Born rule

$$\langle A \rangle_\rho = \sum_i \lambda_i \text{Prob}(\lambda_i) = \sum_i \lambda_i \text{tr}[\rho P_i] = \text{tr}[\rho A], \quad \text{with} \quad \text{Prob}(\lambda_i) = \text{tr}[\rho P_i], \quad (1)$$

where A has spectral decomposition $A = \sum_i \lambda_i P_i$. A remarkable property of quantum mechanics (QM) is that not all physical observable can be jointly measured, the most famous example being the position and momentum of a particle. Such observables are said to be *incompatible* [19, 40, 43]. In the most general experiment on a physical system, then, each round consists in the joint measurement of a set of compatible observables. For each set of compatible observables, a quantum

C. Budroni (✉)
Dipartimento di Fisica "E. Fermi", Università di Pisa, Pisa, Italy
e-mail: costantino.budroni@unipi.it

A. Cintio, A. Michelangeli (Eds.), *Trails in Modern Theoretical and Mathematical Physics*, https://doi.org/10.1007/978-3-031-44988-8_7

state defines a classical probability distribution over all possible outcomes of their measurement. For projective measurements, the textbook QM observables, this is a direct consequence of the spectral theorem applied to a set of mutually commuting observables. The same structure arises for set of jointly measurement generalized measurements, i.e., positive operator-valued measures (POVMs); see e.g., [44]. In the following, when we speak about observables we refer to projective measurements, sometimes called *sharp measurements*, unless explicitly stated otherwise.

Following the original criticism of QM by Einstein, Podolsky, and Rosen [33], several attempts have been made to interpret such a collection of classical probability distributions in terms of a global probability distribution over all physical observables of a system. This is the so-called *hidden-variable program*: to complete QM with additional variables, in order to recover a classical description of the system analogous to the one obtained in classical statistical mechanics. The fact that each set commutative subalgebra of observables allows for a classical description, corresponding to the distribution of outcomes in an experiment where they are jointly measured, motivates the study of partial classical structure in quantum mechanics. For subalgebras of projectors, these commutative structures corresponds to partial Boolean algebras, to which a quantum state assigns a probability measure. These are the central objects in the work of Gleason [38], Kochen and Specker [52], and Bell [10]. In particular, the results of Kochen–Specker (KS) and of Bell significantly constrain the possible hidden-variable models to be, respectively, *contextual* and *nonlocal*. More precisely, KS show that any classical description of the statistics of a collection of measurements must depend on the *measurement context*, i.e., which set of compatible measurements are jointly performed. In contrast, Bell considered a special type of measurement context, namely, a collection of local measurements each performed on a different particle such that all the spacetime regions associated with the measurement events are spacelike separated.

A crucial difference in these results is the fact that KS dealt with the algebraic relations among the quantum mechanical projectors, whereas Bell focused more on the existence of a probability distribution, as the algebraic problem is straightforwardly solved in his scenario. The less constrained notion of measurement context used by KS allows them to prove results that are state-independent: they depend only on the (partial Boolean) algebraic relations between the observables. The two problems are of course related, as KS discussed in detail, since in order to reproduce the probabilistic predictions of QM, one first needs to reproduce (some of) its algebraic structures. For many years, however, these result were treated very differently, with some authors even claiming the impossibility of experimentally test Kochen–Specker contextuality, the so-called *nullification of KS theorem*; see [8, 29, 50, 60] and the reply by [2–5, 20, 42, 56, 59, 62, 67]; see also the discussion in [16].

The work with Gianni on the classical representability of partial Boolean structures in QM addresses the problem of the difference between the Bell and Kochen–Specker approaches to classical representability [18]. In particular, it shows that we can restrict ourselves to the problem of just reproducing the probabilistic structure of the theory, ignoring the problem of representability of the associated algebraic structure, since such a representation is then automatically recovered once the probability representation is obtained. Finally, this work also explored the conditions for the existence of classical representations based only on the compatibility relations among the considered observables. In particular, it was proven that whenever compatibility relations have the structure of a tree graph, in a sense that will be clarified below, a classical representation always exist. This result turned out to be a special case of a general theorem formulated by Vorob'ev [74, 76]. This result was rediscovered several times, within and outside the quantum information community. In recent years, however, it became relatively well known, especially in relation with quantum contextuality; see the discussion in [16].

The formalism used in [18] is that of Boolean algebras and their measures. This allowed us to present the results in a mathematically rigorous form, at the cost, however, of decreasing the accessibility of the results for readers not familiar with this formalism. In this contribution to the volume in memory of Gianni, I would like to review our work from a more modern perspective on classical representations of quantum predictions, and in particular in the framework of Kochen–Specker contextuality, also with the goal of making it accessible to a broader audience. There is a vast literature on Kochen–Specker contextuality and classical representation of QM predictions, but the goal here is to provide a self-contained introduction that provides the reader with the framework and the motivation needed to understand the results that follows and their implications. The rigorous formulation provided in [18] is accompanied by examples and informal discussions.

This chapter is structured as follows. In Sect. 2, we recall some basic facts about the problem of a hidden variable description of QM predictions: From their definition to the celebrated results of Bell and Kochen–Specker. In Sect. 3, we introduce the notion of partial Boolean algebras and formulate the two approaches, that we name Bell-like and Kochen–Specker-like, in this language showing their equivalence for the problem of HV models. In Sect. 4, we address the problem of the existence of classical representations for sets of incompatible observables, which can be derived solely from the structure, more precisely, the graph, of their compatibility relations. In Sect. 5, we present our conclusions and final remarks.

2 Hidden Variable Models and the Problem of Classical Representability for Quantum Mechanics

In this section we provide an introduction to the problem of hidden variable (HV) models and explain its connection with Bell and KS theorems. The goal of the hidden variable program is to reproduce all QM predictions, described above as a collection of probability distributions each associated with a compatible set of measurements, in terms of a single probability distribution. Notice that this is the minimal requirement for a classical model of QM predictions, in line with the approach formulated by Gleason [38], and in contrast with previous attempts to no-go theorems on HV models such as the one of von Neumann [72, 73], which consider relations among incompatible observables.

One construction of a HV model is straightforward and it was pointed out already by KS in their seminal paper [52] one can take as a global probability the product distribution of all the distributions over compatible sets of observables. Let us discuss a simple example to clarify this point. Consider three observables A, B, and C, such that A and B can be measured together and the same holds for B and C. QM then provides two distributions $p_{AB}(a,b)$ and $p_{BC}(b,c)$. We could construct a global distribution as $p_{ABC}(a,b,b',c) := p_{AB}(a,b)p_{BC}(b',c)$. The problem with this construction is that now B is represented by two variables, b and b', depending on whether it is jointly measured with A or with C. We say that the representation of B is *contextual*, i.e., it depends on which other observables are jointly measured with it. As we will see, these observables could be even spatially distributed in such a way that this context-dependence at the level of the classical model requires faster-than-light communication to be described [10, 14]. The notion of *context-independence* or *noncontextuality* formulated above in general terms, then, amounts to *locality* for spatially distributed observables.

2.1 General Formulation of Noncontextual Hidden Variable Models

The question of the existence of a noncontextual hidden variable (NCHV) model can be formulated as the so-called *marginal problem*. Given a set of observables $\mathcal{G} = \{A_1, \ldots, A_n\}$, we denote by $\mathcal{M}$, with $\mathcal{M} \subset 2^{\mathcal{G}}$, the collection of sets of observables that can be jointly measured, i.e., a collection of contexts, where $2^{\mathcal{G}}$ denotes the power set of $\mathcal{G}$. $\mathcal{M}$ is called the *marginal scenario* [26]. The observed data from measurements in each context is interpreted as a marginal of a global probability distribution on all observables. For each context $\{A_i\}_{i\in C} \in \mathcal{M}$, i.e., with $C \subset \{1,\ldots,n\}$, we have a distribution p_C of the outcomes over it. A necessary but not sufficient condition for the existence of a global distribution is for these marginals to be locally consistent. In other words, for each C and C' we have that

$$p_{C|C\cap C'} = p_{C'|C\cap C'}, \tag{2}$$

where $|C \cap C'$ denotes the restriction of the distribution to observables in the intersection of the two contexts, obtained simply by marginalization, i.e., by summing over the variables not in $C \cap C'$. This condition is satisfied in QM as a consequence of the Born rule; for instance, one can think about a pair of commuting observables.

There exist two main formulations of a NCHV model. One is the marginal problem discussed above, the other one is more related to the original idea of a "hidden variable". In a NCHV, we assume the existence of a hidden variable, let us denote it by λ, that determines the outcomes of each observable regardless of the context. For each context, given by the observables $\{A_i\}_{i\in C}$ and $C \subset \{1,\ldots,n\}$ and the outcomes

$\{a_i\}_{i\in C}$, this corresponds to

$$p_C(\{a_i\}_{i\in C}) = \sum_{\lambda} p(\lambda) \prod_{i\in C} p(a_i|\lambda), \tag{3}$$

with $p(\lambda) \geq 0, \sum_{\lambda} p(\lambda) = 1, p(a_i|\lambda) \geq 0, \sum_{a_i} p(a_i|\lambda) = 1$, for $i \in C$. The outcomes a_s are arbitrary, but to simplify the exposition we often consider the case of two-valued measurements, i.e., $a_s \in \{0, 1\}$ or $a_s \in \{-1, 1\}$.

One can show that (3) is equivalent to the existence of a global probability distribution over all observables $A_1, \ldots, A_n$, such that $P(\{a_i\}_{i\in C})$ is obtained by summing over all possible outcomes for the other observables, namely,

$$p_C(\{a_i\}_{i\in C}) = \sum_{a_s : s\notin C} p_{\mathcal{G}}(a_1, \ldots, a_n). \tag{4}$$

In other words, the formulation in terms of the hidden variable is equivalent to the one in terms of the marginals of a global distribution. Intuitively, from (3) one can construct a global distribution just by multiplying together the $p(a_i|\lambda)$ for all possible $\{A_i\}_i$ such that the distribution over all contexts are recovered as marginal. Conversely, every global distribution can be written as a convex mixture of deterministic assignments, which factorizes similarly to (3) since $\{0, 1\}$-valued measures are multiplicative. This general argument is known in the literature as Fine's theorem [34], although Fine stated it only in a special scenario and a complete proof appeared elsewhere [1, 36, 71], see the discussion in [16].

This is the modern formulation of the NCHV models. Notice that the problem here is formulated solely in terms of the reproducibility of the statistics, observed separately in each context, in terms of a global distribution. No question about the algebraic structure of the observables is addressed here. One can say that, implicitly, the algebraic structure is that of a free Boolean algebra [41], i.e., no algebraic relations are assumed between the observables.

From the definition of (4), one can derive conditions for the existence of such a global probability distribution $p_{\mathcal{G}}$. Geometrically, i.e., interpreting each probability as an entry of a vector $\vec{p}$, we have that $\vec{p}$ lies within the convex hull of a finite number of vectors, which represent the deterministic assignments of values to each variable a_i. The problem of reproducibility of a set of observed statistics $\vec{p}$, then, amounts to the membership problem for that vector with respect to the convex hull of the vectors of deterministic assignments. This problem can be solved for a given vector $\vec{p}$ as a linear program [13]. Alternatively, a finite set of necessary and sufficient conditions can be derived by computing the boundaries of such a convex hull. A convex hull of a finite number of points, in fact, is a *polytope* [39], which is also described by a finite number of hyperplanes, i.e., the *facets* of the polytope. Each hyperplane defines an inequality, which, when violated, certifies that a point is outside of the polytope. Such inequalities corresponds to Bell or noncontextuality inequalities, depending on the scenario considered. A general characterization of the set of classical probability distributions, in the sense of distribution admitting a NCHV model description, can be obtained via the *correlation polytope* method [36, 37, 63]; see also the book [64]. In the following, we present examples of Bell inequalities and more general noncontextuality inequalities.

2.2 Bell's Theorem

The simplest Bell scenario is arguably the Clauser–Horne–Shimony–Holt (CHSH) [28] Bell scenario. It consists of two experimenters, let us name them Alice and Bob, making local measurements on their half of a bipartite system, e.g., a pair of photons or spin-1/2 particles. They perform local measurements A_1 and A_2 (Alice) and B_1 and B_2 (Bob), e.g., they measure the spin component of their particle along a certain direction, with possible outcomes ± 1. In the language introduced above, Bell's theorem can be formulated as the impossibility of a NCHV model for the set of observables $\{A_1, A_2, B_1, B_2\}$ with contexts $\{(A_i, B_j)\}_{i,j=1,2}$. In this case, the experimental setting involves two distant parties. In particular, it is assumed that the choice of the measurement settings, i.e., 1 or 2, by Alice cannot influence the measurement outcome of Bob, since the events are located in space-like separated regions. The context-independence, then, amounts to *locality*, the NCHV model is, thus, called *local hidden variable* (LHV) model. From the condition in (4), one can obtain the so-called CHSH inequality

$$\langle A_1 B_1 \rangle + \langle A_1 B_2 \rangle + \langle A_2 B_1 \rangle - \langle A_2 B_2 \rangle \stackrel{\text{LHV}}{\leq} 2, \tag{5}$$

where $\langle A_i B_j \rangle = \sum_{a_i, b_j = \pm 1} a_i b_j p(a_i, b_j)$ and $\stackrel{\text{LHV}}{\leq}$ denotes the fact that the bound holds for all LHV models. Using (4) one can easily show that the bound holds for any LHV model, since the probability is a convex mixture of deterministic assignments to all the variables, the bound 2 can be computed simply by trying all possible ± 1 assignments to a_1, a_2, b_1, b_2.

In contrast, it is well known that in QM for a pair of particles prepared in the entangled state

$$|\psi\rangle = \frac{1}{\sqrt{2}}(|00\rangle + |11\rangle), \tag{6}$$

where $\{|0\rangle, |1\rangle\}$ is the eigenbasis of σ_z, and the measurements

$$A_1 = \sigma_x \otimes \mathbb{1},\ A_2 = \sigma_z \otimes \mathbb{1},\ B_1 = \mathbb{1} \otimes \frac{\sigma_x + \sigma_z}{\sqrt{2}},\ B_2 = \mathbb{1} \otimes \frac{\sigma_x - \sigma_z}{\sqrt{2}}, \tag{7}$$

one can compute the mean values as $\langle A_i B_j \rangle_\psi = \langle \psi | A_i B_j | \psi \rangle$ and obtains

$$\langle A_1 B_1 \rangle_\psi + \langle A_1 B_2 \rangle_\psi + \langle A_2 B_1 \rangle_\psi - \langle A_2 B_2 \rangle_\psi = 2\sqrt{2} > 2, \tag{8}$$

which proves the impossibility of reproducing the QM statistics via a LHV model.

Equation (5) is an example of a Bell inequality: when experimentally violated it certifies the impossibility of explaining the observed correlations in terms of LHV model. Satisfying a Bell inequality is, therefore, a necessary condition for the observed statistic to be reproducible via a LHV model. Interestingly, if "enough" inequalities, always in a finite number, are satisfied, they provide also sufficient condition for the existence of the LHV model, as discussed in the previous subsection.

Notice how in the above discussion no algebraic relations among the observables appear, and the discussion is based solely on the reproducibility of the probability distribution via a LHV model. On the one hand, one can easily see that there are no specific algebraic relations among the compatible pair. If we consider the projectors associated with the observables $\{A_i, B_j\}$, denoting them as $A_i = P_i^+ - P_i^-$ and $B_j = Q_j^+ - Q_j^-$, we have that there are no algebraic relations among compatible ones, i.e., $[P_i^\pm, Q_j^\pm] = 0$ and $P_i^\pm Q_j^\pm \neq 0$. We have that for each pair, e.g., $\{P_1^+, Q_1^-\}$ the Boolean algebra generated by them is a free algebra [41]. On the other hand, one can argue that the minimal requirement for the existence of a LHV (or NCHV) model is the reproducibility of the observed statistics, irrespectively of what algebraic relations among the observables appear in the QM formalism. We come back to this point in Sect. 4.

2.3 Kochen–Specker Theorem

The Kochen–Specker theorem was originally formulated in terms of the impossibility of a context-independent value assignment to a set of QM projectors that respects the algebraic relations among the sets of compatible ones. As shown by KS [52], this was a prerequisite for the existence of any representation of QM predictions via a NCHV model. Intuitively, such a value assignments are the extreme probability distributions, from which all other distributions are obtained as convex mixtures, as discussed above. Thus, at least some of them must exist.

Let us fix the basic objects of the theorem. Consider a Hilbert space $\mathcal{H}$ of dimension d and consider d rank-1 projectors $P_1, P_2, \ldots, P_d$ associated with d orthogonal vectors in $\mathcal{H}$. They satisfy the following relations

(**O**) $P_i P_j = 0$ for any $i \neq j$ (orthogonality).
(**C**) $\sum_{i=1}^d P_i = \mathbb{1}$ or, equivalently, using (**O**), $\prod_{i=1}^d (\mathbb{1} - P_i) = 0$ (completeness).

Such relations can be interpreted in terms of yes-no questions (or truth assignments) $Q_1, \ldots, Q_d$ as follows:

(**O′**) Q_i and Q_j are exclusive; i.e., they cannot be simultaneously "true" for $i \neq j$.
(**C′**) $Q_1, \ldots, Q_d$ cannot be simultaneously "false"; one of them has to be true.

For the case of rank-1 projectors commutativity corresponds to orthogonality, and each collection of commuting projectors generates a Boolean algebra. This means that maximal sets of compatible projectors are in a one-to-one correspondence with orthogonal bases of $\mathcal{H}$. Value assignments, in this case yes/no or $\{0, 1\}$,

Table 1 Table of vectors and contexts for the 18-vector KS sets of [24]. The vectors are unnormalized and, for better alignment, $\bar{1}$ denotes -1. Each row represents a context and each vector appears in exactly two contexts. As a consequence, assigning a noncontextual value of "+1" (or green) to some vectors, one obtains always an even number of "+1", whereas by construction one should get exactly nine "+1", one for each context

$v_{12} = (1, 0, 0, 0)$	$v_{16} = (0, 0, 1, \bar{1})$	$v_{17} = (0, 0, 1, 1)$	$v_{18} = (0, 1, 0, 0)$
$v_{12} = (1, 0, 0, 0)$	$v_{23} = (0, 1, \bar{1}, 0)$	$v_{28} = (0, 0, 0, 1)$	$v_{29} = (0, 1, 1, 0)$
$v_{23} = (0, 1, \bar{1}, 0)$	$v_{34} = (\bar{1}, 1, 1, 1)$	$v_{37} = (1, 1, 1, \bar{1})$	$v_{39} = (1, 0, 0, 1)$
$v_{34} = (\bar{1}, 1, 1, 1)$	$v_{45} = (0, 1, 0, \bar{1})$	$v_{47} = (1, 1, \bar{1}, 1)$	$v_{48} = (1, 0, 1, 0)$
$v_{45} = (0, 1, 0, \bar{1})$	$v_{56} = (1, 1, 1, 1)$	$v_{58} = (1, 0, \bar{1}, 0)$	$v_{59} = (1, \bar{1}, 1, \bar{1})$
$v_{16} = (0, 0, 1, \bar{1})$	$v_{56} = (1, 1, 1, 1)$	$v_{67} = (1, \bar{1}, 0, 0)$	$v_{69} = (1, 1, \bar{1}, \bar{1})$
$v_{17} = (0, 0, 1, 1)$	$v_{37} = (1, 1, 1, \bar{1})$	$v_{47} = (1, 1, \bar{1}, 1)$	$v_{67} = (1, \bar{1}, 0, 0)$
$v_{18} = (0, 1, 0, 0)$	$v_{28} = (0, 0, 0, 1)$	$v_{48} = (1, 0, 1, 0)$	$v_{58} = (1, 0, \bar{1}, 0)$
$v_{29} = (0, 1, 1, 0)$	$v_{39} = (1, 0, 0, 1)$	$v_{59} = (1, \bar{1}, 1, \bar{1})$	$v_{69} = (1, 1, \bar{1}, \bar{1})$

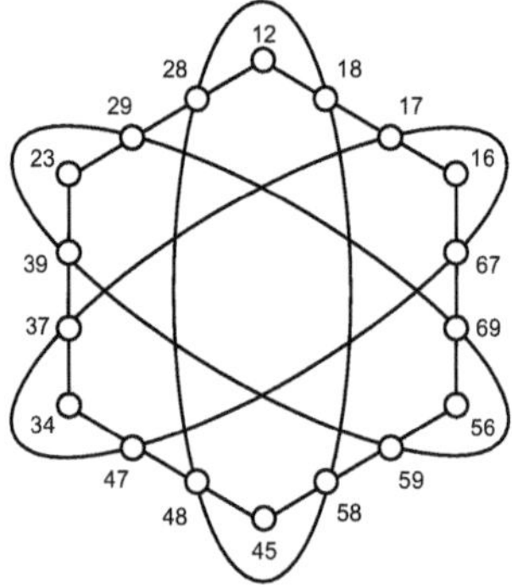

Fig. 1 Graphical representation for the contexts in the 18-vector KS set of Table 1 from [24]. Each smooth line, i.e., straight or ellipse, represents a context, vectors in each context are mutually orthogonal

must then satisfy the conditions **O**′ and **C**′. Orthogonality relations can be represented as a graph, as depicted in Fig. 1: Each node represents a vector, orthogonal vectors are connected by an edge (condition **O**), and complete sets of vectors lie on a smooth line (straight line or an ellipse, condition **C**). The goal is to provide an assignment of $\{1, 0\}$, equivalently (true, false), to these vectors such that conditions **O**′ and **C**′ are satisfied. This problem has also been formulated in terms of an assignment of colors, e.g., green for true and red for false. Referring to Fig. 1, the constraints are therefore that in each smooth line exactly one vertex is colored green and all the others red. Sets of vectors that do not admit any assignment satisfying these rules are called *KS sets*. The example in Fig. 1 has been found by [24] in $d = 4$ and has been proven to be the KS set with the minimal number of vectors [79]. Consider the list of vectors in Table 1. There are 9 contexts, corresponding to the rows of the table, and each vector belongs to two contexts, as can be seen from Fig. 1. If we color green some nodes and we count the total number of green nodes appearing in all contexts, i.e., with repetitions since nodes appear in multiple contexts, we get an even number. However, the contexts are 9 so it should be exactly 9 assignments of green, which give us the contradiction.

The fact that there exists no value assignment that respects condition $\mathbf{O}'$ and $\mathbf{C}'$ implies that there exist no NCHV reproducing the probability associated to these projectors by any quantum state. This is the phenomenon of state-independent contextuality (SI-C) [7, 21, 82]. In particular, this phenomenon can be certified by the violation of noncontextuality inequalities. Similarly to the Bell-CHSH inequality in (5), these are inequalities satisfied by the predictions of any NCHV model, but violated by QM predictions. Interestingly, these phenomenon appears also for sets of projectors that are not KS sets, i.e., they admit some value assignments [12, 80, 81]. As we see in Sect. 3, this possibility is implicit in the discussion of KS vs Bell approaches, namely, there could exist sets of projectors that admit some assignments, but "not enough" to provide an embedding into a Boolean algebra. This should also be a possibility for SI-C, but curiously sets such as [81] are not of this form, and no such example is known. Finally, we remark that necessary and sufficient conditions for SI-C have been investigated, see [25, 65].

3 Unified Approach to the Extension of the Partial Boolean Algebraic and Probabilistic Structures

In the previous section, we briefly recalled the main concepts involved in the Bell and Kochen–Specker theorems. The main difference is that Bell's theorem concerns the extension of a collection of classical distribution (one for each set of local measurements), whereas Kochen–Specker theorem involves the extension of a collection of Boolean algebras (one for each set of commuting projectors). We also discussed how the impossibility of an extension of the algebraic structure implies an impossibility of the extension of all the probabilistic structures associated with a quantum state, i.e., state-independent contextuality. Here, we present the connection between these two problems formulated in terms of the notion of *partial Boolean algebra* and *partial probability theory* and following the presentation of [18].

3.1 Partial Probability Theories

We start with the definition of partial Boolean algebra. This notion goes back to the original paper of Kochen and Specker [52]. However, they consider a special class of partial Boolean algebras that satisfy a property, indicated in the following as *(K–S)* property, which we do not include in our definition. See below for more details.

A *partial Boolean algebra (PBA)* is a set X together with a non-empty family $\mathcal{F}$ of Boolean algebras, $\mathcal{F} \equiv \{\mathfrak{B}_i\}_{i\in I}$, such that $\bigcup_i \mathfrak{B}_i = X$, for which Boolean operations coincide on the intersections, namely

(P_1) for every $\mathfrak{B}_i, \mathfrak{B}_j \in \mathcal{F}$, $\mathfrak{B}_i \cap \mathfrak{B}_j \in \mathcal{F}$ and the Boolean operations $(\cap_i, \cup_i, ^{c_i})$, $(\cap_j, \cup_j, ^{c_j})$ of, respectively, $\mathfrak{B}_i$ and $\mathfrak{B}_j$ coincide on it.

Without loss of generality we can also assume the property

(P_2) for all $\mathfrak{B}_i \in \mathcal{F}$, each Boolean subalgebra of $\mathfrak{B}_i$ belongs to $\mathcal{F}$.

By (P_1), Boolean operations, when defined, are unique and will be denoted by $(\cap, \cup, ^c)$; we denote a partial Boolean algebra by $(X, \{\mathfrak{B}_i\}_{i\in I})$, or simply by $\{\mathfrak{B}_i\}_{i\in I}$. In the following we consider only *finite* partial Boolean algebras. We call their elements *observables*. Given a PBA $(X, \{\mathfrak{B}_i\})$, a *state* is defined as a map $f: X \longrightarrow [0,1]$, such that $f_{|\mathfrak{B}_i}$ is a normalized measure on the Boolean algebra $\mathfrak{B}_i$ for all i. Equivalently, a state is given by a collection of *compatible probability measures* $\{\mu_i\}$, i.e. measures coinciding on intersections of Boolean algebras, one for each $\mathfrak{B}_i$.

A *partial probability theory (PPT)* is a pair $((X, \{\mathfrak{B}_i\}); f)$, where $(X, \{\mathfrak{B}_i\})$ is a partial Boolean algebra and f is a state defined on it. Equivalently, a PPT can be denoted with $((X, \{\mathfrak{B}_i\}); \{\mu_i\})$, where $\mu_i = f_{|\mathfrak{B}_i}$, or simply by $(\{\mathfrak{B}_i\}; \{\mu_i\})$. It can be easily checked that the above properties are satisfied by the set of all orthogonal projections in a Hilbert space of arbitrary dimension, with Boolean operations defined by

$$P \cap Q \equiv PQ, \quad P \cup Q \equiv P + Q - PQ, \quad P^c \equiv 1 - P, \tag{9}$$

for all pairs P, Q of commuting projections. If one considers a finite set of projections, the result of the iteration of the above Boolean operations (on commuting projections) is still a finite set and a partial Boolean algebra.

Moreover, given a set of projections, the corresponding predictions given by a QM state define a PPT on the generated PBA. In fact, given a PBA of projections on a Hilbert space $\mathcal{H}$, by the spectral theorem, a quantum mechanical state ψ defines a state f_ψ on it, given by $f_\psi(P) = (\psi, P\psi)$. The generalization to density matrices is obvious. We name such PPTs *projection algebra partial probability theories*. We see in Sect. 3.2 that they are not the only PPTs that can be associated to QM predictions, other choices being implicit in different approaches to contextuality in QM. As mentioned above, in QM, PBAs of projections also satisfy the following property

(K–S) if $A_1, \dots, A_n$ are elements of X such that any two of them belong to a common algebra $\mathfrak{B}_i$, then there is a $\mathfrak{B}_k \in \mathcal{F}$ such that $A_1, \dots, A_n \in \mathfrak{B}_k$.

In the following, we do not assume *(K–S)*. On the one hand, this is not necessary for the result that we present. We note that in a general theory of measurements, it makes perfect sense to consider, for instance, three measurements such that every pair can be performed jointly, but it is impossible to perform jointly all the three. This is indeed what happens when we move from projective measurements to general measurements, i.e. positive operator-valued measures (POVMs). Such a property, also called *Specker's principle* [22, 52, 70], plays a central role in quantum mechanics, in particular the discussion of physical principle that recovers the bounds for the set of quantum correlations; see [22, 23, 35, 45] for more details.

We now introduce several definitions that are helpful later on. Some of them are simply a reformulation in algebraic terms of concepts that we already encountered, such as the notion of context. Given a PBA $(X, \{\mathfrak{B}_i\})$, we call a *context* each maximal, with respect to inclusion, Boolean algebra of $\{\mathfrak{B}_i\}$. Moreover, given $A, B \in X$, we say that A and B are *compatible* if they belong to a common context. Given a subset $\mathcal{G} \subset X$, we say that $\mathcal{G}$ *generates*, or that $\mathcal{G}$ is a set of *generators* for $(X, \{\mathfrak{B}_i\})$, if each maximal Boolean algebra of $\{\mathfrak{B}_i\}$ is generated by a subset of $\mathcal{G}$. Given two PBAs $(X, \{\mathfrak{B}_i\})$ and $(X', \{\mathfrak{B}'_j\})$; we say that a function $\varphi \colon X \to X'$ is a *homomorphism* if for each $\mathfrak{B}_i$ the image $\varphi(\mathfrak{B}_i)$ belongs to $\{\mathfrak{B}'_j\}$ and $\varphi_{|\mathfrak{B}_i}$ is a homomorphism of Boolean algebras; moreover, if φ is invertible, we say that φ is an *isomorphism*. If $(X', \{\mathfrak{B}'_j\})$ is a Boolean algebra (notice that a Boolean algebra is also a PBA) and the homomorphism φ is an injection, we say that φ is an *embedding*. Homomorphisms of $(X, \{\mathfrak{B}_i\})$ into the Boolean algebra $\{0, 1\}$ define *multiplicative* states, corresponding to the deterministic assignments previously encountered.

The following definitions concerns the possibility of extending a PPT to additional algebras, therefore enlarging the contexts and reducing their total number. We say that $(X', \{\mathfrak{B}'_j\})$ *contains* $(X, \{\mathfrak{B}_i\})$ if $X \subset X'$ and $\{\mathfrak{B}_i\} \subset \{\mathfrak{B}'_j\}$. We say that $(X', \{\mathfrak{B}'_j\})$ *extends* $(X, \{\mathfrak{B}_i\})$ if $(X', \{\mathfrak{B}'_j\})$ contains $(X, \{\mathfrak{B}_i\})$ and X generates $(X', \{\mathfrak{B}'_j\})$. Similar notions apply to states. Given two PPTs $C = ((X, \{\mathfrak{B}_i\}); \{\mu_i\})$ and $C' = ((X', \{\mathfrak{B}'_j\}); \{\mu'_j\})$, we say that C' *contains* C if $(X', \{\mathfrak{B}'_j\})$ contains $(X, \{\mathfrak{B}_i\})$ and $\{\mu_i\} \subset \{\mu'_j\}$; we say that C *extends* C' if $(X', \{\mathfrak{B}'_j\})$ extends $(X, \{\mathfrak{B}_i\})$ and C' contains C. By *classical representation* of a PPT $C = ((X, \{\mathfrak{B}_i\}); \{\mu_i\})$ we mean a Boolean algebra $\mathfrak{B}$ and a (normalized) measure μ such that $(\mathfrak{B}; \mu)$ extends C.

The central point of the KS theorem is precisely the impossibility of an embedding of a PBA of projectors into a Boolean algebra, where this impossibility is shown by the impossibility of constructing an homomorphism from the PBA to the Boolean algebra $\{0, 1\}$, i.e., a deterministic, assignment. The notion of extension presented here concerns the minimal extension, in contrast to the one of KS. The two approaches, however, are equivalent since a PBA is extendable to a Boolean algebra if and only if it is embeddable into a Boolean algebra.

When a PBA $\{\mathfrak{B}_i\}$ extends to a Boolean algebra $\mathfrak{B}$, the problem of a classical representation of a PPT reduces to the extension problem of a function, induced by the corresponding state, defined on a subset of $\mathfrak{B}$; the solution of this extension problem (with necessary and sufficient conditions) is then implicit in the work of Horn and Tarski [47]. Their approach turned out, see [31], to be equivalent to the correlation polytope approach of [36, 37, 63, 64]. For this reason, we call a PPT $C = (\{\mathfrak{B}_i\}; \{\mu_i\})$ such that $\{\mathfrak{B}_i\}$ extends to a Boolean algebra a *Horn–Tarski (H–T) partial probability theory*.

3.2 Reduction to Horn–Tarski PPTs

Empirical Quotients of Partial Probability Theories

Our goal is to provide a unification of Bell-type and KS-type approaches to classical representability. In particular, we want to show that it is enough to focus on the representability of the probabilistic structure ignoring the algebraic one, or more precisely, substituting the PBA of projectors with a a free Boolean algebra and investigating the probability measures induced on it by quantum states. The idea is that, if a representation for the probability measures on such a free algebra exists, the original algebraic structure, and the associated extension to a Boolean algebra, can always be recovered as a quotient with respect to some equivalence relation induced by the probability measures.

The first notion we introduce is that of *empirical quotient*; we briefly discuss it in classical probability theory and then we generalize it to PPTs. Consider a classical probability theory defined by a finite Boolean algebra $\mathfrak{B}$ and a probability measure μ. If for two elements $A, B \in \mathfrak{B}$ it holds $\mu(A \cap B^c) = \mu(A^c \cap B) = 0$, equivalently $\mu(A) = \mu(B) = \mu(A \cap B)$, it follows that every time A happens also B happens and conversely. In terms of conditional probabilities, this can be written as $Pr(A|B) = Pr(B|A) = 1$. It thus makes sense to identify the events A, B and $A \cap B$ with a single event since they cannot be distinguished by any experiment. This procedure induces an equivalence relation $\sim_{\mathcal{I}}$ on $\mathfrak{B}$, given by the ideal $\mathcal{I} = \{A \in \mathfrak{B} | \mu(A) = 0\}$, giving rise to the *empirical quotient algebra* $\widetilde{\mathfrak{B}} \equiv \mathfrak{B}/_{\sim_{\mathcal{I}}}$. The measure μ induces a normalized measure $\tilde{\mu}$ on $\widetilde{\mathfrak{B}}$. Similar notions, with identical interpretation, apply to the case of a finite Boolean algebra $\mathfrak{B}$ and a collection of normalized measures $\{\mu_k\}_{k \in K}$, where K may be any set of indices, through the ideal $\mathcal{I} = \{A \in \mathfrak{B} | \mu_k(A) = 0 \text{ for all } k \in K\}$ (any K being admissible since $\mathfrak{B}$ is finite).

The extension of the above notions to the case of PPTs is not automatic and requires further conditions. Given two collections of PPTs $\{C_k\}_{k \in K} = \{(\{\mathfrak{B}_i\}_{i \in I}; f_k)\}_{k \in K}$ and $\{\widetilde{C}_k\}_{k \in K} = \{(\{\widetilde{\mathfrak{B}}_j\}_{j \in J}; \tilde{f}_k)\}_{k \in K}$, we say that $\{\widetilde{C}_k\}_{k \in K}$ is an *empirical quotient* of $\{C_k\}_{k \in K}$ if there exists an *equivalence relation* $\sim$ on $X = \bigcup_i \mathfrak{B}_i$ such that

(*i*) when restricted to each Boolean algebra $\mathfrak{B}_i$, $\sim$ coincides with the equivalence relation induced by the ideal $\mathcal{I}_i \equiv \{A \in \mathfrak{B}_i \,|\, f_k(A) = 0 \text{ for all } k \in K\}$;

(*ii*) given $A \in \mathfrak{B}_i$ and $B \in \mathfrak{B}_l$, with $\mathfrak{B}_i$ and $\mathfrak{B}_l$ maximal, if $A \sim B$, then there exists $C \in \mathfrak{B}_i \cap \mathfrak{B}_l$ such that $A \sim C$ (and $B \sim C$ by transitivity);

(*iii*) the quotient set $X/_\sim$ is a PBA isomorphic to the PBA $\widetilde{X} = \bigcup_j \widetilde{\mathfrak{B}}_j$; by (*i*), this implies that the quotient preserves Boolean operations, namely for all $A, B \in X$, with A and B compatible, it holds $[A] \cap [B] = [A \cap B]$, where $[A]$ denotes the equivalence class of A with respect to $\sim$, and analogous properties hold for $\cup$ and ^{c};

(*iv*) denoted with $\varphi\colon X/_\sim \longrightarrow \widetilde{X}$ the isomorphism in (*iii*), it holds $f_k(A) = \tilde{f}_k(\varphi([A]))$, for all $k \in K$ and for all $A \in X$.

The above definition clearly applies in the classical case, i.e. when both X and $\widetilde{X}$ are Boolean algebras; we provide below less trivial examples. We remark that, unlike the classical case, an equivalence relation on a PPT satisfying (i) and (iv) does not in general give rise to an empirical quotient; a counterexample can be constructed by considering a PPT given by the PBA consisting of three maximal Boolean algebras, generated respectively by the pairs of observables $\{A, B\}, \{B, C\}$ and $\{A, C\}$, together with the corresponding subalgebras, and a state f that induces in the above Boolean algebras the identification $A \sim B$, $B \sim C$ and $C \sim A^c$. In fact, if an empirical quotient exists, then by transitivity A is identified with A^c and therefore, by (i), both are identified with $\emptyset$; this contradicts $\tilde{f}(\varphi([\mathbf{1}])) = 1$. This example is a version of the the triangle scenario introduced by Specker [70].

The above notion of quotient may look too restrictive; on the contrary, it will turn out that *all* PPTs with a PBA admitting a complete set of states (see below) can be *identified with quotients* of PPTs associated to a collection of freely generated Boolean algebras, which are automatically embeddable into a Boolean algebra. This will imply that *all extension problems* in QM can be put in the H–T form.

Classical Representations of Partial Probability Theories and of Their Empirical Quotients

We want to connect the classical representation of a PPT with that of its empirical quotient with respect to a set of states. Intuitively, this means that we can recover the embedding of the original algebra of projectors (KS approach), from the classical representation of the unconstrained (i.e., free) algebra that we considered instead (Bell approach). An intermediate step is provided by Prop. 1 below.

Given a PBA $\{\mathfrak{B}_i\}_{i\in I}$ and a collection of states $\{f_k\}_{k\in K}$, we say that $\{f_k\}_{k\in K}$ is *complete* with respect to $\{\mathfrak{B}_i\}_{i\in I}$ if for all $A \in X = \bigcup_i \mathfrak{B}_i$, with $A \neq \emptyset$ there exists f_k such that $f_k(A) \neq 0$. If, in addition, for all $A \neq B$, with $A, B \in X$, there exists f_k such that $f_k(A) \neq f_k(B)$ then $\{f_k\}_{k\in K}$ is said to be *separating* for $\{\mathfrak{B}_i\}_{i\in I}$. Notice that for an empirical quotient $\{\widetilde{C}_k\}_{k\in K} = \{(\{\widetilde{\mathfrak{B}}_j\}_{j\in J}; \tilde{f}_k)\}_{k\in K}$, by (i) and (iv), $\{\tilde{f}_k\}_{k\in K}$ is always complete with respect to $\{\widetilde{\mathfrak{B}}_j\}_{j\in J}$. The following result relates classical representations of PPTs with embeddings of PBAs associated to empirical quotients.

Proposition 1 Given $\{C_k\}_{k\in K} = \{(\{\mathfrak{B}_i\}_{i\in I}; f_k)\}_{k\in K}$ and $\{\widetilde{C}_k\}_{k\in K} = \{(\{\widetilde{\mathfrak{B}}_j\}_{j\in J}; \tilde{f}_k)\}_{k\in K}$, with $\{\widetilde{C}_k\}_{k\in K}$ an empirical quotient of $\{C_k\}_{k\in K}$, if there exists $k_0 \in K$ such that C_{k_0} admits a classical representation, then there exists a multiplicative state on $\{\widetilde{\mathfrak{B}}_i\}$, i.e. a homomorphism $\delta_0\colon \widetilde{X} = \bigcup_j \widetilde{\mathfrak{B}}_j \longrightarrow \{0, 1\}$.

Moreover, if there exists $K' \subset K$ such that $\{\tilde{f}_k\}_{k\in K'}$ is separating for $\{\widetilde{\mathfrak{B}}_j\}_{j\in J}$ and C_k admits a classical representation for every $k \in K'$, then $\{\widetilde{\mathfrak{B}}_j\}_{j\in J}$ is embeddable into the Boolean algebra 2^N, the power set of a N-element set, where N is the number of multiplicative states induced by classical representations of the states $\{f_k\}_{k\in K'}$.

Proof Let the Boolean algebra $\mathfrak{B}$ together with the normalized measure μ be a classical representation for C_{k_0}, then μ can be written as a convex combination of multiplicative measures, namely

$$\mu = \sum_i \lambda_i \delta_i, \tag{10}$$

where the δ_i's are multiplicative measures and $\{\lambda_i\}_i$ satisfy $\lambda_i > 0$ and $\sum_i \lambda_i = 1$. It follows that $\mu(A \cap B^c) = \mu(A^c \cap B) = 0$ for all $A, B \in X$ such that $A \sim B$ and A and B belong to a common algebra $\mathfrak{B}_{i_0} \in \{\mathfrak{B}_i\}$; therefore $\delta_i(A \cap B^c) = \delta_i(A^c \cap B) = 0$ for each δ_i that appears in (10). Actually, the same holds even if A and B do not belong to a common maximal algebra of $\{\mathfrak{B}_i\}$. In fact, by (ii), there exists an element C in the intersection of the two maximal algebras containing A and B such that $A \sim C \sim B$ and the above statement follows from $A \cap B^c = (A \cap B^c \cap C) \cup (A \cap B^c \cap C^c)$.

It follows that $\delta_i(A) = \delta_i(B)$ for all $A, B \in X$ such that $A \sim B$ and for all δ_i appearing in (10); therefore each δ_i induces a well defined $\{0, 1\}$-valued function on $\widetilde{X}$. To conclude, we prove that such functions are homomorphisms when restricted to each algebra of $\{\widetilde{\mathfrak{B}}_j\}$. This follows from the isomorphism between $\widetilde{X}$ and $X/_\sim$ and the fact that each δ_i defines a multiplicative measure on $\mathfrak{B}_i/_\sim$ for all $\mathfrak{B}_i$. In fact, given $A, B \in \mathfrak{B}_i$, $[A] \cap [B] = [\emptyset]$ implies $\delta_i(A \cap B) = 0$ and therefore $\delta_i(A) + \delta_i(B) = \delta_i(A \cup B)$; each δ_i defines, therefore, a $\{0, 1\}$-valued function on $\mathfrak{B}_i/_\sim$ which is additive on disjoint elements, i.e. a multiplicative measure, which is a homomorphism with the Boolean algebra $\{0, 1\}$. The proof of the second part follows easily from the first part together with Theorem 0 of [52]. □

Partial Probability Theories as Empirical Quotients of Free H–T Theories

We now show that any complete set of states on a PBA can be regarded as an empirical quotient of a collection of PPTs on a PBA which is embeddable in a (free) Boolean algebra, i.e a collection of H–T PPTs.

Consider a collection of PPTs $\{\widetilde{C}_k\}_{k\in K} = \{(\{\widetilde{\mathfrak{B}}_j\}_{j\in J}; \tilde{f}_k)\}_{k\in K}$ such that $\{\tilde{f}_k\}_{k\in K}$ is complete, and take a subset $\widetilde{\mathcal{G}} = \{\widetilde{A}_1, \ldots, \widetilde{A}_n\} \subset \widetilde{X} = \bigcup_j \widetilde{\mathfrak{B}}_j$ of generators of $\{\widetilde{\mathfrak{B}}_j\}_{j\in J}$ satisfying the following property

(G) given $k \geq 1$ maximal Boolean algebras $\widetilde{\mathfrak{B}}_{i_1}, \ldots, \widetilde{\mathfrak{B}}_{i_k}$, generated respectively by maximal subsets of compatible generators $\widetilde{\mathcal{G}}_{i_1}, \ldots, \widetilde{\mathcal{G}}_{i_k} \subset \widetilde{\mathcal{G}}$, such that $\widetilde{\mathfrak{B}}_{i_1} \cap \ldots \cap \widetilde{\mathfrak{B}}_{i_k} \neq \{\emptyset, \mathbf{1}\}$, the set $\widetilde{\mathcal{G}}_{i_1 \ldots i_k} \equiv \widetilde{\mathcal{G}}_{i_1} \cap \ldots \cap \widetilde{\mathcal{G}}_{i_k}$ is not empty and it generates the Boolean algebra $\widetilde{\mathfrak{B}}_{i_1} \cap \ldots \cap \widetilde{\mathfrak{B}}_{i_k}$;

notice that each maximal algebra is generated by a maximal subset of compatible generators and that the above choice is always possible since one can take $\widetilde{\mathcal{G}} = \widetilde{X}$. The role of this property is clarified below. Denote with $\{\widetilde{\mathcal{G}}_l\}$ the collection of subsets of compatible observables of $\widetilde{\mathcal{G}}$, $\widetilde{\mathcal{G}}_l = \{\tilde{A}_{s_1} \ldots \tilde{A}_{s_{n_l}}\}$. Now consider the PBA $\{\mathfrak{B}_i\}_{i\in I}$ consisting of Boolean algebras freely generated by subsets $\mathcal{G}_l \equiv \{A_{s_1} \dots A_{s_{n_l}}\}$.

The role of such generators is to define a state f_k on the the free Boolean algebra $\{\mathfrak{B}_i\}_{i\in I}$ from a state $\tilde{f}_k$, such that the original PPTs $\{\widetilde{C}_k\}_{k\in K}$ become their empirical quotient. To do so, we start by showing how each state $\tilde{f}_k$ induces a state f_k on $\{\mathfrak{B}_i\}_{i\in I}$. First, notice that, since each state on a PBA is a collection of normalized measures, it is sufficient to define it as measures on maximal Boolean algebras. Each measure on a maximal algebra $\mathfrak{B}_l$ of $\{\mathfrak{B}_i\}_{i\in I}$, generated by a set $\mathcal{G}_l = \{A_{s_1}, \dots, A_{s_{n_l}}\}$, is completely determined by its values on elements of the form $(-1)^{1-\varepsilon_1} A_{s_1} \cap \dots \cap (-1)^{1-\varepsilon_{n_l}} A_{s_{n_l}}$, where $-A \equiv A^c$ and $\varepsilon_i \in \{0, 1\}$, since each element of the algebra can be written as a disjoint union of elements of that form. Now, f_k is defined as $f_k((-1)^{1-\varepsilon_1} A_{s_1} \cap \dots \cap (-1)^{1-\varepsilon_{n_l}} A_{s_{n_l}}) \equiv \tilde{f}_k((-1)^{1-\varepsilon_1} \widetilde{A}_{s_1} \cap \dots \cap (-1)^{1-\varepsilon_{n_l}} \widetilde{A}_{s_{n_l}})$ for all maximal subsets of compatible observables $\widetilde{\mathcal{G}}_l$ of $\widetilde{\mathcal{G}}$, and extended as a measure on each maximal algebra. It can be verified that such measures are normalized and they coincide on intersection of Boolean algebras; therefore, they define a state.

In this way, we obtain a collection of PPTs $\{C_k\}_{k\in K} \equiv \{(\{\mathfrak{B}_i\}_{i\in I}; f_k)\}_{k\in K}$ such that the initial collection $\{\widetilde{C}_k\}_{k\in K} = \{(\{\widetilde{\mathfrak{B}}_j\}_{j\in J}; \tilde{f}_k)\}_{k\in K}$ is an empirical quotient of it. The equivalence relation $\sim$ can be, in fact, defined as follows: to each element A of X, generated by a subset of compatible generators $\mathcal{G}_l \subset \mathcal{G}$ there corresponds, via the correspondence $A_i \mapsto \widetilde{A}_i$, a unique element $\widetilde{A}$ of $\widetilde{X}$, defined as the element generated by $\widetilde{\mathcal{G}}_l \subset \widetilde{\mathcal{G}}$ by means of the same operations that generate A from $\mathcal{G}_l$; then an equivalence relation $\sim$ can be defined on X as $A \sim B$ iff $\widetilde{A} = \widetilde{B}$.

It can be easily verified that $\sim$ is an equivalence relation and that it defines an empirical quotient:

(*i*) it is sufficient to consider each Boolean algebras $\mathfrak{B}_l$, generated by $\mathcal{G}_l = \{A_{l_1}, \dots, A_{l_s}\}$, and notice that, there, $\sim$ coincides with the equivalence relation induced by the ideal $\mathcal{I} \equiv \{B \in \mathfrak{B}_l \mid \bigcup_{\varepsilon\in H_B} (-1)^{1-\varepsilon_1} \widetilde{A}_{l_1} \cap \dots \cap (-1)^{1-\varepsilon_s} \widetilde{A}_{l_s} = \emptyset\}$ with $H_B \equiv \{\varepsilon = (\varepsilon_1, \dots, \varepsilon_n) \in \{0, 1\}^n \mid (-1)^{1-\varepsilon_1} A_{l_1} \cap \dots \cap (-1)^{1-\varepsilon_s} A_{l_s} \subset B\}$; now, since $\{\tilde{f}_k\}_{k\in K}$ is complete and by construction of $\{f_k\}_{k\in K}$, $\mathcal{I}$ coincides with the set $\{B \in \mathfrak{B}_l \mid f_k(B) = 0 \text{ for all } k \in K\}$.

(*ii*) given $A, B \in X$, belonging respectively to maximal algebras $\mathfrak{B}_{l_1}$, generated by $\mathcal{G}_{l_1}$, and $\mathfrak{B}_{l_2}$, generated by $\mathcal{G}_{l_2}$, with $\mathcal{G}_{l_1}$ and $\mathcal{G}_{l_2}$ maximal, if $A \sim B$, then there exists $C \in \mathfrak{B}_{l_1} \cap \mathfrak{B}_{l_1}$, which is the Boolean algebra generated by $\mathcal{G}_{l_1} \cap \mathcal{G}_{l_2}$, such that $A \sim C \sim B$. In fact, $A \sim B$ implies, with the same notation as above, $\widetilde{A} = \widetilde{B}$; therefore the two maximal algebras generated respectively by $\widetilde{\mathcal{G}}_{l_1}$ and $\widetilde{\mathcal{G}}_{l_2}$ have a non-empty intersection containing $\widetilde{A}$, then, by (G), $\mathcal{G}_{l_1} \cap \mathcal{G}_{l_2} \neq \emptyset$ and an element C satisfying the above conditions exists.

(*iii*) by construction, $X/_\sim$ is in a one-to-one correspondence with $\widetilde{X}$; that such a bijection is also an isomorphism follows from the coincidence, within each Boolean algebra, of $\sim$ with the equivalence relation induced by the ideal $\mathcal{I}$ discussed above.

(*iv*) it follows by construction of $\{f_k\}_{k\in K}$.

The above partial Boolean algebra $\{\mathfrak{B}_i\}_{i\in I}$ is embeddable into the Boolean algebra freely generated by the set $\mathcal{G}$. The PPTs $\{C_k\}_{k\in K}$ are therefore of the Horn–Tarski type and we name $\{C_k\}_{k\in K}$ *the collection of free H–T partial probability theories associated to* $\{\widetilde{C}_k\}_{k\in K}$ *and* $\widetilde{\mathcal{G}}$.

Classical Representations and Free H–T Theories

We can finally prove the main result of this section.

Theorem 1 Given a collection of PPTs $\{\widetilde{C}_k\}_{k\in K} = \{(\{\widetilde{\mathfrak{B}}_j\}_{j\in J}; \tilde{f}_k)\}_{k\in K}$ with $\{\tilde{f}_k\}_{k\in K}$ complete with respect to $\{\widetilde{\mathfrak{B}}_j\}_{j\in J}$, a set of generators $\widetilde{\mathcal{G}} = \{\tilde{A}_1, \dots, \tilde{A}_n\}$ satisfying property (G) and the associated collection of free H–T PPTs $\{C_k\}_{k\in K} = \{(\{\mathfrak{B}_i\}_{i\in I}; f_k)\}_{k\in K}$, then

(a) if, for a given $k \in K$, $\widetilde{C}_k$ admits a classical representation, then C_k admits a classical representation;

(b) if there exists $K' \subset K$ such that $\{\tilde{f}_k\}_{k\in K'}$ is separating for $\{\widetilde{\mathfrak{B}}_j\}_{j\in J}$ and C_k admits a classical representation for all $k \in K'$, then $\widetilde{C}_k$ admits a classical representation for all $k \in K'$.

Proof (a) Let the Boolean algebra $\widetilde{\mathfrak{B}}$ together with the normalized measure $\tilde{\mu}_k$ be a classical representation for $\widetilde{C}_k$. By the definition of extension, the set $\widetilde{\mathcal{G}}$ is a set of generators for $\widetilde{\mathfrak{B}}$; therefore the Boolean algebra $\widetilde{\mathfrak{B}}$ is isomorphic to the quotient algebra $\mathfrak{B}/_{\sim}$, where $\mathfrak{B}$ is the Boolean algebra freely generated by n generators $\{A_1, \dots, A_n\}$ and the equivalence relation $\sim$ is that induced by the ideal $\mathcal{I} \equiv \{B \in \mathfrak{B} | \bigcup_{\varepsilon\in H_B} (-1)^{1-\varepsilon_1}\widetilde{A}_1 \cap \dots \cap (-1)^{1-\varepsilon_n}\widetilde{A}_n = \emptyset\}$ with $H_B \equiv \{\varepsilon = (\varepsilon_1, \dots, \varepsilon_n) \in \{0,1\}^n | (-1)^{1-\varepsilon_1} A_1 \cap \dots \cap (-1)^{1-\varepsilon_n} A_n \subset B\}$. Then, denoted with φ the isomorphism between $\mathfrak{B}/_{\sim}$ and $\widetilde{\mathfrak{B}}$, a measure μ_k extending the state f_k on $\mathfrak{B}$ can be defined as $\mu_k(A) \equiv \tilde{\mu}_k(\varphi([A]))$ for all $A \in \mathfrak{B}$, where $[A]$ is the equivalence class of A with respect to $\sim$. It can be easily verified that $(\mathfrak{B}; \mu_k)$ is a classical representation for C_k.

(b) Let the free Boolean algebra $\mathfrak{B}$, defined as above, together with a normalized measure μ_k be a classical representation for C_k, for all $k \in K'$. By Proposition 1, $\{\widetilde{\mathfrak{B}}_j\}_{j\in J}$ is embeddable into the Boolean algebra 2^N, N as in Proposition 1; let us denote with $\widetilde{\mathfrak{B}}$ the subalgebra of 2^N generated by $\widetilde{\mathcal{G}}$ and with S the set of all homomorphism $\delta \colon \widetilde{X} \longrightarrow \{0,1\}$ induced by the normalized measures μ_k, $k \in K'$ (see Proposition 1). Such homomorphisms are, by construction (see Theorem 0 in [52]), in a one-to-one correspondence with the multiplicative measures of 2^N and can be extended to multiplicative measures on $\widetilde{\mathfrak{B}}$ in a way uniquely determined by the values assumed on the set of generators $\widetilde{\mathcal{G}}$. It follows that each element $\bigcup_{\varepsilon\in H}(-1)^{1-\varepsilon_1}\widetilde{A}_1 \cap \dots \cap (-1)^{1-\varepsilon_n}\widetilde{A}_n$, with $H \subset \{0,1\}^n$, generated by $\widetilde{\mathcal{G}}$ is the zero element if and only if $\sum_{\varepsilon\in H}\prod_{i=1}^n \varepsilon_i\delta(\widetilde{A}_i) + (1-\varepsilon_i)(1-\delta(\widetilde{A}_i)) = 0$, i.e. the extension of δ is zero on such an element, for all $\delta \in S$. Since the homomorphisms in S are induced by multiplicative measures associated, see (10), to the normalized measures μ_k, $k \in K'$, it follows that the ideal $\mathcal{I}$ defined as in (a) coincides with the ideal $\mathcal{I}' \equiv \{B \in \mathfrak{B} | \mu_k(B) = 0 \text{ for all } k \in K'\}$. This implies, as in the proof of Proposition 1, that μ_k induces a normalized measure on $\mathfrak{B}/_\sim$, and consequently a normalized measure $\tilde{\mu}_k$ on $\widetilde{\mathfrak{B}}$, for all $k \in K'$. It can be easily checked that $(\widetilde{\mathfrak{B}}; \tilde{\mu}_k)$ is a classical representation for $\widetilde{C}_k$ for all $k \in K'$. □

In simple terms, we may say that the problem of classical representability can be solved with or without taking into account the algebraic relations between the observables. The latter is the path implicitly taken in several approaches to Kochen–Specker contextuality [7, 21], but not all. For instance, some authors derived non-contextuality inequalities that assume some of the algebraic properties of the QM projectors, often as an alternative version of an inequality that do not assume that. A typical example is the Klyachko–Can–Binicioğlu–Shumovsky (KCBS) [51] inequality. It was provided both in the form

$$\langle A_0 A_1 \rangle + \langle A_1 A_2 \rangle + \langle A_2 A_3 \rangle + \langle A_3 A_4 \rangle + \langle A_4 A_0 \rangle \geq -3, \tag{11}$$

in which the observables A_i are $\{+1, -1\}$-valued and the possible contexts are $\{A_i, A_{i+1}\}_i$ (with sum mod 5) and in the form

$$\sum_{i=0}^{4} p(a_i = 1) \leq 2, \tag{12}$$

which assumes the exclusivity of the events $a_i = 1$ and $a_{i+1} = 1$, from the orthogonality relations of the projectors $\{P_i\}_{i=0}^4$, on the eigenvalue $+1$ for A_i, i.e., $A_i = 2P_i - \mathbb{1}$ and $P_i P_{i+1} = 0$. These two variants of noncontextuality inequalities appear in parallel several other works, see, e.g., [12, 81] and in the original proposals of noncontextuality inequalities, called Kochen–Specker inequalities [56, 67], which explicitly included in the definition of the NCHV model conditions such as $\mathbf{O}'$ and $\mathbf{C}'$, discussed in Sect. 2.3 above. From the perspective of experimental tests of NCHV models and noncontextuality inequalities, it is clear that the ideal approach is to minimize the assumptions involved in the interpretation of the experiment, so that the largest set of NCHV can be disproven; see [16] for a detailed discussion. Theorem 1, then, proves that there is no loss of generality in considering just the probabilistic approach.

4 Automatic Extensions Based on Compatibility Relations

We discussed the problem of classical representability of quantum mechanical predictions from the perspective of the algebraic and probabilistic structures. A natural question is whether there exist properties of these structures that guarantees that the classical representation always exists. Two trivial cases come immediately to mind. The first is the case in which all observables are compatible. In this case, QM already provides a classical representation, i.e., a global probability distribution over all considered observables. In the language of the previous section, we have a Boolean algebra and a measure on it. The extreme opposite, namely, the case of all observables being incompatible, also admits a classical representation, as already noted by Kochen–Specker [52] in the general case and by Bell for two-level quantum systems [11]. In fact, since everything is incompatible there are no algebraic or probabilistic relations to be satisfied, one can thus take as a global distribution the product of the distribution of all observables. These two examples are specials cases of a more general result that connects the graph-theoretic properties of the compatibility relations with the existence of a classical representation for the associated QM predictions. We start by introducing some basic definition and then we present the general result.

We fist need some basic definitions from graph theory. See [9, 32] for more details. A *graph* is a pair $G = (V, E)$ where V is the set of vertices, or nodes, and E is the set of edges, i.e., unordered pairs (i, j) for some $i, j \in V$. Two vertices $i, j \in V$ of a graph are *adjacent*, or *connected*, if $(i, j) \in E$. A set of mutually connected vertices is called a *clique* of the graph. A *path* is a sequence of distinct vertices $v_0, \ldots, v_n$ such that v_i is connected to v_{i+1}, for $i = 0, \ldots, n-1$. A *cycle* is defined in the same way, but with $v_0 = v_n$. A graph is an *acyclic*, or a *tree*, graph, if it contains no cycle. A graph is *triangulated*, or *chordal*, if every cycle of length $n \geq 4$ contains a *chord*, i.e., an edge connecting (v_i, v_{i+2}).

A *hypergraph* is a generalization of the above idea obtained by allowing edges to connect more than two vertices, namely, a pair $H = (V, E)$, where V is the set vertices and E the set of hyperedges, i.e., $E \subset 2^V$, with 2^V the power set of V. Hypergraphs can also arise from graphs; for instance, the clique hypergraph H of a graph G is defined by the same set of vertices and has as hyperedges the cliques of G. If a hypergraph contains only maximal hyperedges, i.e., for each hyperedge E there is no hyperedge E' such that $E' \subset E$, the graph is said to be *reduced*. Given a hypergraph H, we say that H' is the *reduced hypergraph* of H if it is obtained from H by removing all nonmaximal hyperedges. The notion of acylicity for hypergraphs that is relevant for us (many are possible) is given by the following two equivalent definitions. A hypergraph is *acyclic* [9] if it has the *running intersection property*, i.e., if there exists an ordering of the hyperedges, $E_1, \ldots, E_n$, such that

$$E_i \cap (E_1 \cup \cdots \cup E_{i-1}) \subset E_j, \text{ with } j < i, \text{ for all } i. \tag{13}$$

An equivalent definition is that a hypergraph is acyclic if it is the clique hypergraph of a triangulated graph.

We defined a marginal scenario $\mathcal{M}$ as the set of of all contexts for a given set of observables $A_1, \ldots, A_n$. A natural representation of a marginal scenario is given by a hypergraph $H = (V, E)$: Each vertex $v \in V = \{A_1, \ldots, A_n\}$ represents an observable, whereas hyperedges $h \in E = \mathcal{M} \subset 2^{\mathcal{G}}$ represent contexts. Given its relevance, we often discuss the specific case of sharp measurements. For projective measurements in quantum mechanics the K-S property, or Specker's principle,

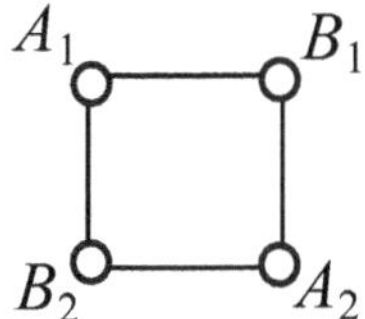

Fig. 2 Compatibility graph associated with the observables of the CHSH scenario corresponding to the marginal scenario $\{(A_i, B_k)\}_{i,j=1,2}$. This graph can also be used to illustrate the basic notions of path and cycles. A path is given by any sequence of sequentially connected vertices, such as (A_1, B_1, A_2). A cycle is a closed path such as (A_1, B_1, A_2, B_2). Notice that this graph is not triangulated, as there exists a cycle of length 4 without a chord

defined in Sect. 3 holds, hence, pairwise compatibility is equivalent to global compatibility. As a consequence, for the case of projective measurements, it is enough to represent the marginal scenario as a graph, interpreting edges as pairwise compatibility relations and cliques as contexts. For the case of sharp measurements, we call such graphs *compatibility graphs*. Interestingly, any graph can be interpreted as a compatibility graph, i.e., there always exists a set of sharp observables that realizes it [46], and the a similar result holds when considering hypergraphs and POVMs [54]. An example of a compatibility graph is given in Fig. 2 for the CHSH scenario. Each vertex represents an observable A_1, A_2, B_1, B_2, and edges connect vertices corresponding to the joint measurements, or contexts, $\langle A_i B_j \rangle$ appearing in (5).

The connection between acyclicity properties of the compatibility graph and the existence of classical representation for the associated marginal scenario was investigated in [18] and independently in [55, 66]. It was shown that for any set of observables such that their compatibility graph is acyclic, i.e., a tree graph, a classical representation always exists independently of the assigned probabilities, i.e., for any quantum state. Moreover, in [18] an extra condition was considered, namely the possibility of a graph representation where nodes represent larger contexts instead of single observables. These represent special cases of a general result for marginal scenario hypergraphs that follows from a theorem by Vorob'ev, originally developed in the context of coalition games [76]. In our terminology, the result can be stated as follows:

Theorem 2 ([74]) Any marginal scenario represented by an acyclic hypergraph admits a joint probability distribution.

Sketch of the Proof An elementary proof of the theorem can be obtained as follows. Let $\mathcal{M}$ be the marginal scenario hypergraph, with hyperedges (contexts) $C_1, \ldots, C_n$. To each of them is associated a probability distribution, let us denote it by $p_i(C_i)$, such that the distributions coincide on their intersection. Since $\mathcal{M}$ is acyclic, we can assume, up to a relabelling of the contexts, that the ordering $C_1, \ldots, C_n$ respects the running intersection property. To prove that a global distribution exists, we proceed by induction on n. For $n = 1$, $p_1(C_1)$ is a valid probability distribution. We then apply the inductive hypothesis. Let us assume that for $n-1$ $p_{(1,n-1)}(C_1 \cup \ldots \cup C_{n-1})$ is a valid probability distribution extending the marginals $p_i(C_i)$ for $1 \leq i \leq n-1$. We want to extend it to $P(C_1 \cup \ldots \cup C_{n-1} \cup C_n)$. By the running intersection property, $C_n \cap (C_1 \cup \ldots \cup C_{n-1}) =: S_n \subset C_j$ for $j < n$. We define $R_n := C_n \setminus S_n$ and

$$p(R_n|S_n) := \frac{p_n(C_n)}{p_j(S_n)}, \tag{14}$$

defining $p(R_n|S_n) = 0$ when $p_n(C_n) = p_j(S_n) = 0$, and we define the joint distribution as

$$p_{(1,n)}(C_1 \cup \ldots \cup C_{n-1} \cup C_n) := p(R_n|S_n)p_{(1,n-1)}(C_1 \cup \ldots \cup C_{n-1} \setminus S_n|S_n)p_j(S_n). \tag{15}$$

It can be straightforwardly verified that this is a valid probability distribution and its marginals coincide with $p_i(C_i)$ for $1 \leq i \leq n$, so it is an extension of the marginal scenario. □

The theorem was originally stated in [76] [translate into English as [77]] and later proven in [74]; see also [75]. The same result was also independently proven in [48, 49, 57]. Intuitively, Vorob'ev's result can be understood as the construction of a global probability by "gluing together" probability distributions on their intersection, the so-called "adhesivity" property [58]. Due to the acyclicity property, i.e., the running intersection property, such a construction can always be made in a consistent way.

In recent year, Vorob'ev's result has been rediscovered in different areas of quantum information and quantum foundations, from contextuality [68, 69, 78] and causal discovery methods [17], and entanglement theory [61]. This result has implication for the computation of correlation polytopes and entropic cones associated with noncontextuality scenarios [6, 15, 53] and more general causal structures [17, 27]. The problem of classical representability for sharp observables, then, can be discussed simply in terms of compatibility graphs. In this case, it is sufficient to verify that the graph is triangulated, since this corresponds to an acyclic hypergraph of the marginal scenario; see the discussion given by [78] for additional details.

We can now comment on the examples we have seen so far. First, the two examples discussed at the beginning: the case of all observables being compatible and all observable being incompatible. In one case, we have a fully connected graph, i.e., consisting of a single clique, which is triangulated, and in the other we have a fully disconnected graph, which is, again, triangulated since there are no paths. In the latter case, Vorob'ev's theorem also tells us that the global distribution is just the product distribution. Another interesting example is the case of a marginal scenario consisting only of nondegenerate observables. In this case, compatibility becomes a transitive relation [30], giving rise only to a graph consisting of disconnected cliques, again admitting a classical representation by Vorob'ev's theorem.

It is not a coincidence that the two simplest example of scenarios not admitting a classical representation form a cycle of length greater than 3, namely, the CHSH scenario, whose compatibility graph is a square (see (5) and Fig. 2), and the KCBS scenario, whose compatibility graph is a pentagon (see (11)).

5 Conclusions

We provided a review of the main results of [18], presented in the modern perspective of Kochen–Specker contextuality and with connections with recent progress in the field. First, we showed that two approaches of classical representability, what we called the KS-type, i.e., classical representation of the algebraic structure and then the probabilistic one, and the Bell-type, i.e., classical representation of the probabilistic structure without algebraic constraints, are fundamentally equivalent. This implies that the general problem can be solved by considering only the problem for the probabilistic structure on a relaxed algebraic structure, i.e., a free algebra. The classical representation for the original algebra can, then, be recovered as a quotient. Over the years, this approach became the standard one in the discussion of noncontextuality, especially in relation with experimental tests, due to the minimality of the assumptions involved in the definition of the associated NCHV model, since no algebraic relations need to be assumed. The results presented here show that there is no loss of generality in this choice. Second, we discussed how the compatibility relations, more precisely, their graph-theoretic structure, guarantee the existence of a classical representation independently of the assigned probability. The result presented in [18] turned out to be a special case of a theorem by Vorob'ev with several implications both for the identification of interesting experimental scenarios, leading to the violation of classical constraints, and the explicit computation of the boundaries of the classical set of correlations (Bell and noncontextuality inequalities), as well as implications even beyond the field of quantum foundations.

References

1. Abramsky, S., Brandenburger, A.: The sheaf-theoretic structure of non-locality and contextuality. New J. Phys. **13**, 113036 (2011)
2. Appleby, D.M.: Contextuality of approximate measurements. arXiv:quant-ph/0005010 [quant-ph] (2000)
3. Appleby, D.M.: Nullification of the nullification. arXiv:quant-ph/0109034 [quant-ph] (2001)
4. Appleby, D.M.: Existential contextuality and the models of Meyer, Kent, and Clifton. Phys. Rev. A **65**, 22105 (2002)
5. Appleby, D.M.: The Bell–Kochen–Specker theorem. Stud. Hist. Philos. Sci. B **36**, 1–28 (2005)
6. Araújo, M., Quintino, M.T., Budroni, C., Cunha, M.T., Cabello, A.: All noncontextuality inequalities for the n-cycle scenario. Phys. Rev. A **88**, 22118 (2013)
7. Badziağ, P., Bengtsson, I., Cabello, A., Pitowsky, I.: Universality of state-independent violation of correlation inequalities for noncontextual theories. Phys. Rev. Lett. **103**, 50401 (2009)

8. Barrett, J., Kent, A.: Non-contextuality, finite precision measurement and the Kochen-Specker theorem. Stud. Hist. Philos. Sci. B **35**, 151–176 (2004)
9. Beeri, C., Fagin, R., Maier, D., Yannakakis, M.: On the desirability of acyclic database schemes. J. ACM **30**, 479–513 (1983)
10. Bell, J.S.: On the Einstein Podolsky Rosen paradox. Physics **1**, 195–200 (1964)
11. Bell, J.S.: On the problem of hidden variables in quantum mechanics. Rev. Mod. Phys. **38**, 447–452 (1966)
12. Bengtsson, I., Blanchfield, K., Cabello, A.: A Kochen–Specker inequality from a SIC. Phys. Lett. A. **376**, 374–376 (2012)
13. Boyd, S., Vandenberghe, L.: Convex optimization. Cambridge University Press, Cambridge (2004)
14. Brunner, N., Cavalcanti, D., Pironio, S., Scarani, V., Wehner, S.: Bell nonlocality. Rev. Mod. Phys. **86**, 419–478 (2014)
15. Budroni, C., Cabello, A.: Bell inequalities from variable-elimination methods. J. Phys. A. Math. Theor. **45**, 385304 (2012)
16. Budroni, C., Cabello, A., Gühne, O., Kleinmann, M., Larsson, J.-Å.: Kochen-Specker contextuality. Rev. Mod. Phys. **94**, 45007 (2022)
17. Budroni, C., Miklin, N., Chaves, R.: Indistinguishability of causal relations from limited marginals. Phys. Rev. A **94**, 42127 (2016)
18. Budroni, C., Morchio, G.: The extension problem for partial Boolean structures in quantum mechanics. J. Math. Phys. **51**, 122205 (2010)
19. Busch, P., Lahti, P., Werner, R.F.: Colloquium: Quantum root-mean-square error and measurement uncertainty relations. Rev. Mod. Phys. **86**, 1261–1281 (2014)
20. Cabello, A.: Comment on "Non-contextual hidden variables and physical measurements". arXiv:quant-ph/9911024 [quant-ph] (1999)
21. Cabello, A.: Experimentally testable state-independent quantum contextuality. Phys. Rev. Lett. **101**, 210401 (2008)
22. Cabello, A.: Specker's fundamental principle of quantum mechanics. arXiv:1212.1756 [quant-ph] (2012)
23. Cabello, A.: Simple explanation of the quantum violation of a fundamental inequality. Phys. Rev. Lett. **110**, 60402 (2013)
24. Cabello, A., Estebaranz, J.M., García-Alcaine, G.: Bell-Kochen-Specker theorem: A proof with 18 vectors. Phys. Lett. A. **212**, 183–187 (1996)
25. Cabello, A., Kleinmann, M., Budroni, C.: Necessary and sufficient condition for quantum state-independent contextuality. Phys. Rev. Lett. **114**, 250402 (2015)
26. Chaves, R., Fritz, T.: Entropic approach to local realism and noncontextuality. Phys. Rev. A **85**, 32113 (2012)
27. Chaves, R., Luft, L., Gross, D.: Causal structures from entropic information: Geometry and novel scenarios. New J. Phys. **16**, 43001 (2014)
28. Clauser, J.F., Horne, M.A., Shimony, A., Holt, R.A.: Proposed experiment to test local hidden-variable theories. Phys. Rev. Lett. **23**, 880–884 (1969)
29. Clifton, R.K., Kent, A.: Simulating quantum mechanics by non-contextual hidden variables. Proc. R. Soc. Lond. A **456**, 2101–2114 (2000)
30. Correggi, M., Morchio, G.: Quantum mechanics and stochastic mechanics for compatible observables at different times. Ann. Phys. **296**, 371–389 (2002)
31. De Simone, A., Pták, P.: On the Farkas lemma and the Horn–Tarski measure-extension theorem. Linear Algebra Appl **481**, 243–248 (2015)
32. Diestel, R.: Graph theory. Springer Publishing Company, Incorporated (2018)
33. Einstein, A., Podolsky, B., Rosen, N.: Can quantum-mechanical description of physical reality be considered complete? Phys. Rev. **47**, 777–780 (1935)
34. Fine, A.: Hidden variables, joint probability, and the Bell inequalities. Phys. Rev. Lett. **48**, 291–295 (1982)

35. Fritz, T., Sainz, A.B., Augusiak, R., Brask, J.B., Chaves, R., Leverrier, A., Acín, A.: Local orthogonality as a multipartite principle for quantum correlations. Nat Commun **4**, 1–7 (2013)
36. Froissart, M.: Constructive generalization of Bell's inequalities. Nuovo Cimento B **64**, 241–251 (1981)
37. Garg, A., Mermin, N.D.: Farkas's lemma and the nature of reality: Statistical implications of quantum correlations. Found Phys **14**, 1–29 (1984)
38. Gleason, A.M.: Measures on the closed subspaces of a Hilbert space. J. Math. Mech. **6**, 885 (1957)
39. Grünbaum, B.: Convex polytopes, 2nd edn. Springer, New York (2003)
40. Gühne, O., Haapasalo, E., Kraft, T., Pellonpää, J.-P., Uola Colloquium, R.: Incompatible measurements in quantum information science. Rev. Mod. Phys. **95**, 11003 (2023)
41. Halmos, P., Givant, S.: Introduction to boolean algebras. Springer (2009)
42. Havlicek, H., Krenn, G., Summhammer, J., Svozil, K.: Colouring the rational quantum sphere and the Kochen-Specker theorem. J. Phys. A **34**, 3071–3077 (2001)
43. Heinosaari, T., Miyadera, T., Ziman, M.: An invitation to quantum incompatibility. J. Phys. A. Math. Theor. **49**, 123001 (2016)
44. Heinosaari, T., Ziman, M.: The mathematical language of quantum theory: from uncertainty to entanglement. Cambridge University Press, Cambridge, New York (2012)
45. Henson, J.: Quantum contextuality from a simple principle? arXiv:1210.5978 [quant-ph] (2012)
46. Heunen, C., Fritz, T., Reyes, M.L.: Quantum theory realizes all joint measurability graphs. Phys. Rev. A **89**, 32121 (2014)
47. Horn, A., Tarski, A.: Measures in boolean algebras. Trans. Amer. Math. Soc. **64**, 467–497 (1948)
48. Kellerer, H.G.: Maßtheoretische marginalprobleme. Math. Ann. **153**, 168–198 (1964)
49. Kellerer, H.G.: Verteilungsfunktionen mit gegebenen marginalverteilungen. Z. Wahrscheinlichkeitstheorie Verwandte Gebiete **3**, 247–270 (1964)
50. Kent, A.: Noncontextual hidden variables and physical measurements. Phys. Rev. Lett. **83**, 3755–3757 (1999)
51. Klyachko, A.A., Can, M.A., Binicioğlu, S., Shumovsky, A.S.: Simple test for hidden variables in spin-1 systems. Phys. Rev. Lett. **101**, 20403 (2008)
52. Kochen, S., Specker, E.P.: The problem of hidden variables in quantum mechanics. J. Math. Mech. **17**, 59–87 (1967)
53. Kujala, J.V., Dzhafarov, E.N., Larsson, J.-Å.: Necessary and sufficient conditions for an extended noncontextuality in a broad class of quantum mechanical systems. Phys. Rev. Lett. **115**, 150401 (2015)
54. Kunjwal, R., Heunen, C., Fritz, T.: Quantum realization of arbitrary joint measurability structures. Phys. Rev. A **89**, 52126 (2014)
55. Kurzyński, P., Ramanathan, R., Kaszlikowski, D.: Entropic test of quantum contextuality. Phys. Rev. Lett. **109**, 20404 (2012)
56. Larsson, J.-Å.: A Kochen-Specker inequality. Europhys Lett **58**, 799–805 (2002)
57. Malvestuto, F.M.: Existence of extensions and product extensions for discrete probability distributions. Discrete. Math. **69**, 61–77 (1988)
58. Matúš, F.: Adhesivity of polymatroids. Discrete. Math. **307**, 2464–2477 (2007)
59. Mermin, N.D.: A Kochen-Specker theorem for imprecisely specified measurement. arXiv:quant-ph/9912081 [quant-ph] (1999)
60. Meyer, D.A.: Finite precision measurement nullifies the Kochen-Specker theorem. Phys. Rev. Lett. **83**, 3751–3754 (1999)
61. Navascués, M., Baccari, F., Acín, A.: Entanglement marginal problems. Quantum **5**, 589 (2021)
62. Peres, A.: Finite precision measurement nullifies Euclid's postulates. arXiv:quant-ph/0310035 [quant-ph] (2003)
63. Pitowsky, I.: The range of quantum probability. J. Math. Phys. **27**, 1556–1565 (1986)

64. Pitowsky, I.: Quantum probability-quantum logic. Lecture Notes in Physics, vol. 321. Springer, Berlin (1989)
65. Ramanathan, R., Horodecki, P.: Necessary and sufficient condition for state-independent contextual measurement scenarios. Phys. Rev. Lett. **112**, 40404 (2014)
66. Ramanathan, R., Soeda, A., Kurzyński, P., Kaszlikowski, D.: Generalized monogamy of contextual inequalities from the no-disturbance principle. Phys. Rev. Lett. **109**, 50404 (2012)
67. Simon, C., Brukner, Č., Zeilinger, A.: Hidden-variable theorems for real experiments. Phys. Rev. Lett. **86**, 4427–4430 (2001)
68. Soares Barbosa, R.: On monogamy of non-locality and macroscopic averages: examples and preliminary results. EPTCS, vol. 172., pp. 36–55 (2014)
69. Soares Barbosa, R.: Contextuality in quantum mechanics and beyond. PhD thesis. University of Oxford (2015)
70. Specker, E.: Die Logik nicht gleichzeitig entscheidbarer Aussagen. Dialectica **14**, 239–246 (1960)
71. Suppes, P., Zanotti, M.: When are probabilistic explanations possible? Synthèse **48**, 191–199 (1981)
72. von Neumann, J.: Über Funktionen von Funktionaloperatoren. Ann. Math. **32**, 191–226 (1931)
73. von Neumann, J.: Mathematische Grundlagen der Quantenmechanik. Springer, Berlin (1932)
74. Vorob'ev, N.: Consistent families of measures and their extensions. Theory Probab. Appl. **7**, 147–163 (1962)
75. Vorob'ev, N.: Markov measures and markov extensions. Theory Probab. Appl. **8**, 420–429 (1963)
76. Vorob'ev, N.N.: On coalition games (in russian). Dokl. Akad. Nauk. SSSR **124**, 253–256 (1959)
77. Vorob'yev, N.N.: Coalition games. Theory Probab. Appl. **12**, 251–266 (1967)
78. Xu, Z.-P., Cabello, A.: Necessary and sufficient condition for contextuality from incompatibility. Phys. Rev. A **99**, 20103 (2019)
79. Xu, Z.-P., Chen, J.-L., Gühne, O.: Proof of the Peres conjecture for contextuality. Phys. Rev. Lett. **124**, 230401 (2020)
80. Xu, Z.-P., Chen, J.-L., Su, H.-Y.: State-independent contextuality sets for a qutrit. Phys. Lett. A. **379**, 1868–1870 (2015)
81. Yu, S., Oh, C.H.: State-independent proof of Kochen-Specker theorem with 13 rays. Phys. Rev. Lett. **108**, 30402 (2012)
82. Yu, X.-D., Guo, Y.-Q., Tong, D.M.: A proof of the Kochen–Specker theorem can always be converted to a state-independent noncontextuality inequality. New J. Phys. **17**, 93001 (2015)

Relativistic Newtonian Gravitation

P. Christillin and G. Morchio

Contents

Abstract The physical principles at the basis of an "elementary derivation" of the General Relativity (GR) effects in a static centrally symmetric field are reexamined. We propose a theoretical framework in which all the GR results follow from the EP, local SR and Newton law in intrinsic coordinates.

1 Introduction

The deviations from Newtonian gravity [1] in the centrally symmetric static case, i.e. slowing of time, light deflection and precession of orbits have played a fundamental role for the confirmation of General Relativity (GR) [2].

It is not clear to which extent the full GR theory is necessary for the prediction of such effects. In particular, the possibility that the above three crucial tests could be derived only or mainly from Special Relativity (SR) was first explored in [3], but its conclusions have been criticized in [10, 11]. The problem of a derivation of

P. Christillin (✉) · G. Morchio
Dipartimento di Fisica "E. Fermi", Università di Pisa, Pisa, Italy
e-mail: christ.pao@gmail.com

G. Morchio
e-mail: morchio@df.unipi.it

A. Cintio, A. Michelangeli (Eds.), *Trails in Modern Theoretical and Mathematical Physics*, https://doi.org/10.1007/978-3-031-44988-8_8

the above effects from a restricted number of assumptions has been reconsidered in many other investigations [4–9] and ref.s therein.

While Schiff's analysis consists substantially in the attempt of a direct intepretation of the metric of the popular Schwarzschild (Ss) solution, the subsequent work has been directed to a similar interpretation of the (less widespread) Painlevé–Gullstrand (PG) solution.

For the Ss solution, as we shall see, the attempt to identify the correponding terms as Special Relativity (SR) effects [3] is only tenable for time, whereas the factor given by SR for the radial coordinate is the inverse of that. The simplicity of the Ss solution, in particular the separation between time and space effects in the invariant interval, both given by Lorentz factors, is in this sense misleading.

The PG metric (originally interpreted as a *physically different solution of the GR equations*) has also been obtained without recurring to Eintein equations, by arguments which seemed so "elementary" to be considered as purely heuristic even by the authors themselves. This sounds indeed a bit paradoxical since the "elementary" and GR approaches lead to the same results, so that the different ingredients of the first cannot be discarded as fortuitous because they are not complicated enough. This judgement is rather the result, in our opinion, of an uncomplete control of the proposed theoretical framework, related to the ambiguities in the identification of the components of the metric tensor as physical objects (see the discussion in Sect. 4.3).

The purpose of the present paper is to reexamine the problem and clarify the ingredients substantiating an elementary derivation of the GR effects in the static centrally symmetric case.

While Schiff's considerations, based on the Ss solution, try to reconstruct the GR results as SR effects, in the approach based on the PG solution the Equivalence Principle (EP) plays the most important role, the metric being the result of an argument involving radial free fall trajectories, provided by a velocity field $v(r)$ associated to gravity and common to all particles in free fall from infinity.

Free fall trajectories can be interpreted as defining a "modified inertial law". They have therefore nothing to do with accelerations in Minkowski space and they should rather be regarded as transporting to all the space-time points the metric of local Minkovski structure assumed at infinity.

This amounts to use the EP in the form that, to the first order in the displacements, the metric remains Minkowski for a free falling observer, in the space and time variables defined by free falling clocks and rods.

The determination of the metric then depends on the free fall law and on a parameter describing a possible space curvature, and the essential question is whether they are affected by the velocity of light. One might indeed conceive a relativistic modification of Newton's law; as we shall see, the result is sensitive to such corrections in the case of Mercury's precession, which disproves their presence.

On the basis of the above interpretation of free fall as a modified inertial law, the basic idea is that only the Newton constant must be relevant for the velocity of radial free fall and for the space geometry parameter [6], in the intrinsic variables mentioned above.

This fixes the free fall law, which holds at all distances and even provides a common description of the outer an inner regions of the Schwarzschild black hole. The velocity of light of course enters, but only in the construction of the invariant interval, which is at the basis of local relativistic physics.

We will show that indeed, in the static centrally symmetric case, the ordinary Minkowski structure at space infinity and the Newton constant uniquely identify radial trajectories and a metric on space-time, determined in physically constructed ("intrinsic") coordinates and coinciding with the one given by the Painlevé–Gullstrand [16, 17] solution of the Einstein equations.

The presentation will be elementary, starting from the modifications of Newton's treatment required by an "operational" construction of the space and time coordinates, which will be done on the basis of the EP for radial free fall trajectories (Sect. 2). The local Minkowski structure will be determined in such coordinates in Sect. 3. To better clarify the elementary character of our arguments, in Sects. 4 and 5 the predictions of the main GR effects will be derived directly from our approach.

The results will be compared to other approaches, and a discussion of the Sagnac effect along similar lines will follow.

2 Newtonian Time and the Equivalence Principle

2.1 Absolute Time from the EP

Newtonian space-time consists of Euclidean space and "absolute" time, a notion which clearly conflicts with the basics of Special Relativity. This justifies the general consensus about the fact that the inclusion of SR into Newton's theory be forbidden from the beginning, leaving as the only solution a complete reformulation of the entire problem of gravity, which is in fact the case of the General Relativity approach. However, it is our aim to show that there is a simple and direct way to give a meaning to "relativistic corrections to Newton gravity", on the basis of the Equivalence Principle and Special Relativity. This can be done in a *substantially unique way*, and reproduces all the General Relativity results for static central gravitational fields.

The basic form of the EP is that feathers and lead balls experience the same gravitational effects. All objects acquire the same free fall velocity: *their inertial and gravitational masses are equal.*

In general, the purpose of the EP is to use frames associated to free falling observers to "eliminate" gravity locally, to the first order in the space-time displacements from a given point.

In the following, we will use the EP to include gravity effects in the Newtonian notion of time. *We assume central symmetry and the validity at large distances of the Newtonian description: space is Euclidean and time satisfies, in that limit, all the synchronization properties defining Minkowski frames in the absence of gravity.*

We also assume stationarity with respect to the time at infinity. Then, the EP, which asserts that gravity effects are not felt by free falling observers, suggests to define time by clocks in free fall from infinity.

We assume therefore that clocks can be arranged to fall freely from infinity, starting at all times, along radial trajectories, with zero initial velocity; they provide a unique notion of time, defined for all space-time points. We will adopt such a notion of time as the **"EP absolute time" t** with a similar role as Newton's absolute time.

Let us also remark that, by the above stationarity assumption, the time needed for free falling clocks to reach a given point in space from infinity is independent from the starting time and that therefore in this sense time intervals at any space point "coincide with time intervals at ∞".

We assume invariance of all the physical laws with respect to the translations of the EP absolute time.

Notice that *no velocity parameter appears in such a construction*, due to the null velocity of the clocks at infinity. Other possible constructions, with a non-zero velocity at infinity, would require the use of SR to account for the initial motion near infinity, preventing a clear separation of roles between SR and EP.

Clearly, the introduction of the above notion of time has important consequences, even on the description of space alone, since the very identification of the space variables and of space geometry concerns, by definition, space-time points at the same time. This affects in particular the notion of space distances, which will be defined as measured by sequences of small rods, with their ends taken at the same time.

The same construction can be performed for radial trajectories reaching infinity (at time $+\infty$) with zero velocity. Our results will be independent of the choice between the corresponding (alternative) notions of time. To be definite we will consider in the following the case of infalling velocities i.e. $v(r) < 0$.

2.2 Newton Laws

Adopting the above reformulation for time, we now endorse the Newton principles of gravitation, for a centrally symmetric static gravitational field:

1. *space is assumed to be Euclidean*. This amounts, due to central symmetry, to the Euclidean relation between radial and angular distances

$$\oint d\,l = 2\pi r \tag{1}$$

2. *radial free fall is asserted to be given by the Newtonian velocity law*

$$v^2(r) = \frac{2GM}{r} = -2\,\Phi(r) \tag{2}$$

As well known, since the same law applies to all bodies, (1) includes, for the case of radial free fall from infinity, the basic form of the EP, i.e. the uniqueness of free fall trajectories for given initial position and velocity.

We emphasize that all the above notions refer to measured space and time intervals; as a consequence of assumption 1), space distances are given by the Euclidean expression in Cartesian coordinates x; the velocity in (2) is defined by the above measurements of space distances and by time intervals given by free falling clocks.

Opposite to the ordinary GR point of view, we do not start from coordinate independence, but rather identify coordinates allowing for a description in the spirit of Newton gravitation. We have therefore

1. *chosen* to discuss the property of space and time in presence of gravity in "intrinsic coordinates", obtained in terms of Euclidean coordinates near infinity, extended by using times and distances measured by radially free falling observers.
2. *assumed* Euclidean space and Newton's free fall law in such coordinates.

The velocity of light does not appear in the above considerations: only the Newton constant G and the mass M enter in the above description of space and time. On such a basis (1) and (2) are forced by dimensional analysis and space flatness at infinity, which determines in particular the value 2π in (1).

2.3 Free Falling Frames and the EP

So far the EP has only been used to derive a notion of time, in which Newton laws have been expressed. Let us now formulate the complete EP. To this purpose, the essential step is to introduce, around each space-time point, local variables associated to Free Falling Frames.

Observers falling along radii starting at infinity (at time minus infinity) with zero velocity employ, around a trajectory $X(t)$, time intervals measured by free falling clocks, and space distances from $X(t)$, measured by (small) rods with ends at the same time and therefore given by Euclidean expressions in $x - X(t)$. Using $dX/dt = v(X(t))$, $v(r)$ given by (2), the corresponding differentials at a space-time point r_0, t_0 are (see Fig. 1)

$$\begin{aligned} dt' &= dt \\ dr' &= dr - v(r_0)dt \\ dx'_\perp &= dx_\perp \end{aligned} \tag{3}$$

with $dx'_\perp$ the space displacements in the directions orthogonal to r, given by the differential $dx_\perp$ of local Newton cartesian coordinates orthogonal to the radius.

Even if (3) have the form of Galilei transformations, they have a very different nature, since they describe "small" (infinitesimal) displacements in Free Falling Frames, on the l.h.s., in terms of global coordinates in the r.h.s.

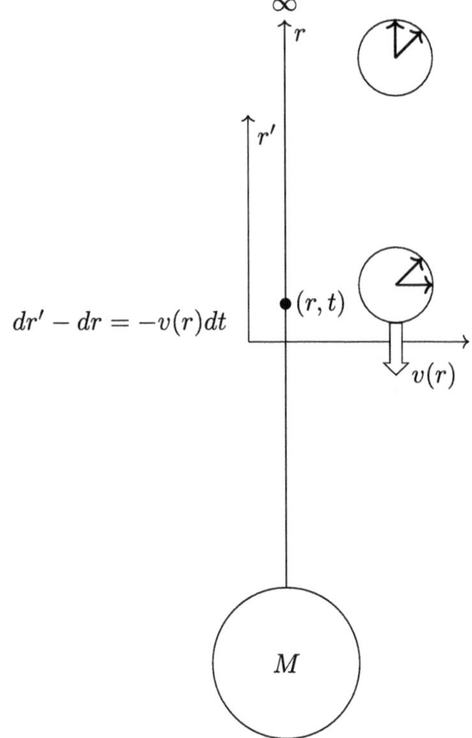

Fig. 1 The clocks of the infinitesimal Equivalence Principle Inertial Frames (EPIFs), starting at any time, associate to each point the time $t' = t$. The relativistic space-time effects of gravitation are determined by the infinitesimal invariant Minkowski interval $ds^2 = c^2 dt'^2 - dr'^2$ in the EPIFs coordinates, which gives the metric in the global coordinates (r, t) through (3)

It is a fundamental fact that the above relations are not *given by Lorentz transformations. They have the form of Galilei transformations because they arise from the use of a common, "absolute", notion of time.*

Equation (3) should not be interpreted therefore as a low velocity approximation of Lorentz transformations (as in [7]). They are exact in our approach, for all values of the free fall velocity. By the above derivation, they hold independently of (2), which only fixes the value of v; morever, only the third equation should be modified (by an r dependent factor) in the absence of the Euclidean relation, (1).

In fact, in the present Section we are only discussing the modifications to the *inertia principle*, which holds both in relativistic and non relativistic physics and has little to do with the velocity of light, which was in fact never mentioned in the above discussion.

Clearly, the use of a Lorentz transformation in (3) would lead to a trivial local Minkowski structure, excluding gravity effects on clocks and light deflection.

The differentials dx' and dt' representing the description of space and time in Free Falling Frames, *to the first order in the displacements* from a free fall trajectory, are the substitute of coordinates satisfying the inertia principle. They will be denoted as (infinitesimal) Equivalence Principle Inertial Frames (EPIFs) and are the object of the following form of the *Einstein's Equivalence Principle*:

All the physical laws which can be written, in the absence of gravity, in inertial frames in terms of local variables and their first order variations around each point, hold true in the presence of gravity in terms of the same variables in EPIFs.

It is important to notice that the *differentials* defining EPIFs *do not in general define coordinates, even locally*. In fact the differential form

$$dr' = dr - v(r)dt$$

is not integrable, unless the velocity field $v(r)$ is constant (the trivial inertial case), since

$$\frac{\partial v}{\partial r} = -\frac{\partial 1}{\partial t} = 0$$

is precisely its integrability condition.

On the contrary for the time variable the restriction to the infinitesimal interval is not essential and in fact $t = t'$ is the "gravity free" time measured by clocks on EPIFs, which does not suffer from the limitations produced by gravity on the space variables.

Clearly, even if the formulation of the EP only uses EPIFs, its implications crucially depend on the relation between the above differentials at different points, given by (3). In other terms, the introduction of global coordinates and the expression of the EPIF differentials in terms of them is an essential step for an effective use of the EP.

3 Relativistic Physics

3.1 Basic Gravitational Effects

a) Newton's Mechanics from the Principle of Least Action in EPIFs

Let us first show how the use of the EP reproduces classical non-relativistic mechanics for a particle in the gravitational field of a mass M.

Classical Mechanics can be indeed formulated in terms of the principle of least action in inertial frames. The implementation of the EP in the non relativistic free Lagrangian $\mathcal{L} = m/2\,\dot{x}^2$ is immediate. Following (3), it is enough to express the velocity in

$$\dot{r} \to \dot{r} - v(r)$$

The principle of least action in EPIFs for a free particle in the presence of gravity is thus given by

$$\delta \int dt \left(\frac{m}{2}[(\dot{r} - v(r))^2 + r^2\dot{\theta}^2]\right) \tag{4}$$

The Lagrange equations yield for the radial coordinate

$$\frac{d}{dt}(\dot{r} - v(r)) + (\dot{r} - v(r))\frac{dv}{dr} - r\,\dot{\theta}^2$$
$$= \ddot{r} - \frac{d}{dr}\left(\frac{v^2(r)}{2}\right) - r\,\dot{\theta}^2 = 0 \qquad (5)$$

Since $1/2\,v^2(r) = GM/r$ the Newton's radial equation of motion is obtained, and the same holds also for the angular variables. A constant v would only amount to a change of inertial frame.

b) The Invariant Minkowski Interval

Relativistic physics is governed by the "infinitesimal" invariant interval. By the EP, the invariant interval has the standard Minkowski form in EPIFs, in which the ordinary inertial frame laws hold and light propagates isotropically and always with velocity c, to first order in space-time displacements:

$$ds^2 = c^2 dt'^2 - dr'^2 - dx_\perp'^2 \qquad (6)$$

The local (first order) validity of the principles of SR implies that *ds^2 is still given by the same expression*, (6), in the coordinates employed by any observer around the given space-time point, to the first order, independently of his motion. All the SR results hold locally, *for all observers (on arbitrary trajectories) to first order in the coordinates defined by their clocks and rods*, with the Minkowski interval given by the ordinary expression.

The above expression of the EPIF differentials in terms of globally defined variables allows to write the Minkowski intervals, all of the same form in their EPIF variables around different points, in global coordinates:

$$ds^2 = c^2 dt^2 - (dr - v(r)dt)^2 - dx_\perp^2$$
$$= c^2(1 - v^2(r)/c^2)dt^2 + 2v(r)\,dt\,dr - dr^2 - dx_\perp^2 \qquad (7)$$

(7) gives nothing else than *the Painlevé* [16]–*Gullstrand* [17] *metric (P–G), a solution of the GR equations in a central field*, obtained here (as a solution of no whatsoever equation other than Newton's law) *on the pure basis of Euclidean space and absolute "free fall" time*. Even if built on Euclidean space and absolute time, it represents a non-Minkowskian space-time, due to the crossed term. The P–G metric is equivalent to the Schwarzschild [18] metric, the relation being given by a change of the time variable (see Sect. 4.2).

Clearly, in our approach, SR enters at a different stage with respect to the description of radial free fall. The latter is given by the Newton free fall velocity in global coordinates; the velocity of light enters in the local relativistic structure of space time, which is trivial in EPIFs and globally determined by (6) and (3).

c) Clocks Ticking and Red Shift

How time flows in a gravitational field for observers at rest, in the above (P–G) coordinates, is immediately got from the P–G metric. Actually, the notion of rest is independent of coordinate transformations preserving stationarity of the metric tensor. By setting $dr = 0$,

$$d\tau^2 = (1 - v^2(r)/c^2)dt^2 = (1 - \epsilon(r))dt^2 \tag{8}$$

relates the (P–G) Newtonian free fall absolute time t to the relativistic invariant interval $d\tau$ measured by observers at rest in the P–G coordinates, thus defining their proper time.

The parameter

$$\epsilon(r) \equiv 2GM/rc^2 = v^2(r)/c^2 \equiv -2\Phi(r)/c^2 \tag{9}$$

does not appear for non relativistic mechanics and enters in our approach only through the invariant interval.

By time translation invariance and linearity of propagation, the frequency of light propagating from infinity remains constant in (the above, P–G) time, and therefore frequencies observed by observers at rest are given by (the inverse of) the above relation. Because the velocity of light remains the same for all observers, *this can also expressed in terms of wavelengths*, relating the one at ∞, λ_∞ to the one at r λ_r,

$$\lambda_r = (1 - \epsilon(r))^{1/2}\, \lambda_\infty \simeq (1 - \epsilon(r)/2)\lambda_\infty$$

For small radial distances h,

$$\Delta\omega/\omega \simeq \epsilon(r)/2\ (h/r)$$

i.e. the well known red shift. For moving sources, one has to add the usual Doppler effect.

d) Light Cones

Light velocity is obtained by setting to zero the invariant interval (7). The velocity in directions orthogonal to the radius takes the value

$$c_\perp = rd\theta/dt = c(1 - v^2/c^2)1/2\,. \tag{10}$$

Along the radius, the velocity is

$$c_r = dr/dt = \pm c + v(r)\,, \tag{11}$$

$v = v(r)$, given by (2) ($v < 0$). *Both equations directly follow from* (3) *by ordinary (Galilean) vector composition of the (isotropic) velocity c in EPIFs with the EPIF velocity* v,

$$\mathbf{c}_{PG} = \mathbf{c} + \mathbf{v}(r) \,. \tag{12}$$

For comparison, we recall the results for the Schwarzschild metric

$$ds^2 = (1 - v^2/c^2)c^2\, dt_S^2 - dr_S^2/(1 - v^2/c^2) - r^2 d\Omega^2 \,. \tag{13}$$

There the radial velocity is

$$dr/dt = \pm c\,(1 - v^2/c^{)} \tag{14}$$

whereas the tangential one is the same as in the P–G coordinates.

Thus the Schwarzschild light cones shrink for decreasing distances down to a pseudo singularity at $R_S = 2MG/c^2$. In the P–G coordinates, they rotate, without any singularity. If one believes in the validity of the extrapolation of Newtonian dynamics to such extremes (to be discussed later), a physical effect emerges from the SR constraints in EPIFs.

Indeed, when $v(r) < -c$ light cannot propagate outwards for positive times. This happens below R S, where the P–G metric describes a black hole. Reversing the sign of time is equivalent to reversing the free fall velocity in the above construction. In this case, light cannot propagate inwards (since this time $v(r) - c > 0$), for positive times and the corresponding P–G metric describes a "white hole".

Notice that the result only depends on the free fall velocity (in intrinsic coordinates) exceeding c at some radius, and has therefore nothing to do with any "interior dynamics" below such a radius.

e) Relativistic Mechanics

The formulation of relativistic mechanics is immediate via the EP, which only amounts to substitute in the variational principle the ordinary Minkowski intervals with the Minkovski intervals in EPIFs. The corresponding action is therefore obtained, as in the non-relativistic case, by the substitution $\dot{r} \to \dot{r} - v(r)$

$$\delta A = dt = \delta \int ds = \delta \int \mathcal{L}\, \delta \int dt\,(mc^2)\sqrt{(1 - 1/c^2[(\dot{r} - v(r))^2 + r^2\dot{\theta}^2])} \tag{15}$$

The equation of motion are given by the corresponding Euler Lagrange equations. They are equivalent to the geodesic equations in the metric defined by the above invariant interval, i.e. in the P–G metric.

Both solve in fact the same variational problem, the ordinary geodesic equations being obtained from a parametrization of trajectories with the proper time and the substitution of the Lagrangian with its square (which is allowed since the Lagrangian associated to the proper time parametrization takes the value 1 on the solution of the stationarity problem). Let us also mention that the use of proper times is inappropriate in the many body case.

3.2 Light Deflection

Since our invariant interval coincides with the one of the Painlevé–Gullstrand solution of the Einstein equations and the principle of stationary action amounts to geodesic motion in the corresponding metric, all the results of GR for the dynamics of particles and light follow.

We show below that these results can also be derived directly in a rather elementary way.

The problem of light bending has been paramount in assessing the view of space distortion associated to the Schwarzschild solution. Indeed, classically, (see e.g. [19]) it is well known that Newtonian mechanics can account for light deflection, at variance however by a factor of 2 from the GR result and from experimental data. This is explained in Schwarzschild coordinates by saying that Newton just reproduces the time part (g_{00}) of the metric tensor and that the space part g_{ii} is the new fundamental contribution of GR.

Let us first recall the classical treatment and then discuss the contributions arising from the EP. We restrict to first order in the (relativistic) parameter $\epsilon(R)$, which completely covers the experimental situation.

Take a luminous ray grazing the sun, coming from infinity and calculate the light deflection observed at large distances (on the earth, practically at infinity).

a) Newtonian Light Deflection

To first order, the bending angle of light associated to the unperturbed trajectory ($x = ct$, $y = R$), is given by $\theta(x) = -c_y/c$, c_y the y component of the light velocity, c the unperturbed velocity. If light is assumed to accelerate according to Newton's law

$$\theta(x) \simeq -c_y(x)/c = \int_{-\infty}^{t} \frac{\partial}{\partial R} \Phi(ct',\, R)\, dt'/c = \int_{-\infty}^{x} \frac{\partial}{\partial R} \Phi(x',\, R)\, dx'/c^2 \qquad (16)$$

Φ the Newton potential, (2). The deflection angle is then obtained by integration over the whole x axis in terms of the relativistic weak field parameter $2GM/c^2R = \epsilon(R)$ as

$$\theta = \theta(\infty) = \epsilon(R) \qquad (17)$$

b) Wave Fronts and Light Velocity

In order to discuss light deflection as a refraction effect produced by position dependent velocities, let us see how it can be obtained in general, for small angles, in terms of wave fronts. The Newtonian result will be shown to follow from the light

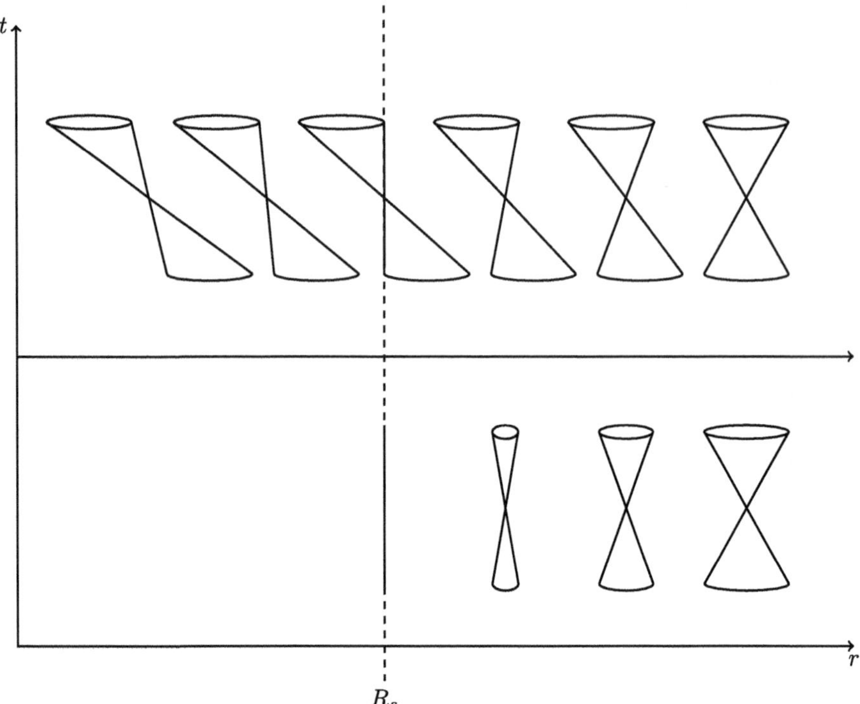

Fig. 2 Light cones as a function of r in the Painlevé–Gullstrand and Schwarzschild metric respectively. A singularity at R_S only arises for the second metric; in the P–G coordinates, because of the distorted light cone, nothing can simply get out of the hypothetical black hole

velocity given by the pure time component of the Schwarzschild metric, while the full GR result will follow from the above (very elementary) P–G light velocity.

We will consider the propagation of light in a first approximation along straight lines (see Fig. 2)) at different heights, with velocity $dx/dt = c(x, y)$ and calculate the orientation of wave fronts, at $y \simeq R$.

Since, as motivated before, light frequency ω remains constant in the P–G time t (and also for the Schwarzschild time, see below), the phase φ of the wave changes with the time it takes a wavefront to travel in the x direction

$$d\varphi = dx/d\lambda(x) = \omega dx/c_x(x, y) .$$

c_x the velocity of propagation along the x axis. Thus, to first order, the difference of the wavefronts at different heights

$$\frac{\partial}{\partial y}\varphi(x, y) \simeq -\int_{-\infty}^{x} c_x(x', y)^{-1}\omega \, dx' \tag{18}$$

determines the bending. With the same sign convention as above the bending angle θ of the wavefront at $y \simeq R$ is given by

$$\theta(x) = -\frac{\partial}{\partial R}\varphi(x, R)\,\lambda(x) \simeq -\int_{-\infty}^{x} (\frac{\partial}{\partial R} c_x(x', R)^{-1}) c_x(x, R) dx' \quad (19)$$

and the deflection angle is $\theta = \theta(\infty)$.

c) Schwarzschild

The modifications of the light velocity given by the "pure time component" of the Schwarzschild metric, $ds^2 = c^2(1 - v(r)^2/c^2)dt^2 - dx^2$, are independent of the direction and given by

$$c(x, y) = c(1 - v^2(r)/c^2)^{1/2} \simeq c(1 - \epsilon/2) \quad (20)$$

and therefore (19) gives the deflection angle

$$\theta \simeq \int_{-\infty}^{\infty} \frac{\partial}{\partial R} \Phi(x, R)/c^2 \, dx \, , \quad (21)$$

which coincides with the Newtonian expression, (16) and (17). The Newtonian equation of motion is in fact equivalent to the stationarity principle for the optical length in such a metric.

For the complete Schwarzschild metric, the velocity of propagation of light along the x axis, with $y = R = r\cos\alpha$, is given by

$$c^2 dx^2((1 - v^2/c^2)^{-1} \sin^2(\alpha) + \cos^2\alpha) = (1 - v^2/c^2)dt^2$$

which gives

$$c_x(x, R)^{-1} \simeq c^{-1}(1 + \epsilon(r)(1 - R^2/(x^2 + R^2)/2)) \, . \quad (22)$$

The integral of the last term is finite and independent of R, so that the result changes by the well known factor of 2.

d) Deflection from EPIF Galilean Light Velocity Composition

In our approach light velocity is given by the Galilei formula, (12).

By imposing to first order its propagation along the x axis, the y components cancel in (12) and therefore the x component is given by (see Fig. 3)

$$c_x = (c^2 - v_y^2(r))^{1/2} + v_x(r)$$

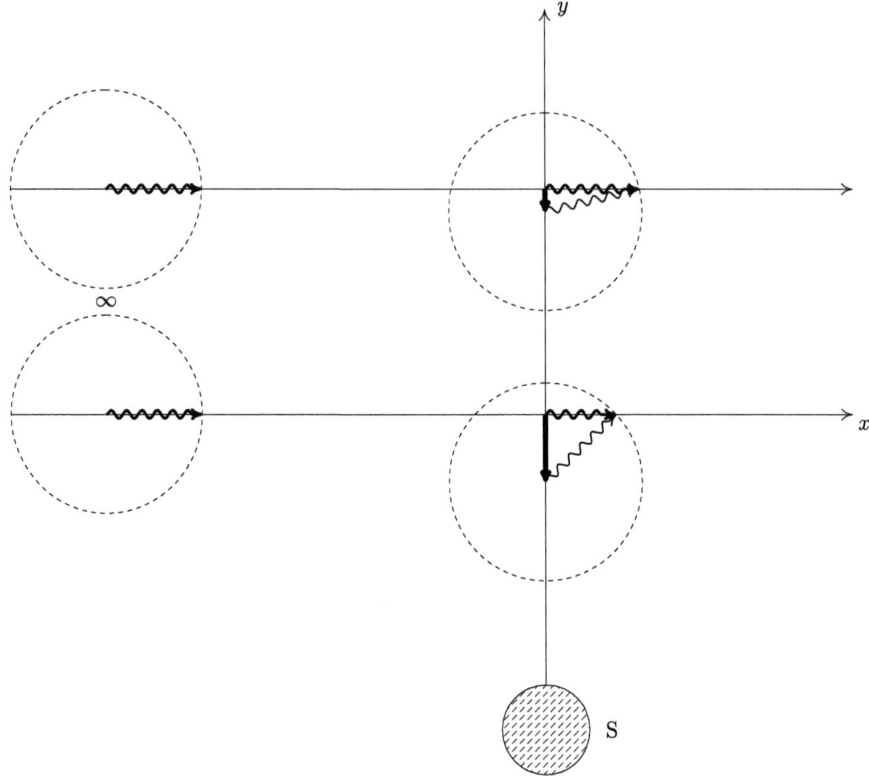

Fig. 3 Spherically symmetric light propagation in the EPIFs is vectorially composed with the free fall velocity $v(r)$ to yield a resultant c_x along the unperturbed trajectory. The phase variation and hence the bending of the wave front as a mirage effect comes from the dependence of the x velocity c_x on the height y

Since $v(r)/c$ is of order $\epsilon(R)^{1/2}$, to first order in ϵ

$$\begin{aligned} c_x^{-1} &= c^{-1}((1 - v_y^2(r)/c^2)^{1/2} + v_x(r)/c)^{-1} \\ &\simeq c^{-1}(1 + v_y^2/2c^2 - v_x(r)/c + v_x^2(r)/c^2) \\ &= c^{-1}(1 - v_y^2/2c^2 - v_x(r)/c + \epsilon(r)) \end{aligned} \tag{23}$$

The second term is proportional to $\Phi(r)R^2/(x^2 + R^2)$ and its integral is independent of R, as above. The third is antisymmetric in x and its integral vanishes; the last term gives the GR result,

$$\theta = \theta(\infty) = 2\epsilon(R)$$

The factor 2 has emerged from *the second order* expansion of the inverse of the velocity in the parameter $\epsilon(r)^{1/2}$.

Notice that in the P–G coordinates light velocity is different on its way towards and away from the source of gravitation, due to the linear dependence on the free fall velocity. As we have seen, light deflection arises as a second order effect in the free fall velocity. Linear terms are present in the deflection angles at finite distances from the source; they depends crucially on the notion of simultaneity implicit in the definition of wave fronts, which is different in different coordinate systems.

3.3 Perihelion Precession

Let us show how the perihelion precession can be calculated directly, **for motion close to a circular orbit and to first order in** ϵ, from the above equations of motion.

The relativistic Lagrangian is given by (15). The angular equation of motion is

$$d/dt\, \partial\mathcal{L}/\partial\dot{\theta} = -d/dt(\mathcal{L}^{-1}r^2\dot{\theta}) = 0\,, \tag{24}$$

i.e.,

$$L \equiv r^2\, d\theta/dt\ \mathcal{L}^{-1} = r^2 d\theta/ds = \text{const} \tag{25}$$

The radial motion is given by

$$d/dt\,(\partial\mathcal{L}/\partial\dot{r}) - \partial L/\partial r = 0\,. \tag{26}$$

By using the proper time, here denoted by s, $ds/dt = \mathcal{L}$, we obtain

$$d^2r/ds^2 = v(r)(d/ds\mathcal{L}^{-1}) - \mathcal{L}^{-2}d\Phi/dr + L^2/r^3 = 0\,.$$

This equation is equivalent to the Schwarzschild equation

$$d^2r/ds^2 = -d\Phi/dr + L^2/r^3\,(1 - 3/2\epsilon) \tag{27}$$

since they solve the same variational problem in the same variables, and therefore implies the GR result for the perihelion. Even if the P–G radial equation is more involved (a dependence on r and dr/ds being hidden in the terms involving the Lagrangian), a calculation of the precession effect in the above approximations is straightforward.

Let us derive it directly from the equations of motion in our (P–G) time variable. The radial equation (26) reads

$$-d/dt(\mathcal{L}^{-1}\dot{r}) + (d/dt\mathcal{L}^{-1})v(r) - \mathcal{L}^{-1}d\Phi/dr + \mathcal{L}\,L^2/r^3 = 0\,.$$

Multiplying by $\mathcal{L}$,

$$d^2r/dt^2 = -(\dot{r} - v)\mathcal{L}d/dt\mathcal{L}^{-1} - d\Phi/dr + \mathcal{L}^2\,L^2/r^3 \tag{28}$$

Circular orbits are given by

$$-\,d\Phi/dr + \mathcal{L}^2\,L^2/r^3 = 0 \tag{29}$$

and their frequency is

$$\omega_\theta^2 \equiv \dot{\theta}^2 = \mathcal{L}^2\,L^2/r^4 = GM/r^3\,, \tag{30}$$

the same as in the Newton case.

(29) differs from Newton's equation by the term

$$\mathcal{L}^2 = 1 + 3\epsilon$$

For small oscillations around a circular orbit the term $\dot{r}\mathcal{L}\dot{\mathcal{L}}^{-1}$, quadratic in $\dot{r}$, can be dropped; using $v^2 = -2\Phi$,

$$\mathcal{L} \sim 1 + (\Phi + v\dot{r} - L^2/2r^2)/c^2$$

and the circular orbit constraint, the terms linear in $\dot{r}$ of (28) are readily seen to cancel, corresponding to the absence of damping. As a result, the only contribution of the first term in the r.h.s of (28) is

$$-v^2/c^2 d^2r/dt^2 = -\epsilon d^2r/dt^2\,.$$

(28) therefore reduces to

$$\begin{aligned}(1+\epsilon)d^2r/dt^2 &= -d\Phi/dr + \mathcal{L}^2\,L^2/r^3\\ &= -d\Phi/dr + (1-\epsilon-L^2/r^2c^2)L^2/r^3\,.\end{aligned} \tag{31}$$

The frequency for circular orbits is thus given by

$$\begin{aligned}\omega_r^2(1+\epsilon) &= d^2\Phi/dr^2 - \mathcal{L}^2 d/dr L^2/r^3 + L^2/r^3 d/dr(\epsilon(r) + L^2/r^2c^2)\\ &\sim \omega_\theta^2 + d\Phi/dr\;2d\epsilon/dr\end{aligned}$$

so that

$$\omega_r^2 \sim \omega_\theta^2(1-3\epsilon) = GM/r^3(1-3\epsilon)$$

i.e.

$$\omega_r/\omega_\theta \simeq (1 - 3/2\epsilon)$$

and the precession angle is therefore

$$\frac{\Delta\phi}{2\pi} = \frac{3}{2}\epsilon = 3\frac{GM}{rc^2} \tag{32}$$

Since $GM/rc^2 = v^2/c^2$, v the Newtonian velocity for circular orbits, the result can be interpreted as a correction given by the relativistic equation of motion, in a gravity field which is described by Newton law (in intrinsic coordinates).

A "relativistic" ($O(GM/rc^2)$) correction to Newton law would result in an additional term, of the same order, in (32). On the contrary, such a correction would give a second order contribution to time ticking and light bending; therefore, the perihelion precession, usually interpreted as the test of GR (see e.g. Schiff [3]), can be equally seen as the test of Newton law in intrinsic coordinates.

Notice, in connection to MOND [20], that the relativistic corrections appearing in the above equations have nothing to do with effects of order $O(v^2/r)$. In other words comparable velocities (e.g. Earth and orbiting HI lines), even at very different radii, have the same sort of relativistic corrections (with negligible effect in the second case).

4 Dynamics, Metrics, Observables and all that

4.1 The Newtonian Fall Velocity and the Mass

So far our treatment has relied on a somewhat abstract framework, assuming that the free fall velocity is given by Newton's law. Here we want to ascertain to which extent this assumption follows from dynamical considerations in Minkovski space.

To start with, assuming that energy is the source of gravitation, the mass in the potential term in Newton formula should be corrected both by the self energy and by the kinetic term.

Energy conservation for our free falling particle would thus read

$$m_0 v^2/2 = \frac{GMm_0}{r}(1 - GM/c^2 r + v^2/2c^2) \tag{33}$$

or

$$v^2/2 - \frac{GM}{r} = \frac{GM}{c^2 r}(v^2/2 - GM/r) \tag{34}$$

whence

$$v^2 = 2GM/r \tag{35}$$

The result only uses conservation of energy. *This puts Newton's law on a somewhat safer ground in the sense that the above energy corrections cancel out.* It is paramount to underline that *the self energy correction to the mass, which embodies the fact that the graviton is itself a source of gravity, a relativistic effect, is cancelled for radial trajectories by the relativistic kinetic corrections of the gravitational mass.*

This has to be contrasted with what happens in the PPN parametrization (see e.g. [21]), where the non linearity of Einstein equations appears in the form (destitute of any measurement prescription)

$$h_{00} = 2GM/c^2 x(1 - GM/c^2 x)$$

Thus one concludes that, remarkably, the non linearity of gravitation may depend on the formulation.

We also notice that the Newtonian $1/r^2$ form of the force is crucial in canceling possible contributions from external masses, within a reasonable schematization of the outer world as a homogeneous sphere.

One might inquire about other relativistic corrections to the preceding expression. A possible relativistic extension of (35) is

$$\frac{m_0}{\sqrt{1-v^2/c^2}} - \frac{GMm_0}{c^2 r}(1/\sqrt{1-v^2/c^2} - GM/c^2 r) = m_0 \qquad (36)$$

The terms in brackets sum up to 1 as in (33)-35 so that the only possible correction is given by the l.h.s. It is important to notice that the inertial mass in the l.h.s. cannot be gravitationally corrected by an additional factor $-GM/c^2 r$ since one would get in this case for the escape velocity $v^2 = 4GM/r$

Additional higher order terms in v^2/c^2 in the l.h.s. of (36) are therefore the only possible modification of the free fall law for radial trajectories. They would result in first order corrections to Mercury's precession, which are ruled out.

In that respect the LLR experiments [22] of the free falling Earth–Moon system in the gravitational field of the Sun should also exclude such relativistic corrections with higher accuracy. Gravitational quantum interference experiments [23] are very far from providing possible additional information.

The above arguments imply that free fall is determined only by GM without kinetic and self energy corrections and that gravity does not contribute to the inertial mass m_0. Thus all GR corrections to Newtonian dynamics simply follows from the treatment of *free fall where one mass is enough, and cancels out in the motion in a given field.* This can be summarized as

$$m_I = m_0 = m_0(1 - GM/c^2 r + v^2/2c^2) = m_0 = m_G$$

I standing for inertial.

In conclusion our treatment based on Newton's law and the free falling frames gets validated beyond expectations and all the geometrical relations of GR are direct consequences. Of course the possible distinction between different substances (WEP), ruled out by the terrestrial experiments of Eöt–Wash [24], is automatically implemented here. Let us finally comment on the role of moving frames which play such a fundamental role in SR and GR, although sometimes with a misleading interpretation.

In the former case they represent a physical entity: e.g., in the flying muon frame the atmosphere thickness is shorter than the one measured on earth and the time needed to reach it is correspondingly shorter.

In the latter the accelerating falling frame is the basis for a construction which describes gravity in terms of local "inertial" frames. Thus the popular expression that GR dilates spacial distance has no relation with the physical fact that in SR distances for moving particles are shorter. That statement should be supplemented by the phrase: in the Ss metric, whereas in the P–G it does not happen as also stressed by Mizony [25]. In that sense the P–G metric, with its clear and physically founded combination of local SR with a global "Galilean" free fall law (also preserving at infinity the gravity free absolute time), has an interpretation which is "closer to reality".

4.2 From the P–G to the Ss Metric

In this paragraph the relation of the Ss metric the P–G one will be elucidated by explicitly considering the combination of Galilean transformations for the free fall with the local space time Minkowski structure which yield the physical time and length measured at rest. In short:

$$(dr, dt)_{\text{global coord.}} \Longrightarrow_{\text{Galilei}} (dr_0, dt_0)_{\text{EPIF}} \Longrightarrow_{\text{Lorentz Tranf.}} (d\rho, d\tau)_{\text{at rest}} \tag{37}$$

i.e., on the basis of the globally defined coordinates (r, t), extending the Newtonian coordinates at infinity, EPIF differentials are given by (3), (eliminating gravity through free fall) and then a SR transformation yields the frame at rest at a given point. Time and space coordinates at rest, $d\tau$, $d\rho$, are thus obtained as

$$d\tau = \gamma(v)(dt_0 + v dr_0) = \gamma(v)(dt + v(dr - vdt)) = dt/\gamma(v) + \gamma(v)vdr \tag{38}$$

$$d\rho = \gamma(dr_0 + vdt_0) = \gamma(dr - vdt + vdt) = \gamma dr \tag{39}$$

where $\gamma(v) = \gamma[v(r)] = \frac{1}{\sqrt{1-v^2/c^2}} = \frac{1}{\sqrt{1-2GM/c^2r}}$

We stress the role of the Galilei transformation in the above derivation. The last equation shows a dilation of lengths, contrary to the ordinary SR effect.

According to the second to last, times are indeed shortened, as in SR, were it not for a space dependent term. Since that term *does not alter the time rate at a given space point, it can be dropped through a redefinition of the global time*:

$$dt \to dt + v/c^2/(1 - v^2/c^2)dr = \gamma d\tau \equiv dt_S \tag{40}$$

Together with $dr_S \equiv dr$, this gives the Schwarzschild coordinates and the Schwarzschild metric

$$ds^2 = c^2 d\tau^2 - d\rho^2 = (1 - v^2/c^2)c^2 dt_S^2 - dr_S^2/(1 - v^2/c^2) - r^2 d\Omega^2 \tag{41}$$

The difference between the two notions of time is particularly evident in the treatment of light deflection, where in the P–G metric a linear factor in the velocity shows up, with a presumed huge effect when observed half way (e.g. on earth in the measurement of parallaxes). The point is that one must not confuse the P–G notion of simultane- ity with the one defined by clocks at rest (and therefore on earth apart from an easy relativistic correction for its motion).

It should also be noticed that the Ss metric has spurious singularities not only at the S radius $r = 2GM/c$, as well known, but also at $r = \infty$ where the Newton time t and t_S differ by $\simeq \sqrt{r}$.

In conclusion the P–G coordinates have the advantage of the underlined physical foundation, the lack of singularities, no necessity of an equation of motion beyond Newton's law and can be directly and simply applied to all processes, apart from the discussion of equal time geometrical effects, as the parallax, where the Ss coordinates give a notion of simultaneity which coincides with the one at rest.

Let us recall that the requirement to eliminate the off diagonal term of the P–G metric is generally accomplished just by redefining time in an ad hoc way, as in (40), without any discussion about its physical meaning, nor about its effects in the interpretation of experiments.

Finally, let us underline an inherent "paradox" of GR. The pretension that coordinate independence of the formulation is fundamental backfires, in the sense that Newton's absolute time not only has the right of citizenship, but gives rise to an independent description based on fundamental physical motivations.

4.3 On Alternative Derivation

Ever since the appearance of GR, the endeavor to find other solutions than Schwarzschild's, to "derive" it from SR and to eventually propose alternative theories has been paramount.

To start with, let us recall that Einstein's rebuttal of the Painlevé–Gullstrand *solution* has led to an ostracism (their metric is not even mentioned in most textbooks) which has lasted till almost the end of the last century. Only recently the P–G metric has been reevaluated as a singularity-free solution. In addition, it has been realized that it could be obtained directly from basic principles, without recursion to GR.

This possibility has provoked a heap of warnings: that it could be only heuristic, that it only may apply to the weak field case, accompanied as well by the (trivial) argument that it cannot reproduce all of GR results. In connection with the first points we want to comment on some of the most relevant and cited articles.

Schiff's [3] work had already been criticized by Schild [10]. The usual result for time had been obtained by using a SR argument, comparing local time with that of gravity free infinity via a flying, time-shortened, clock. However his (incomplete) argument about space cannot be correct since in the end, contrary to SR, the velocity of light is not constant nor isotropic. His statement about Mercury's perihelion being the crucial test of GR has already been commented upon above.

Kassner's [9] work is relevant in the present context because of his discussion of the necessity of supplementary assumptions in order to derive the Ss metric on the basis of pre-general-relativistic physics alone, i.e., SR, the Einstein EP and the "Newtonian limit".

This is not contradictory with our findings. As a matter of fact, the Newton law is used by us globally, not only to first order at infinity, supplemented by the two (almost unescapable because of our motivations) subsidiary conditions on space (length of the circumference) and absolute time. They can be seen as substitutes of Kassner's two additional "postulates", which serve the same aim but which are, in our opinion, less transparent and motivated.

Czerniawsky's point of view [4, 5] is the closest to ours. Our assumption on the Euclidean properties of space, at equal "free fall" times, is somewhat hidden in his considerations about the EP. As a result, his treatment does not include the dependence on two functions of the radius ((1) and (2)), a general fact already recognized in [6]. On the other hand we agree with Czerniawsky's considerations on the difference in the notion of simultaneity between the Ss and PG metrics and on the physical significance of the time-reversed PG metric.

Finally Visser [7] and Padmanabhan [8] have strived to maintain the inadequacy of the free fall approach for the following reasons: to be only a weak field approximation of a more general theory and to be heuristic since it does not reproduce the Kerr metric. The first point has already been commented upon. The second is irrelevant in the present context. For rotating masses results have been reproduced successfully via gravitomagnetism in a parameter free way just from SR without the need to invoke GR [26].

In general, it is important to underline the peculiarity of proper time effects, with respect to those involving space. is is clear already to first order, since time contraction can be obtained as a SR effect, while the treatment of space depends on the overall analysis, see (37).

Their effect to first order has been evaluated by Einstein using only mass energy equivalence and is as follows:

Consider an atom at $B = r_0 = r + h$ and an identical one at $A = r$. Then the photon emitted by B reaches A, because of the coupling between its energy and the gravitational field, with a greater energy due to the effect of the gravitational field. The photon frequencies at the two places are related by

$$\hbar\omega(1 - GM/c^2r) = \hbar\omega'(1 - GM/c^2r') \tag{42}$$

from which trivially follows

$$\omega = \omega' \frac{1 - GM/c^2r'}{1 - GM/c^2r} \simeq \omega'(1 + gh/c^2) \tag{43}$$

This implies the reverse relatation for times

$$t' = t\frac{1 - GM/c^2r'}{1 - GM/c^2r} \tag{44}$$

i.e. that time runs quicker in regions of smaller gravitational field. When the comparison is made with respect to ∞, where gravity is absent, one gets the proper time at r denoted by τ

$$t' = t_\infty = \tau/(1 - GM/c^2 r) \tag{45}$$

and this agrees to first order with the above result from the invariant interval.

Notice that a basic form of principle of equivalence has been tacitly assumed: atoms are the same (as locally measured) in different points of a gravitational field. Otherwise a correction factor would arise.

This goes along with the parallel argument about atomic energy levels. The mass m **at rest** in a gravitational field of M at the height R_T has an energy

$$E_0 = m_0 c^2 (1 - GM/c^2 R_T) \tag{46}$$

and at $R_T + h$

$$E' = m_0 c^2 (1 - GM/c^2 (R_T + h)) \tag{47}$$

It follows that at the earth surface

$$E' - E_0 \simeq m_0 c^2 GM/c^2 R_T^2 h = m_0 g h \tag{48}$$

This energy difference exactly corresponds in classical terms to the gravitational potential energy difference or, in other words, to the work done against the standard Newtonian force $F = mg$. The two arguments are consistent because of *local* energy conservation of the atom-photon systems. They also show that the use of the gravitational interaction energy is consistent, to first order, with the dynamical treatment based on the elimination of gravity in free fall motion.

Independently of the first order approximation, the peculiarity of pure time effects is that they are physical, i.e. coordinate independent. In fact, time effects are given by the invariant interval at *a fixed point in space*, and, as observed above, such a notion is independent of coordinate transformations preserving stationarity of the metric tensor. This does not happen (not even to first order) for the other components of the metric tensor, which are in fact different in the Ss and PG formulation.

5 Rotating Frames and the Sagnac Effect

We pass now to a subject, not directly related to gravitation, whose treatment may help in shedding some more light on the use of metrics and of synchronization. The Sagnac effect has a long history, remarkable practical applications and has caused a considerable amount of discussions concerning its connection with SR and GR. In its standard form two counter propagating photon beams in a circular waveguide mounted on a disk are made to interfere after having traveled one circumference. When the disk is put in rotation with angular velocity Ω, the interference figure is seen to shift by an amount proportional to Ω.

Let us consider the problem from the point of view of the external observer (inertial frame). For him, light propagates of course with velocity c; however when the disk rotates the interference of the two light waves is observed at a moving angle, $\theta = \Omega t$. The lenghts $l_{1,2}$ of the two paths satisfy

$$l_1 - l_2 = 2\theta r$$

where θ is the shift of the angle in the traveling time and r the radius of the disk (with time and distances measured in the fixed frame). This corresponds, for light of frequency ω, to a shift in phase

$$\Delta\varphi = 2\theta r\omega/c$$

To first order in Ω, the traveling time of the two rays is $t \simeq 2\pi r/c$ and therefore

$$\Delta\varphi \simeq \frac{4\pi r^2 \omega\Omega}{c^2} = \frac{4\omega\Omega S}{c^2} \tag{49}$$

S standing for the area perpendicular to the rotation axis enclosed by the given contour.

This is all, since the effect is frame independent. However for the sake of the argument and in order to make contact with the previous treatment of gravity, let us consider it from another point of view.

On the Invariant Interval in Rotating Frames

The above kinematical constraint about the meeting of two rays at a moving point can of course be written as the condition of meeting at *the same point in a rotating coordinate system* and can be therefore discussed in terms of light propagation *in such coordinates*. This does *not* mean that quantities *measured* "on a rotating body" enter the discussion and in fact the introduction of such "physical frames", in particular of local frames associated to observers at each point on the disk, is not necessary.

Consider then a uniformly rotating reference system, whose local cylindrical coordinates are denoted by (t, r, z, ϕ_R) connected to those of the fixed inertial one (t, r, z, ϕ) by

$$\phi_R = \phi - \Omega t\,; \tag{50}$$

the invariant interval in the rotating system reads

$$ds^2 = c^2 dt^2 - (r d\phi_R + \Omega r dt)^2 - dz^2 - dr^2\,, \tag{51}$$

in complete analogy with gravity in our Newtonian coordinates. This immediately yields that light propagates tangentially with velocity

$$c_\perp = \pm c - \Omega r$$

as in the composition of light velocity with that of the free fall frame in gravity.

The previous equation can be rewritten in terms of $v = \Omega r$ and $dy = r d\phi_R$

$$ds^2 = c^2(1 - v^2/c^2)dt^2 - 2v dt dy - dy^2 - dr^2 - dz^2 \tag{52}$$

The similarity with the P–G formula is once more apparent. The essential difference is that now v is independent of y, and in fact the differential $r d\phi = dy + v dt$ from (50), corresponding to the EPIF differential $dr' = dr - v(r)dt$, is now exact.

The above difference between light velocities in the two directions easily leads to the same result as before. We emphasize that the analysis applies to first order in Ω, that no relativistic effect appear to that order and that the above discussion of the relativistic interval has nothing to do with Lorentz transformations, rather expressing the interval in the inertial frame in terms of different coordinates (intervals in such coordinates coinciding with measured intervals "on the moving disk" only in the non-relativistic limit).

It is also of some interest to write the above relativistic interval in "Schwarzschild coordinates": the off-diagonal term can be disposed of along the previous lines via the transformation

$$\begin{aligned} dt_S &= dt + v/(1 - v^2/c^2)dy \\ dy_S &= dy \end{aligned} \tag{53}$$

and for the relevant part (i.e. apart from dr^2 and dz^2 terms) the invariant interval takes the "Schwarzschild form"

$$ds^2 = c^2(1 - v^2/c^2)dt_S^2 - dy_S^2/(1 - v^2/c^2) \tag{54}$$

In both forms, the invariant interval is **not** Minkovskian, and in fact light velocity is different from c both in the "P–G" and in the "Schwarzschild" coordinates, where the tangential velocity is direction independent:

$$c_S = c(1 - v^2/c^2)$$

This is compatible with the Sagnac effect because (53) only gives rise to a local notion of time and global, topological, effects are hidden in the angular nature of the y variable. In Ss coordinates the Sagnac effect comes in fact from the difference in time coordinates obtained after following closed paths. The time difference between two path enclosing the circle in opposite directions is

$$\Delta t_S = 2/c^2 \oint \frac{\Omega r^2}{1 - \Omega^2 r^2/c^2}$$

which for low angular velocities yields

$$\Delta T = \frac{4\Omega S}{c^2}$$

corresponding to the result obtained via elementary considerations at the beginning of the paragraph.

6 Conclusions

In the present work Einstein's equations have been shown to be unnecessary in the static symmetric case, where all GR results have been obtained via EP, Newton's law and local SR.th The result is the PG metric, which realizes the Einstein program of "eliminating gravity locally" through free fall motion.

The calculational ingredient has been the usual variational principle (like the Fermat one for light) applied to the infinitesimal invariant non Minkovskian interval and the associated Euler–Lagrange equations realizing what usually appears as the result of a rather complicated and formal description of gravitation.

The connection with the Ss GR solution, which is superfluous given the explicit calculations of the present approach (and which had been considered as a necessary endorsement for the Painlevé–Gullstrand solution) only helps in clarifying the arbitrariness of attaching physical significance to the metric. In particula, spacetime curvature is perfectly compatible with Newtonian absolute time and Euclidean space.

The treatment of rotating frames, which played a role in the genesis of GR, leads, at each point, to a metric of the same form, with the same role of local frames as in the case of gravitational free fall. The only difference is here the existence of a global Minkovski frame.

Acknowledgements We would like to thank Dr. E. Cataldo for having brought to our attention a relevant piece of literature, prof.s F. Strocchi and P. Menotti for many discussions and Dr. A. Rucci for the drawings.

References

1. Newton, I.: Philosophiae Naturalis Principia Mathematica (Mathematical Principles of Natural Philosophy. Jussu Societatis Regiae ac Typis Josephi Streater, London (1687). Cambridge (1713). London (1726)
2. Einstein, A.: Die Grundlage der allgemeinen Relativitätstheorie. Ann. Phys. **49**, 769–822 (1916)
3. Schiff, L.I.: Am. J. Phys. **28**, 340 (1960)
4. Czerniawski, J.: What is wrong with Schwarzschild's coordinates? (2004). https://doi.org/10.48550/arXiv.gr-qc/0201037
5. Czerniawski, J.: The possibility of a simple derivation of the Schwarzschild metric (2016). https://doi.org/10.48550/arXiv.gr-qc/0611104
6. Gruber, R.P., et al.: Am. J. Phys. **56**, 265 (1988)
7. Visser, M.: Heuristic approach to the Schwarzschild geometry (2004). https://doi.org/10.48550/arXiv.gr-qc/0309072
8. Padmanabhan, T.: Snippets of physics. Reson. **14**, 1060–1070 (2009). https://doi.org/10.1007/s12045-009-0101-x
9. Kassner, K.: Eur. J. Phys. **36**, 65031 (2015). arXiv:1502.00149 [pdf, ps, other]
10. Schild, A.: Am J Phys **28**, 778 (1960)
11. Rindler, W.: Am J Phys **36**, 240–245 (1968)
12. Strauman, N.: Reflections on gravity. arXiv:astro-ph/0006423 (2000)

13. Ehlers, J., Rindler, W.: Gen. Rel. Grav. **29**, 519 (1997)
14. Pirani, F.A.E.: Acta Phys. Polonica **15**, 389 (1956)
15. Feynman, R.: Lectures on gravitation. Penguin Books (1999)
16. Painlevé, P.: La mécanique classique et la theorie de la relativité. C. R. Acad. Sci. (Paris) **173**, 677–680 (1921)
17. Gullstrand, A.: Allgemeine Lösung des statischen Einkörper-Problems in der Einsteinschen Gravitationstheorie. Arkiv. Mat. Astron. Fys **16**(8), 1–15 (1922)
18. Schwarzschild, K.: Über das Gravitationsfeld eines Massenpunktes nach der Einsteinschen Theorie. Sitzungsberichte der Königlich Preussischen Akademie der Wissenschaften, vol. 7., pp. 189–196 (1916)
19. Berkeley Physics Course, Mechanics, McGraw-Hill (1962)
20. Milgrom, M.: Astrophys. J. **270**, 365–370 (1983)
21. Damour, T.: La Matematica, vol. IV. G. Einaudi Editore, p. 453 (2010)
22. Williams, J.G., Turyshev, S.G., Boggs, D.H.: Int. J. Mod. Phys. D **1129**(2009), 18 (2004). https://doi.org/10.1142/S021827180901500X
23. Colella, R., Overhauser, A.W., Werner, S.A.: Phys. Rev. Lett. **34**, 1472 (1975)
24. Eöt-Wash Group: Class. Quantum Grav. **29**, 184002 (2012)
25. Mizony, M.: https://math.univ-lyon1.fr/~mizony/cargese1
26. Christillin, P., Barattini, L.: Gravitomagnetic forces and quadrupole gravitational radiation from special relativity. https://doi.org/10.48550/arXiv.1205.3514 (2013)

Two Results in the Quantum Theory of Measurements

Simone Del Vecchio, Jürg Fröhlich, Alessandro Pizzo, and Alessio Ranallo

Contents

Abstract Two theorems with applications to the quantum theory of measurements are stated and proven. The first one clarifies and amends von Neumann's Measurement Postulate used in the Copenhagen interpretation of quantum mechanics. The second one clarifies the relationship between "events" and "measurements" and the meaning of measurements in the *ETH*-Approach to quantum mechanics.

S. Del Vecchio
Dipartimento di Matematica, Università degli Studi di Bari, Bari, Italy
e-mail: simone.delvecchio@uniba.it

J. Fröhlich (✉)
Institute for Theoretical Physics, ETH Zürich, Zürich, Switzerland
e-mail: juerg@phys.ethz.ch

A. Pizzo · A. Ranallo
Dipartimento di Matematica, Università di Roma Tor Vergata, Roma, Italy
e-mail: pizzo@mat.uniroma2.it

A. Ranallo
e-mail: ranallo@mat.uniroma2.it

A. Cintio, A. Michelangeli (Eds.), *Trails in Modern Theoretical and Mathematical Physics*, https://doi.org/10.1007/978-3-031-44988-8_9

1 Introduction and Summary of Contents

In this paper, we present two mathematical results of relevance to the quantum theory of measurements,[1] which we treat in a spirit close to the Copenhagen interpretation/heuristics of quantum mechanics (QM), as amended in [1–3].

Let $\mathfrak{E}$ be a large ensemble of physical systems identical (isomorphic) to a specific system, S, of finitely many degrees of freedom to be described quantum-mechanically. We are interested in understanding the effect of measurements of a physical quantity, $\widehat{X}$, characteristic of S for all systems in $\mathfrak{E}$. In text-book QM, one tends to invoke *von Neumann's measurement postulate* (see [4]) to predict properties of the resulting state, averaged over all systems in $\mathfrak{E}$, right after a successful completion of the measurements of $\widehat{X}$. The standard formulation of this postulate appears to be afflicted with some problems, which we will discuss and attempt to clarify in the following.

We begin this paper by describing the systems $\simeq S$ we have in mind. A physical quantity, $\widehat{X}$, characteristic of S is represented by a self-adjoint operator, $X = X^*$, acting on a separable Hilbert space, $\mathcal{H}$. An average over $\mathfrak{E}$ of states of these systems is called an *"ensemble state"* and is given by a *density matrix,* i.e., by a positive, trace-class operator, Ω, on $\mathcal{H}$ of trace $\operatorname{tr}\Omega = 1$. To mention an example, a system $S \in \mathfrak{E}$ might consist of a particle, such as an electron, propagating in physical space $\mathbb{E}^3$, $\widehat{X}$ might be a component of the spin or a bounded function of a component of the position- or the momentum of the particle, and

$$\mathcal{H} = L^2(\mathbb{R}^3, d^3x) \otimes \mathbb{C}^{2s+1},$$

where $x \in \mathbb{R}^3$ is the position and s the spin of the particle.

The purpose of this paper is to clarify what is meant by the statement that a measurement of the quantity $\widehat{X}$ has been completed successfully. Since we will try to follow the spirit of the Copenhagen Interpretation/heuristics of QM, where appropriate, we will usually adopt an *ensemble point of view*, emphasizing statements that are obtained by taking averages over all systems in the ensemble $\mathfrak{E}$. But when combined with results in [2, 3], our results have implications relevant for the theory of measurements carried out on individual systems.

Next, we outline the contents of this paper. In Sect. 2, we recall von Neumann's measurement postulate and point out some problems with it. We then formulate a revised version of this postulate and state the main result proven in this paper. In Sect. 3 we sketch how measurements are described in the *ETH*-Approach to quantum mechanics [1–3]. In Sect. 4, we present the proof of our main result.

---

[1] As far as we remember, Gianni Morchio had an interest in the foundations of quantum mechanics; so he would probably have appreciated our results.

2 Von Neumann's Measurement Postulate

We imagine that the initial ensemble state when measurements of $\widehat{X}$ set in, for all systems in $\mathfrak{E}$, is described by a density matrix Ω_{in}. In his book [4] on the foundations of QM, von Neumann postulated that, when averaging over $\mathfrak{E}$, the effect of measuring $\widehat{X}$ for all systems belonging to $\mathfrak{E}$ amounts to replacing the state Ω_{in} by a certain ensemble state, Ω_{out}, describing the average of states of systems belonging to $\mathfrak{E}$ right *after* the measurements of $\widehat{X}$ have been completed, where Ω_{out} satisfies the following postulate.

Von Neumann's Postulate *Let $X = X^*$ be the self-adjoint operator on $\mathcal{H}$ representing the physical quantity $\widehat{X}$, and let*

$$X = \int_{\mathbb{R}} \xi \, d\Pi(\xi) \tag{1}$$

be the spectral decomposition of X, with $\Pi(\Delta)$ its spectral projection associated with an arbitrary Borel set $\Delta \subset \mathbb{R}$. The ensemble state Ω_{out} right after completion of the measurements of $\widehat{X}$ has the properties that

$$\begin{aligned} &[\Omega_{\text{out}}, X] = 0, \quad \textit{and} \\ &\operatorname{tr}\big(\Omega_{\text{in}} \cdot \Pi(\Delta)\big) = \operatorname{tr}\big(\Omega_{\text{out}} \cdot \Pi(\Delta)\big), \ \forall \textit{ Borel sets } \Delta \subset \mathbb{R} \quad \textit{(Born's Rule)} \end{aligned} \tag{2}$$

Remark We will see shortly that this formulation of von Neumann's postulate is inadequate, except if the operator X has pure point spectrum (for which case it was originally formulated)—but even then it is problematic, as will become apparent in Sect. 3.

The spectral decomposition of a density matrix Ω has the form

$$\begin{aligned} &\Omega = \sum_{n=1}^{N} \omega_n \, \pi_n, \quad 1 \geq \omega_1 > \omega_2 > \cdots > \omega_N > 0, \\ &\pi_n = \pi_n^*, \quad \pi_n \cdot \pi_m = \delta_{nm} \pi_n, \quad \forall \, n, m = 1, 2, \ldots, N, \end{aligned} \tag{3}$$

for some $N \leq \infty$. The operators π_n are disjoint orthogonal projections of finite rank (the eigen-projections of Ω), and

$$\operatorname{tr}\big(\Omega\big) = \sum_{n=1}^{N} p_n = 1, \quad \text{where} \quad p_n = \omega_n \cdot \dim \pi_n, \ \ n = 1, 2, \ldots, N \, .$$

We set

$$\pi_\infty := \mathbf{1} - \sum_{n=1}^{N} \pi_n \, .$$

If, as in the formulation of von Neumann's postulate given in (2), the operator X is assumed to commute with Ω_{out}, then it satisfies the identity

$$X = \sum_{n=1}^{N} \pi_n X \pi_n + \pi_\infty X \pi_\infty, \tag{4}$$

where $\pi_n X \pi_n$ is of finite rank, $\forall\, n = 1, 2, \ldots, N$, where the operators π_n are the eigenprojections of Ω_{out}.

We observe that, for every $n = 1, 2, \ldots, N$, $\pi_n X \pi_n$ is a selfadjoint, finite-dimensional matrix; hence its spectrum consists of finitely many (discrete) eigenvalues. Let $\mathcal{H}^+$ be the subspace of $\mathcal{H}$ given by the range of $\mathbf{1} - \pi_\infty$. It follows that if X satisfies (2) then the operator $X\big|_{\mathcal{H}^+}$ has *pure-point spectrum.* (Of course, if the range of π_∞ is infinite-dimensional then $\pi_\infty\, X\, \pi_\infty$ may have continuous spectrum; but this is irrelevant for measurements of $\widehat{X}$ that result in states occupied by the systems in $\mathfrak{E}$ whose average is given by Ω_{out}.) Thus, at best, von Neumann's postulate in the formulation of (2) can only be applied to measurements of physical quantities with pure-point spectrum. However, a component of the position or of the momentum of a quantum particle propagating in physical space $\mathbb{E}^3$ has continuous spectrum occupying the entire real line $\mathbb{R}$.

We conclude that (2) *cannot* be valid verbatim when physical quantities represented by operators with *continuous spectrum* are measured, and we should find out how to modify them in such instances.

2.1 An Amended Form of von Neumann's Postulate

We imagine that measurements of a physical quantity $\widehat{X}$ are carried out for all systems belonging to a large ensemble $\mathfrak{E}$ of systems identical to a system S, with the result that the average over $\mathfrak{E}$ of the final states of these systems after completion of the measurements of $\widehat{X}$ is found to be *close* (but not necessarily equal) to an ensemble state given by a density matrix Ω_{out} with the property that

$$\big|\big|[\Omega_{\text{out}}, X]\big|\big| < \varepsilon\,, \tag{5}$$

for some ε smaller than the error margin of the instrument used to measure $\widehat{X}$. One may add the assumption that, for Ω_{out}, *Born's Rule* holds, as formulated in the second equation of (2). We will establish the following

Main Result If condition (5) holds for a sufficiently small $\varepsilon \ll 1$ then one may replace Ω_{out} by a modified density matrix Ω'_{out} and X by a modified operator X',

$$X' = \sum_{k=1}^{K} \xi_k\, \Pi_k\,, \quad \text{for some} \quad K \leq \infty\,, \tag{6}$$

where $\xi_1 > \xi_2 > \cdots > \xi_K > -\infty$ are the eigenvalues of X' and $\Pi_1, \ldots, \Pi_K$ the corresponding eigen-projections, with the properties that

(i) the operator X' has pure-point spectrum and is close to the operator X representing $\widehat{X}$ in the operator norm;
(ii) the density matrix Ω'_{out} is close to the density matrix Ω_{out} in the trace norm; and
(iii) the operators X' and Ω'_{out} commute, i.e.,

$$\left[\Omega'_{\text{out}}, X'\right] = 0\,. \tag{7}$$

The closeness of X' to X and of Ω'_{out} to Ω_{out} depends on the size of the commutator of X with Ω_{out}: the smaller the norm, $\|[\Omega_{\text{out}}, X]\|$, of this commutator the closer are X' to X and Ω'_{out} to Ω_{out}. The size of $\|[\Omega_{\text{out}}, X]\|$ is thus a measure for the precision of the instrument used to measure $\widehat{X}$—the smaller this norm, the higher the precision of the instrument.

The proof of the *Main Result* stated above is given in Sect. 4. At the end of the present section, we sketch the very easy proof in the special case where $\dim(\mathcal{H}) < \infty$.

Remarks

(1) Another possible amendment of von Neumann's postulate can be formulated as follows. We cover the spectrum, spec(X), of the operator X with small closed intervals $\Delta_k \subset \mathbb{R}, k = 1, 2, \ldots, K$, for some $K < \infty$, with the properties that $\Delta_k \cap \Delta_{k'}$ is empty or consists of a single point (assumed not to be an eigenvalue of X) whenever $k \neq k'$, and $\bigcup_{k=1}^{K} \Delta_k \supseteq \text{spec}(X)$. These intervals are assumed to be determined by properties of the instrument used to measure $\widehat{X}$. One may then assume that the $\mathfrak{E}$-average of the states of the systems after completion of the measurements of $\widehat{X}$ is given by a density matrix Ω_{out} satisfying

$$\Omega_{\text{out}} = \sum_{k=1}^{K} \Pi_k \Omega_{\text{out}} \Pi_k, \quad \text{where} \quad \Pi_k = \Pi(\Delta_k), \;\; \forall\, k = 1, 2, \ldots, K\,. \tag{8}$$

The operator X' is chosen to be given by

$$X' = \sum_{k=1}^{K} \xi_k \, \Pi_k,$$

where ξ_k is the midpoint of the interval $\Delta_k \subset \mathbb{R}$, for all k. Assuming that the length of all the intervals Δ_k is bounded above by 2ε, we conclude that

$$\left\|\left[\Omega_{\text{out}}, X\right]\right\| < \varepsilon\,, \quad \left[\Omega_{\text{out}}, X'\right] = 0\,, \quad \text{and} \quad \|X - X'\| < \varepsilon\,. \tag{9}$$

This amendment of von Neumann's postulate is somewhat arbitrary and involves assumptions on what is meant by a measurement of a physical quantity that are more detailed than condition (5).

(2) The *Main Result* stated above is reminiscent of a theorem that says that if two bounded self-adjoint operators almost commute then there are two operators close in norm to the original ones that *do* commute; see [5–7].

(3) We conjecture that our *Main Result* is a special case of the following more general statement: Let $\mathfrak{A}$ be a von Neumann algebra with unit $\mathbf{1}$, and let ω be a normal state on $\mathfrak{A}$. For an operator $X \in \mathfrak{A}$, we define a bounded linear functional on $\mathfrak{A}$ by

$$\mathrm{ad}_X[\omega](Y) := \omega([Y, X]), \quad \forall\, Y \in \mathfrak{A}. \tag{10}$$

Suppose now that ω and X are such that

$$\left|\mathrm{ad}_X[\omega](Y)\right| < \varepsilon \|Y\|, \quad \forall\, Y \in \mathfrak{A}, \quad \text{for some } \varepsilon \ll 1. \tag{11}$$

Then there exist a normal state ω' on $\mathfrak{A}$ and an operator $X' \in \mathfrak{A}$, with $\|\omega' - \omega\| < \delta(\varepsilon)$ and $\|X' - X\| < \delta(\varepsilon)$, for some $\delta(\varepsilon) \searrow 0$, as $\varepsilon \searrow 0$, such that

$$\mathrm{ad}_{X'}[\omega'] = 0. \tag{12}$$

Our *Main Result* shows that this conjecture holds in the special case where $\mathfrak{A}$ is isomorphic to the algebra of all bounded operators on a separable Hilbert space.

As a warm-up we prove the *Main Result* in the special case of a finite-dimensional Hilbert space $\mathcal{H}$, which is very easy. In items (i) through (iii), one may then set $\Omega'_{\text{out}} = \Omega_{\text{out}}$ and only slightly modify the operator X, or one may set $X' = X$ and only slightly modify Ω_{out}, and end up with (7).

Let $\mathcal{H} = \mathbb{C}^M$, with $M < \infty$. Then

$$X = \sum_{k=1}^{K} \xi_k\, \Pi_k, \quad K \le M, \quad \text{and}$$
$$\Omega = \sum_{n=1}^{N} \omega_n\, \pi_n, \quad N \le M, \tag{13}$$

where $\xi_1 > \xi_2 > \cdots > \xi_K > -\infty$ are the eigenvalues of X and $\Pi_1, \Pi_2, \ldots, \Pi_K$ are the corresponding eigen-projections, and $\omega_1 > \omega_2 > \cdots > \omega_N > 0$ are the non-zero eigenvalues of Ω, with $\pi_1, \pi_2, \ldots, \pi_N$ the corresponding eigen-projections. We define $\pi_{N+1} := \mathbf{1} - \sum_{n=1}^{N} \pi_n$, and

$$\gamma_\Omega := \min_{1 \le n \le N} \left(\omega_n - \omega_{n+1}\right) > 0, \quad \text{with } \omega_{N+1} := 0, \tag{14}$$

to be the smallest gap between distinct eigenvalues of Ω. Let us assume that

$$\left|\left|[X, \Omega]\right|\right| \le \varepsilon, \quad \text{for some } \varepsilon \ll \gamma_\Omega. \tag{15}$$

We define an operator X' by setting

$$X' := \sum_{n=1}^{N+1} \pi_n\, X\, \pi_n. \tag{16}$$

Obviously X' commutes with Ω, and we claim that

$$\|X' - X\| < \text{const.}\, \varepsilon\,. \tag{17}$$

Proof of (17) Clearly

$$X = \sum_{n,n'=1,2,\ldots,N+1} \pi_n\, X\, \pi_{n'}\,. \tag{18}$$

By (15) we have that

$$\|[\pi_n\, X\, \pi_{n'}, \Omega]\| = \|\pi_n[X, \Omega]\pi_{n'}\| \le \varepsilon, \quad \forall\; n, n'\,. \tag{19}$$

Plainly $[\pi_n\, X\, \pi_n, \Omega] = 0\,,\ \forall\; n = 1, 2, \ldots, N+1$. If $n \neq n'$ then

$$[\pi_n\, X\, \pi_{n'}, \Omega] = (\omega_{n'} - \omega_n)\, \pi_n\, X\, \pi_{n'}\,.$$

By (14) and (19), we have that

$$\|\pi_n\, X\, \pi_{n'}\| \le \gamma_\Omega^{-1}\, \varepsilon\,, \quad \text{for } n \neq n'\,.$$

Thus, using (18) we find that

$$\|X - X'\| \le (N+1)N\, \gamma_\Omega^{-1}\, \varepsilon < M^2 \gamma_\Omega^{-1} \varepsilon\,, \tag{20}$$

as claimed in (17).

In the calculations just shown we can obviously exchange the roles of X and Ω. We set

$$\gamma_X := \min_{1 \le k < K} \left(\xi_k - \xi_{k+1}\right) > 0\,,$$

and we then replace the density matrix Ω by

$$\Omega' := \sum_{k=1}^{K} \Pi_k\, \Omega\, \Pi_k\,.$$

Clearly Ω' is a non-negative operator, and $\mathrm{tr}(\Omega') = 1$, because $\sum_{k=1}^{K} \Pi_k = \mathbf{1}$; i.e., Ω' is a density matrix; and it obviously commutes with X. Repeating the arguments shown above, we find that

$$\mathrm{tr}(|\Omega - \Omega'|) \le M\,(K-1)K\, \gamma_X^{-1}\, \varepsilon < M^3\, \gamma_X^{-1} \varepsilon\,. \tag{21}$$

Of course, the problem with the estimates in (20) and (21) is the dependence of the right sides on the dimension, M, of the Hilbert space $\mathcal{H}$. This problem is addressed in Sect. 4, where we state a result that is *uniform* in the dimension of the Hilbert space, but at the price that we have to slightly modify both, X and Ω. This result enables one to modify von Neumann's measurement postulate so as to avoid the shortcomings of the original version, as indicated above.

3 The Description of Measurements in the *ETH*-Approach to QM

In this section we sketch how measurements can be described in the formulation of QM proposed in [1–3] under the name of "*ETH*-Approach to QM" (assuming some familiarity with these papers).

We begin with the obvious observation that a successful measurement of a physical quantity $\widehat{X}$ characteristic of a system S (belonging to an ensemble $\mathfrak{E}$) results in an *event*, namely the event that $\widehat{X}$ takes a—possibly somewhat imprecise—value belonging to some small interval contained in the real line whose length depends on the accuracy of the instrument used to measure $\widehat{X}$. To understand the significance of this statement it is necessary to clarify what, in the *ETH*-Approach to QM, is meant by an *"event"*. We recall the definition proposed in [2, 3]. Abstractly, a *"potential event"*, e, associated with a physical system $S \in \mathfrak{E}$ is a partition of unity, $\mathrm{e} = \{\pi_n\}_{n=1}^{\infty}$, by orthogonal projections satisfying

$$\pi_n = \pi_n^*, \quad \pi_n \cdot \pi_{n'} = \delta_{nn'}\, \pi_n, \ \forall\ n, n' = 1, 2, \ldots, \quad \sum_{n=1}^{\infty} \pi_n = \mathbf{1}. \tag{22}$$

An operator X representing a physical quantity $\widehat{X}$ characteristic of a system $S \in \mathfrak{E}$ at some time $\geq t$ and the projections $\pi \in \mathrm{e}$ of an arbitrary potential event e that may occur in S at a time $\geq t$ are supposed to belong to some algebra $\mathfrak{A} = \mathcal{E}_{\geq t}$, which, in general, depends *non-trivially* on time t. For systems, S, with finitely many degrees of freedom, $\mathfrak{A}$ is the algebra, $B(\mathcal{H})$, of all bounded operators on a separable Hilbert space $\mathcal{H}$ and is independent of t. But, for systems with infinitely many degrees of freedom, including those describing the quantized electromagnetic field,[2] the time-dependence of $\mathfrak{A} = \mathcal{E}_{\geq t}$ tends to be *non-trivial*, and $\mathfrak{A}$ is a more exotic (type-III_1) algebra. Our analysis in this section does not require any specific assumptions on $\mathfrak{A}$. (It is only assumed that the algebra $\mathfrak{A}$ is weakly closed, i.e., that it is a von Neumann algebra; but it need not and usually will *not* be isomorphic to $B(\mathcal{H})$.) States at time t are states on $\mathfrak{A} = \mathcal{E}_{\geq t}$ (i.e., positive, normalized linear functionals on $\mathcal{E}_{\geq t}$). They are denoted by lower-case Greek letters, $\omega, \ldots$.

In the following discussion we fix a time t and suppress explicit reference to time-dependence wherever possible. We suppose that a state, ω, on $\mathfrak{A}$ is an *ensemble state,* i.e., that it has the meaning of being an average over the ensemble $\mathfrak{E}$ of states of individual systems, all $\simeq S$. If a potential event $\mathrm{e} = \{\pi_n\}_{n=1}^{\infty} \subset \mathfrak{A}$ is *actualizing* (i.e., is observed to happen) at some time t then, according to the *ETH*-Approach to QM, the state $\omega = \omega_t$ has the property that

$$\omega(X) = \sum_{\pi \in \mathrm{e}} \omega(\pi \cdot X \cdot \pi), \quad \forall\ X \in \mathfrak{A}, \tag{23}$$

[2] The only systems for which (in our opinion) the "measurement problem" has a satisfactory solution.

i.e., ω is a convex combination of states in the images of the projections $\pi \in \mathfrak{e}$. Potential events actualizing at some time are called *"actualities"*. (For a more precise characterization of actualities, see, e.g., [3].) If $\mathfrak{A} = B(\mathcal{H})$ then

$$\omega(X) = \mathrm{tr}\big(\Omega \cdot X\big), \quad \forall\ X \in \mathfrak{A},$$

for some density matrix Ω on $\mathcal{H}$, and the projections π belonging to the event $\mathfrak{e}$ that actualizes, given the state ω, are the spectral projections of the density matrix Ω.

If $\mathfrak{e}$ is an event actualizing at some time t then the state at time t of an *individual* system in the ensemble $\mathfrak{E}$ is expected to belong to the image of a projection $\pi \in \mathfrak{e}$, with a probabilty, $prob_{\omega}(\pi)$, given by *Born's Rule*, namely

$$prob_{\omega}(\pi) = \omega(\pi),$$

where ω is the ensemble state at time t.

We are interested in characterizing actualities $\mathfrak{e} = \{\pi_n\}_{n=1}^{N} \subset \mathfrak{A}$, $N \leq \infty$, that can be interpreted as corresponding to the completion of the measurement of a certain physical quantity $\widehat{X}$. We thus consider a state ω satisfying (23). Given a non-negative number $\varepsilon \ll 1$, there exists an integer $N_0 < \infty$ such that

$$\sum_{n=1}^{N_0-1} \omega(\pi_n) > 1 - \varepsilon, \text{ i.e.,} \quad \omega\big(\pi^{(N_0)}\big) < \varepsilon, \quad \text{where} \quad \pi^{(N_0)} := \sum_{n=N_0}^{N} \pi_n. \quad (24)$$

It is then very unlikely that an individual system in $\mathfrak{E}$ is found to occupy a state in the range of a projection $\pi \leq \pi^{(N_0)}$. If $\mathfrak{e}$ is the potential event actualizing at a certain time t and ω is the ensemble state at time t satisfying (23) then the slightly coarser event $\mathfrak{e}_0 := \{\pi_1, \pi_2, \dots, \pi_{N_0-1}, \pi^{(N_0)}\}$ can be viewed to be an actuality at time t, too. To avoid irrelevant complications, we henceforth replace $\mathfrak{e}$ by $\mathfrak{e}_0$ throughout the following discussion, and we simplify our notations by writing $\mathfrak{e}$, instead of $\mathfrak{e}_0$, and π_{N_0}, instead of $\pi^{(N_0)}$, with $N_0 < \infty$.

We assume that the operator X representing the physical quanitiy $\widehat{X}$ has the form

$$X = \sum_{k=1}^{K} \xi_k \Pi_k, \quad \text{for some} \quad K < \infty, \quad (25)$$

where the real numbers ξ_k are the eigenvalues of X and the operators Π_k are the corresponding eigen-projections, $k = 1, 2, \dots, K$. (We should mention that the projections Π_k may be given by $\Pi_k = \Pi(\Delta_k)$, where the sets Δ_k are intervals of the real line of length $< 2\varepsilon$ whose union covers $\mathrm{spec}(X)$, and ξ_k may be (e.g.) the midpoint of the interval Δ_k, for all k, as discussed in Remark (1) of Subsect. 2.1.)

If the actuality $\mathfrak{e}$ can be interpreted to correspond to the likely completion of a measurement of $\widehat{X}$, with an accuracy measured by ε, then there must exist a decomposition of $\{1, 2, \ldots, N_0\}$ into disjoint subsets $\mathcal{I}_k$, $k = 1, 2, \ldots K$, such that

$$\|[\pi_n, \Pi_k]\| < \mathcal{O}(N_0^{-2}\varepsilon), \quad \forall\, n \leq N_0, \quad \text{and}$$
$$\sum_{n \notin \mathcal{I}_k, n < N_0} \|\pi_n\, \Pi_k\, \pi_n\| < \mathcal{O}(\varepsilon), \quad \forall\, k = 1, 2, \ldots K\,. \tag{26}$$

The second equation tells us that if a system is found in a state in the range of a projection $\pi_n, n \notin \mathcal{I}_k, n < N_0$, then the quantity $\widehat{X}$ is very unlikely to have the measured value ξ_k. By (24), if the ensemble state is given by ω then it is very unlikely that an individual system in $\mathfrak{E}$ is found in a state belonging to the range of the projection π_{N_0}.

Since $\sum_{n=1}^{N_0} \pi_n = \mathbf{1}$, one obviously has that

$$X = \sum_{n,n'=1,2,\ldots,N_0} \pi_n\, X\, \pi_{n'}\,.$$

Since $\pi_n \cdot \pi_{n'} = 0$, for $n \neq n'$, the first inequality in (26) then implies that the operator X is approximated in norm by

$$X' := \sum_{n=1}^{N_0} \pi_n\, X\, \pi_n\,, \tag{27}$$

up to an error of $\mathcal{O}(\varepsilon)$; and (24) tells us that the Born probability of picking up a correction in determining the outcome of the measurement of $\widehat{X}$ that is due to the operator $\pi_{N_0}\, X\, \pi_{N_0}$ is bounded by $\mathcal{O}(\varepsilon)$, hence very small. One may then wonder whether the actuality e could occur as the result of a measurement of a *slightly different* physical quantity $\simeq \widehat{X}$.

The second inequality in (26) implies that X' is well approximated by the operator

$$X'' := \sum_{k=1}^{K} \sum_{n \in \mathcal{I}_k} \xi_k\, \pi_n\, \Pi_k\, \pi_n + \pi_{N_0}\, X\, \pi_{N_0} \tag{28}$$

with

$$\|X'' - X'\| < \mathcal{O}(K\,\varepsilon)\,. \tag{29}$$

Next, we note that the first inequality in (26) implies that

$$\left|\left|\left(\pi_n\, \Pi_k\, \pi_n\right)^2 - \pi_n\, \Pi_k\, \pi_n\right|\right| < \mathcal{O}\left(N_0^{-2}\varepsilon\right).$$

This estimate enables us to apply the following

Lemma *Let P be a self-adjoint operator in a von Neumann algebra $\mathfrak{A}$, and let $\delta < \frac{1}{2}$. If $\|P^2 - P\| < \delta$ then there exists an orthogonal projection $\widehat{P} \in \mathfrak{A}$ whose image belongs to the range of P such that*

$$\|\widehat{P} - P\| < \delta\,.$$

See Lemma 8 and Appendix C of [8]. This lemma implies that if $N_0^{-2}\varepsilon$ is small enough then there exists an orthogonal projection $\pi_{k,n}$ with the property that the image of $\pi_{k,n}$ is contained in or equal to the image of π_n and such that

$$\|\pi_{k,n} - \pi_n \Pi_k \pi_n\| < \mathcal{O}\big(N_0^{-2}\varepsilon\big).$$

We define

$$X''' := \sum_{k=1}^{K} \xi_k \Big(\sum_{n\in \mathcal{I}_k} \pi_{k,n}\Big) + \pi_{N_0} X \pi_{N_0}, \quad \text{and} \quad X_{\text{fin}} := X''' - \pi_{N_0} X \pi_{N_0}. \tag{30}$$

We are ready to state a result in the theory of measurements, according to the *ETH*-Approach to QM.

Theorem 1 *We assume that the bounds in* (24) *and* (26) *hold for some* $\varepsilon \ll 1$. *Then we have that*

(i) *the Born probability of finding an individual system in the ensemble* $\mathfrak{E}$ *in a state that belongs to the range of the projection* $\pi_{N_0} = \pi^{(N_0)}$ *is bounded above by* ε*;*
(ii) *the operator* X''' *defined in* (30) *is reduced by the projections* π_{N_0} *and* $\mathbf{1} - \pi_{N_0}$*;*
(iii) *the norm of* $X''' - X$ *is bounded by*

$$\|X''' - X\| < \mathcal{O}(\varepsilon),$$

i.e., the physical quantity $\widehat{X}$ *is well approximated by a slightly modified physical quantity represented by the operator* X'''*;*
(iv) *the eigenvalues of* $X_{\text{fin}} = X''' - \pi_{N_0} X \pi_{N_0}$ *are contained in or equal to the spectrum,* $\{\xi_k\}_{k=1}^{K}$*, of the operator* X *and the eigen-projection of* X_{fin} *corresponding to* ξ_k *is given by the projection* $\sum_{n\in\mathcal{I}_k} \pi_{k,n}$ *(which is dominated by the projection* $\sum_{n\in\mathcal{I}_k} \pi_n$*), for* $k = 1, 2, \ldots, K$*; and*

$$\big[X_{\text{fin}}, \pi_n\big] = 0, \quad \forall\ \pi_n \in \mathfrak{e}.$$

We conclude that, under the hypotheses of Theorem 1, one may interpret the actualization of the event $\mathfrak{e}$ as being accompanied by the completion of a measurement of a physical quantity $\widehat{X}''' \approx \widehat{X}$, where $\widehat{X}'''$ is represented by an operator X''' that is a tiny modification of the operator X representing $\widehat{X}$.

In this section, we have not tried to optimize our estimates; we have attempted to outline the basic ideas of how measurements can be interpreted in the *ETH*-Approach described in [1–3].

4 Proof of the Main Result

In this section we prove the *Main Result* announced in Sect. 2. We consider a density matrix Ω on a separable Hilbert space $\mathcal{H}$ with spectral decomposition

$$\Omega = \sum_{n=1}^{\infty} \omega_n \, \pi_n \, , \quad \omega_1 > \omega_2 > \cdots . \tag{31}$$

as in (3) of Sect. 2. We define $p_n := \omega_n \cdot \dim \pi_n$, $n = 1, 2, \ldots$ Given a positive number $\varepsilon \ll 1$, we define Δ_ε by

$$\Delta_\varepsilon := \sum_{n \,:\, \omega_n \leq \varepsilon^{1/4}} p_n \, . \tag{32}$$

Clearly, $\Delta_\varepsilon \searrow 0$, as $\varepsilon \searrow 0$. The *Main Result* is a consequence of the following theorem.

Theorem 2 *Let Ω and Δ_ε be as in* (31) *and* (32), *respectively, and let X be a self-adjoint operator on $\mathcal{H}$, with $\|X\| \leq 1$. We assume that*

$$\big\|[\Omega, X]\big\| \leq \varepsilon \, . \tag{33}$$

Then, for sufficiently small values of ε and Δ_ε, there exist a density matrix Ω' and a self-adjoint operator X' such that $[\Omega', X'] = 0$ and

$$\|X - X'\| \leq \varepsilon^{1/4}, \quad \text{and} \quad \mathrm{tr}\big|\Omega - \Omega'\big| \leq 2\Delta_\varepsilon + \mathcal{O}(\varepsilon^{1/4}) \, . \tag{34}$$

Proof As announced in the theorem, our goal is to construct a density matrix Ω' close to Ω in the trace norm and a self-adjoint operator X' close to X in the operator norm such that $[\Omega', X'] = 0$. We begin with the construction of Ω'.

In the following it is convenient to rewrite the spectral decomposition of Ω as follows:

$$\Omega = \sum_{j=1}^{\infty} \omega_j \, |u_j\rangle\langle u_j| \, , \quad \omega_1 \geq \omega_2 \geq \cdots \geq 0, \quad \sum_{j=1}^{\infty} \omega_j = 1 \, , \tag{35}$$

where $\{u_j\}_{j=1}^{\infty}$ is an orthonormal system of eigenvectors of Ω, and $|u_j\rangle\langle u_j|$ is the orthogonal projection onto u_j, for all j. Then assumption (33) implies that

$$\|[\Omega, X]u_i\|^2 = \sum_{j=1}^{\infty} (\omega_i - \omega_j)^2 |\langle u_i, X u_j\rangle|^2 \leq \varepsilon^2, \quad \forall i \, . \tag{36}$$

In the following steps, we construct a positive trace-class operator $\widetilde{\Omega} \leq \Omega$, (hence $\mathrm{tr}\, \widetilde{\Omega} \leq 1$).

1) We preserve the eigenvectors of the density operator Ω, but—where necessary—modify the corresponding eigenvalues in such a way that the spectrum of the modified operator $\widetilde{\Omega}$ consists of (possibly degenerate) eigenvalues separated by gaps of specified size. To begin with we choose two exponents, δ and β (later set equal to $1/4$ and $3/4$, respectively), with

$$0 < \delta < \beta < 1 \quad \text{and} \quad \beta > 2\delta\,, \tag{37}$$

and we modify the spectrum of $\widetilde{\Omega}$ in such a way that the gaps between the non-coinciding modified eigenvalues, i.e., between the distinct eigenvalues of $\widetilde{\Omega}$, will be larger than ε^{β}.

1-i) We observe that, since $\Omega \geq 0$, with $\operatorname{tr}\Omega = 1$, the dimension of the direct sum of the eigenspaces of Ω corresponding to eigenvalues larger than or equal to ε^{δ} is bounded above by $O(1/\varepsilon^{\delta})$.

1-ii) Next, we define ω_{i_1} to be the smallest eigenvalue of Ω larger than ε^{δ} with the property that its separation from the previous (next larger) eigenvalue is bounded below by ε^{β}. It is not assumed that an eigenvalue with the properties of ω_{i_1} exists.

But if such an eigenvalue ω_{i_1} exists then we denote by $(\omega_{i_1})_-$ its precursor. By construction, we have that $(\omega_{i_1})_- \leq \varepsilon^{\delta} + \mathcal{O}(\varepsilon^{\beta-\delta})$, because there are at most $\mathcal{O}(\varepsilon^{-\delta})$ eigenvalues separated by gaps bounded by $\leq \varepsilon^{\beta}$ in between ε^{δ} and ω_{i_1}, as follows from 1-i).
We define

- an interval I_0 by $I_0 := [0, (\omega_{i_1})_-]$,
- and a subspace $\mathcal{H}_0 \subset \mathcal{H}$ as the direct sum of the eigenspaces of Ω corresponding to eigenvalues contained in the interval I_0.

If an eigenvalue with the properties of ω_{i_1} does not exists then we conclude that the largest eigenvalue, $\omega_{\max}$, of Ω must be smaller than $\mathcal{O}(\varepsilon^{\delta} + \varepsilon^{\beta-\delta})$. In this case, we define $I_0 := [0, \omega_{\max}]$.
We define $\widetilde{\Omega}$ to vanish on the subspace $\mathcal{H}_0$.

1-iii) We next assume that $I_0 \neq [0, \omega_{\max}]$, i.e., that an eigenvalue with the properties of ω_{i_1} exists. Then we consider the smallest eigenvalue of Ω larger than ω_{i_1} with the property that its separation from the previous eigenvalue is larger than ε^{β}.

If such an eigenvalue exists we denote it by ω_{i_2} and its precursor by $(\omega_{i_2})_-$, and we then have that $\omega_{i_2} \leq \omega_{i_1} + \mathcal{O}(\varepsilon^{\beta-\delta})$. We also define

- $I_1 := [\omega_{i_1}, (\omega_{i_2})_-]$;
- $n_1 :=$ number of eigenvalues (with multiplicity) of Ω contained in I_1;
- $\mathcal{H}_1 :=$ direct sum of the corresponding eigenspaces (notice that $\dim\mathcal{H}_1 = n_1$).

If an eigenvalue with the properties of ω_{i_2} does not exist we conclude that the largest eigenvalue, $\omega_{\max}$, of Ω is smaller than $\omega_{i_1} + \mathcal{O}(\varepsilon^{\beta-\delta})$, and we define $I_1 := [\omega_{i_1}, \omega_{\max}]$.

On the subspace $\mathcal{H}_1$ we define

$$\widetilde{\Omega}\big|_{\mathcal{H}_1} := \widetilde{\omega}_1 \cdot \mathbf{1}\big|_{\mathcal{H}_1},$$

where $\widetilde{\omega}_1 := \omega_{i_1}$.

1-iv) We iterate these arguments: If $I_{m-1} \neq [\omega_{i_{m-1}}, \omega_{\max}]$, then, starting from ω_{i_m}, we consider the eigenvalue of Ω with the property that its separation from the previous one is bounded below by ε^{β}.

If an eigenvalue of Ω with these properties exists we denote it by $\omega_{i_{m+1}}$ and the previous one by $(\omega_{i_{m+1}})_- \leq \omega_{i_m} \mathcal{O}(\varepsilon^{\beta-\delta})$. We also define

- $I_m := [\omega_{i_m}, (\omega_{i_{m+1}})_-]$;
- $\mathcal{H}_m :=$ direct sum of eigenspaces of Ω corresponding to eigenvalues contained in the interval I_m; and $n_m := \dim \mathcal{H}_m$.

If this eigenvalue does not exists we conclude that the largest eigenvalue, $\omega_{\max}$, of Ω is bounded above by $\omega_{i_m} + \mathcal{O}(\varepsilon^{\beta-\delta})$, and we define $I_m := [\omega_{i_m}, \omega_{\max}]$
On the subspace $\mathcal{H}_m$ we define the operator $\widetilde{\Omega}$ by $\widetilde{\Omega}\big|_{\mathcal{H}_m} := \widetilde{\omega}_m \cdot \mathbf{1}\big|_{\mathcal{H}_m}$, where $\widetilde{\omega}_m := \omega_{i_m}$.

1-v) The construction described above must necessarily stop at some step $\overline{m} \geq 0$, because Ω is trace-class and $\varepsilon^{\beta} > 0$. The spectrum of the operator $\widetilde{\Omega}$ constructed above consists of the points

$$\{\, \omega_{i_0} := 0\,, \omega_{i_1}\,, \ldots\,, \omega_{i_{\overline{m}}} \}\,. \tag{38}$$

1-vi) We note that $\widetilde{\Omega}$ has been defined as the operator whose eigenspaces are the subspaces $\mathcal{H}_m$ and the corresponding eigenvalues are given by $\widetilde{\omega}_m$. (To avoid possible confusion we stress that the eigenvalues $\widetilde{\omega}_m$ of $\widetilde{\Omega}$ are *increasing* in m whereas the eigenvalues ω_i of Ω are *decreasing* in i.) The operator $\widetilde{\Omega}$ enjoys the property

$$\mathrm{tr}|\Omega - \widetilde{\Omega}| \leq o(1) + \mathcal{O}\big(\varepsilon^{\beta-\delta}(n_1 + \cdots + n_{\overline{m}})\big) \leq o(1) + O(\varepsilon^{\beta-2\delta})\,, \tag{39}$$

which holds, because

$$0 < \sum_{\omega_i \leq \varepsilon^{\delta}} \omega_i \leq o(1)\,; \tag{40}$$

(recall that Ω is trace-class and that, in (32), we have noticed that $\sum_{i:\omega_i \leq \varepsilon^{1/4}} \omega_i =: \Delta_\varepsilon \ll 1$). Moreover, we use the facts that any eigenvalue of Ω corresponding to an eigenvector in $\mathcal{H}_m$ is included in the interval $[\tilde{\omega}_m\,, \tilde{\omega}_m + \epsilon^{\beta-\delta}]$, by construction of $\mathcal{H}_m$, and that

$$n_1 + \cdots + n_{\overline{m}} \leq O(1/\varepsilon^{\delta})\,, \tag{41}$$

as shown in 1-i).

2) Next, we modify the operator X. The modified operator is denoted by X' and is defined by its matrix elements in the basis, $\{u_j\}_{j=1}^{\infty}$, of eigenvectors of Ω, which are given by

$$(X')_{i,j} := \langle u_i\,,\, X\,u_j\rangle\,, \tag{42}$$

provided that u_i and u_j belong to the same subspace $\mathcal{H}_p,\ p \leq \overline{m}$, and

$$(X')_{i,j} := 0\,, \tag{43}$$

if u_i and u_j belong to *different* eigenspaces, $\mathcal{H}_p$, $\mathcal{H}_{p'}$, of $\widetilde{\Omega}$.
We thus have by construction that

$$[\widetilde{\Omega}\,,\, X'] = 0\,. \tag{44}$$

Next, we show that $\|X - X'\| = o(1)$. This follows from the following inequality proven below:

$$\sup_i \sum_{j=1}^{\infty} |\langle u_j\,,\,(X - X')\,u_i\rangle|^2 \leq \varepsilon^{2(1-\beta)}\,, \tag{45}$$

where the summands are non-zero only if u_i and u_j belong to different eigenspaces $\mathcal{H}_{p_i}$, $\mathcal{H}_{p_j}$ of $\widetilde{\Omega}$, so that

$$\sum_{j=1}^{\infty} |\langle u_j,\,(X - X')\,u_i\rangle|^2 = \sum_{j\,:\,u_j \in \mathcal{H}_{p_j}\,,\,p_j \neq p_i} |\langle u_j,\,X\,u_i\rangle|^2\,. \tag{46}$$

But if $p_j \neq p_i$ then $|\omega_i - \omega_j| \geq \varepsilon^{\beta}$, where ω_i and ω_j are the eigenvalues of Ω on the vectors $u_j \in \mathcal{H}_{p_j}$ and $u_i \in \mathcal{H}_{p_i}$, respectively. Next, we exploit the bound assumed in (33), namely

$$\varepsilon^2 \geq \|\,[\,\Omega\,,\,X\,]\,u_i\,\|^2 \tag{47}$$

$$= \sum_{j=1}^{\infty} (\omega_i - \omega_j)^2 |\langle u_i\,,\,X\,u_j\,\rangle|^2 \tag{48}$$

$$\geq \varepsilon^{2\beta} \sum_{j\,:\,u_j \in \mathcal{H}_{p_j}\,,\,p_j \neq p_i} |\langle u_i\,,\,X\,u_j\,\rangle|^2 \tag{49}$$

To conclude the proof of the theorem, we normalize $\widetilde{\Omega}$ by dividing by its trace, defining $\Omega' := \frac{\widetilde{\Omega}}{\operatorname{tr}\widetilde{\Omega}}$. Setting $\delta = \frac{1}{4}$ and $\beta = \frac{3}{4}$, and using that

$$\operatorname{tr}|\Omega - \widetilde{\Omega}| \leq \Delta_{\varepsilon} + \mathcal{O}(\varepsilon^{\frac{1}{4}})\,,$$

we conclude that

$$\operatorname{tr}|\Omega - \Omega'| \leq 2\,\Delta_{\varepsilon} + \mathcal{O}(\varepsilon^{\frac{1}{4}})\,.$$

References

1. Fröhlich, J., Schubnel, B.: Quantum probability theory and the foundations of quantum mechanics. In: Blanchard, P., Fröhlich, J. (eds.) The Message of Quantum Science. Springer, Berlin (2015). arXiv:1310.1484 (see also: Ph. Blanchard, J. Fröhlich and B. Schubnel, A "garden of forking paths" – The quantum mechanics of histories of events, Nucl. Phys. B **912**, 463–484 (2016))
2. Fröhlich, J., Pizzo, A.: The Time-Evolution of States in Quantum Mechanics according to the ETH-Approach. Commun. Math. Phys. **389**, 1673–1715 (2022)
3. Fröhlich, J., Gang, Z., Pizzo, A.: A Tentative Completion of Quantum Mechanics. arXiv:2303.11112v1 [math-ph] (2023)
4. von Neumann, J.: Mathematical foundations of quantum mechanics (with an introduction by Iván Abonyi). Akadémiai Kiadó (Publishing House of the Hungarian Academy of Sciences), Budapest (1980). (translated from the 1964 Russian edition by Ákos Sebestyén) (see also: G. Lüders, Über die Zustandsänderung durch den Messprozess, Annalen der Physik, **443**(5–8) 322–328 (1950))
5. Lin, H.: Almost commuting self-adjoint matrices and applications. Fields Inst. Commun. **13**, 193 (1995)
6. Hastings, M.B.: Making Almost Commuting Matrices Commute. Commun. Math. Phys. **291**, 321–345 (2009)
7. Kachkovskiy, I., Safarov, Y.: Distance to Normal Elements in C^*-Algebras of Real Rank Zero. arXiv:1403.2021v3 [math.OA] (2015)
8. Fröhlich, J., Schubnel, B.: Do we understand quantum mechanics – finally? In: Reiter, W.L., Yngvason, J. (eds.) Erwin Schrödinger – 50 Years After, ESI Lectures in Mathematics and Physics. pp. 37–84. European Mathematical Society Publ (2013)

Why Imaginary Time? QM of the Fermi Oscillator and Path Integrals

Jürgen Löffelholz

Contents

Abstract The path integral for the Fermi oscillator defined by the evolution kernel $U(t|x, y)$, with $x, y = \pm 1$, is analysed in detail. We derive a polar decomposition of the complex-valued cylinder measures $\mathrm{d}W_N$ governing N time steps of length $\tau = T/N$ and show how it survives the continuous time limit, $N \to \infty$. Moreover, we confront the formulae with those for the corresponding imaginary time Markoff process.

1 Introduction

Two level quantum systems are very useful to describe interesting aspects of real physical phenomena like tunneling of the N atom in the Ammonium molecule NH_3 within the lowest energy states [1].

Moreover, they allow to discuss basic ideas of quantum mechanics (QM) without very difficult mathematics. The aim of this paper is to demonstrate how naturally emerge path integrals in "physical" and imaginary time.

J. Löffelholz (✉)
Angewandte Informatik, Fachhochschule Erfurt FHE, Erfurt, Germany
e-mail: jloeffholz@web.de

A. Cintio, A. Michelangeli (Eds.), *Trails in Modern Theoretical and Mathematical Physics*, https://doi.org/10.1007/978-3-031-44988-8_10

Let us briefly review the history. As well known the first formulation of QM in terms of matrices was given by W. Heisenberg 1923. But it was confronted soon with the theory of E. Schrödinger. He introduced a so called wave function $\varphi(t|x)$ which must solve his famous equation.

Using an idea of P. Dirac 1933 concerning the possible role of Lagrangean for quantum theory R. Feynman 1946 gave a recipe for explicit calculation of the propagator $U(t|x, y)$ which allows to get $\varphi(t|x)$. His path integral reformulation of QM meanwhile is standard.

But this was not yet the end of the story. In 1972 E. Nelson showed how for certain QFT models the propagator may be recovered from the transfer matrix $p(t|.,.)$, $t \geq 0$, governing a stochastic process "at imaginary time" [2]. Later the crucial Markoff property could be relaxed, also the requirement of having a positive measure, ergodicity and even regularity [3–5].

In fact the trick of analytic continuation to imaginary time was originally introduced in the opposite direction and described by J. Schwinger 1958 with "Euclidean" fields related to positive metric as rather miraculous objects. It stimulated constructive QFT also including gauge fields.

2 Model

The pure states of our two level QM system are described by vectors $f = (a, b)$ in $\mathcal{H}_{\text{phys}} = \mathbf{C}^2$, often normalized by $|a|^2 + |b|^2 = 1$. We will denote the ground state by $\Omega = (0, 1)$ and the excited state by $h = (1, 0)$ corresponding to energy levels $E_0 = 0$, and $E_1 = \omega > 0$.

The Hamiltonian then reads $H = \omega P_1$, where P_1 is the 2×2-matrix of projection onto h. Introducing the operator A sending h to Ω and annihilating Ω and its adjoint A^* we get $A^* A = P_1$ and hence $H = \omega A^* A$. Moreover, $AA = A^* A^* = 0$, $AA^* = P_0$ and $AA^* + A^* A = I$.

For $Q = (A + A^*)$ it holds $QQ = I$ and hence its eigenvalues are $x = \pm 1$. Let

$$P = \mathrm{i}\,\omega(A^* - A) = \mathrm{i}[H, Q]$$

and

$$U(t) = \exp(-\mathrm{i}\,tH) = P_0 + \exp(-\mathrm{i}\,t\omega)P_1\,, \quad -\infty < t < +\infty\,,$$

denote the group of unitary operators governing time evolution. The resulting formula

$$Q(t) = U^*(t)QU(t) = \cos(t\omega)Q + \sin(t\omega)\frac{P}{\omega}$$

suggests to interpret Q as position, P as momentum and the name "Fermi oscillator" for this quantum system. Observe that $PQ + QP = 0$. Instead of CCR one finds an interesting operator

$$\frac{(QP - PQ)}{2\mathrm{i}\,\omega} = R = P_0 - P_1$$

implementing reflections $RQR = -Q$. Hence R intertwines the eigenstates of Q and also of P. Moreover, it holds $R\Omega = \Omega$ and one finds the representation of the Hamiltonian

$$H = \frac{\omega}{2}(I - R)\,.$$

Explicitly:

$$P_0 = \begin{pmatrix} 0 & 0 \\ 0 & 1 \end{pmatrix}, \qquad P_1 = \begin{pmatrix} 1 & 0 \\ 0 & 0 \end{pmatrix}, \qquad P = \omega \begin{pmatrix} 0 & \mathrm{i} \\ -\mathrm{i} & 0 \end{pmatrix},$$

$$A = \begin{pmatrix} 0 & 0 \\ 1 & 0 \end{pmatrix}, \qquad Q = \begin{pmatrix} 0 & 1 \\ 1 & 0 \end{pmatrix},$$

$$U(t) = \begin{pmatrix} r & 0 \\ 0 & 1 \end{pmatrix}, \qquad R = \begin{pmatrix} -1 & 0 \\ 0 & 1 \end{pmatrix},$$

with $r = \exp(-\mathrm{i}\,t\omega)$.

In fact, H is renormalized. In most textbooks one considers the energy levels $\pm\omega/2$, i.e. the naive Hamiltonian $H_0 = \omega/2(P_1 - P_0) = -\omega R/2$. Hence $H = H_0 + \omega/2$. Also for the momentum operator P using R we find another formula

$$P = -\mathrm{i}\,\omega/2[R, Q] = \mathrm{i}\,\omega QR\,.$$

3 Schrödinger Equation

The matrix algebra generated by Q and P (with $PP = \omega^2 I$) contains the commutative subalgebra generated by the position operator Q with spectrum $\{-1, +1\}$. So we may realize vectors in the physical Hilbert space $\mathcal{H}_{\text{phys}}$ as functions $f(x) = a1 + bx$, where $x = \pm 1$, equipped with scalar product

$$\langle g, f\rangle = \frac{1}{2}\sum g(x)^*\, f(x)\,,$$

summing over x. The factor $1/2$ makes

$$f = (a, b) \to a1 + bx$$

an isometry, in particular $\Omega \to 1$ and $h \to x$. In this representation eigenfunctions of Q are given by

$$\delta_+ = \frac{1 + x}{2}\,,\ \delta_- = R\delta_+ = \frac{1 - x}{2}\,.$$

It holds $\delta_+(1) = +1, \delta_+(-1) = 0$ and $\langle \delta_+, \delta_-\rangle = 0$.

The eigenfunctions of the momentum operator P read $(1-\mathrm{i}\,x)/2$ and $(1+\mathrm{i}\,x)/2$ with the corresponding eigenvalues $\pm\omega$. It holds $P\delta_+ = -\mathrm{i}\,\omega\delta_-$. Moreover, the eigenstates of H are just the symmetric and antisymmetric superpositions of those for Q with $x = \pm 1$

$$\Omega = (\delta_+ + \delta_-)\,, \quad h = (\delta_+ - \delta_-)\,.$$

All operators are defined by their "integral kernels" where integral here means sum. In particular we obtain $A(x, y) = y$, $A^*(x, y) = x$, $Q(x, y) = (x + y)$, $P(x, y) = \mathrm{i}\,\omega(x - y)$ and $R(x, y) = (1 - xy)$. Moreover, $P_0(x, y) = 1$ and $P_1(x, y) = xy$ so that $I(x, y) = (1 + xy)$. Here we did not include the factor $1/2$. This leads to

$$\begin{aligned} Pf(x) &= \mathrm{i}\,\omega x f(-x)\,, \\ Hf(x) &= \frac{\omega}{2}\big(f(x) - f(-x)\big)\,. \end{aligned}$$

Below we will include the factor $1/2$ appearing in the scalar product of $\mathcal{H}_{\mathrm{phys}}$. So in particular for the evolution kernel we obtain the expression

$$U(t|x, y) = \frac{1}{2}\big(1 + \exp(-\mathrm{i}\,t\omega)\,xy\big)\,, \qquad x, y = \pm 1.$$

Given f in $\mathcal{H}_{\mathrm{phys}}$, $\varphi(t|\cdot) = \exp(-\mathrm{i}\,tH)f$ with initial condition $\varphi(0|\cdot) = f$ satisfies the Schrödinger equation $\mathrm{i}\,\mathrm{d}\varphi(t|\cdot)/\,\mathrm{d}t = H\varphi(t|\cdot)$. For $f = \delta_+$ as above we obtain

$$\begin{aligned} \varphi(t|x) &= \exp(-\mathrm{i}\,tH)\delta_+ \\ &= \frac{1 + \exp(-\mathrm{i}\,t\omega)x}{2} = u_+(t)\delta_+ + u_-(t)\delta_-\,, \end{aligned}$$

with the complex coefficients

$$u_+(t) = \frac{\langle\delta_-, \varphi(t|x)\rangle}{\langle\delta_+, \delta_+\rangle} = \frac{1 + \exp(-\mathrm{i}\,t\omega)x}{2}$$

and

$$u_-(t) = \frac{1 - \exp(-\mathrm{i}\,t\omega)x}{2}\,,$$

respectively.

As vectors $u_+(t), u_-(t)$ are orthogonal to each other. In Fig. 1 we have drawn their geometry. The corresponding probabilities (which sum up to 1) are

$$|u_\pm(t)|^2 = \frac{1 \pm \cos(t\omega)}{2}\,.$$

Hence $\varphi(t|\cdot)$, starting at $\varphi(0|\cdot) = \delta_+$, after time $t = \pi/\omega$ is equal to the state vector δ_-, i. e. "the quantum system moves from $x = 1$ to $x = -1$". And after

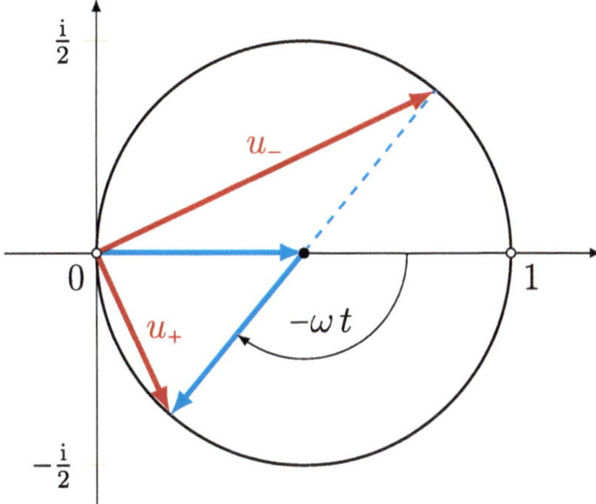

Fig. 1 Complex $u_\pm$-plane

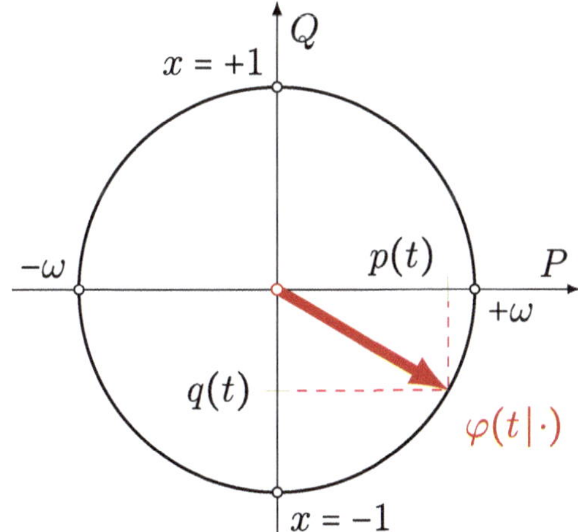

Fig. 2 Circulation of $\varphi(t|\cdot)$ within $0 \leq t \leq 2\pi/\omega$

time $t = T_0 = 2\pi/\omega$ "it returns to $x = +1$" i. e. the original one. We easily check $\varphi(t|\cdot) = (1 - \mathrm{i}\, x)/2$ for $t = \pi/2$ and $\varphi(t|\cdot) = (1 + \mathrm{i}\, x)/2$ for $t = 3\pi/2$.

More precisely, the corresponding vector of the Bloch sphere [6] associated to our QM system moves around its equator, which is a circle in the Q, P/ω-plane of mean energy

$$\langle \varphi(t|\cdot), H\varphi(t|\cdot)\rangle / \langle \varphi(t|\cdot), \varphi(t|\cdot)\rangle = \omega/2\,.$$

For the mean values of Q and P in these states $\varphi(t|\cdot)$ we calculate $q(t) = \cos t\omega$ and $p(t)/\omega = \sin t\omega$, as one may expect from an oscillator, see Fig. 2.

The amplitudes $u_\pm(t)$ tell us that $\varphi(t|\cdot) \to \delta_+, \delta_-$ respectively with probabilities $|u_\pm(t)|^2$, for any time $t > 0$. Naively one would expect that within time $t = kT_0/2\,, k = 0, 1, 2, 3, \ldots$ a number k of flips $x = -1 \leftrightarrow x = +1$ would happen.

These flips are virtual and cannot be observed. But they allow to get an intuitive picture of what happens with the QM system Fermi oscillator. Although

$$\varphi(t|\cdot) \equiv \exp(-itH)\delta_+ = \delta_-, \delta_+$$

for $t = T_0/2$ respectively $t = T_0$ there is no probability for finding it in a state such that $-1 < x < +1$ which contradicts our classical logic and understanding of physics.

4 Feynman Integral

The a priori probability for the quantum system to be in a state with eigenvalue $x = \pm 1$ of Q is equal to $1/2$ and we rewrite

$$\mathcal{H}_{\text{phys}} = L^2(S_0, dP_0), \quad \text{where } S_0 = \{-1, +1\}.$$

Below we consider the N-point vacuum expectation values of time ordered operators $Q(t_n)$, $n = 0, 1, 2, \ldots, N$. Using $U(t)\Omega = \Omega$, for $N = 1$ we obtain $< \Omega, Q(t)\Omega >= 0$ because $Q\Omega = h$ is orthogonal to Ω. Next, for $N = 2$ we have

$$\langle \Omega, Q(t_1)Q(t_2)\Omega \rangle = \langle \Omega, Q \exp(\mathrm{i}\,\tau H)Q\Omega \rangle = \exp(\mathrm{i}\,\tau\omega),$$

where $\tau = (t_2 - t_1) > 0$.

We will always have the variable $x_0 = x(t_0)$ for some starting time t_0 with probability $dP_0(x_0) = 1/2$. Let $t_0 < t_1 < t_2$ and $x_1 = x(t_1)$, $x_2 = x(t_2)$, $\tau_1 = (t_1 - t_0)$, $\tau_2 = (t_2 - t_1)$. The Feynman path integral measure which will be shown to reproduce the above expression is given by

$$dW_2(x_0, x_1, x_2) = dP_0(x_0)w(\tau_1|z_1)w(\tau_2|z_2)$$

with

$$w(\tau|z) = \frac{1 + \exp(\mathrm{i}\,\tau\omega)z}{2} = \frac{U(-\tau|x, y)}{2}, \quad z = xy = \pm 1.$$

One verifies

$$\sum dW_2(x_0, x_1, x_2) = 1,$$

summing over $x_0, x_1, x_2 = \pm 1$.

“Integrating out x_2” we obtain dW_1 equal to $dP_0(x_0)w(\tau_1|z_1)$. By iteration we obtain dW_N defined on points $(x_0, x_1, x_2, \ldots, x_N)$ in $S_0 \times S_N$, where

$$S_N = \bigotimes_{n=1}^{N} \{-1, +1\}.$$

There are exactly $2 \cdot 2^N$ elements in $S_0 \times S_N$ and these can be interpreted as discrete time trajectories $t_n \to x_n = \pm 1$. One easily checks a consistency condition for "summing out" any of the variables and hence $d W_N$, $N = 0, 1, 2, \ldots$, define so called cylinder measures [7] on continuous time trajectories $t \to x(t) = \pm 1$ "controlling their values only sometimes".

Let $0 \le t \le T$, with finite T. Then the trajectories are elements in $S[0, T]$ which is an uncountable set. The idea of R Feynman was to define a complex-valued measure $d W[0, T]$ on such set of "classical" trajectories. For simplicity, in the following we put $t_n = n\tau, n = 0, 1, 2, \ldots, N$ so that all time steps $(t_n - t_{n-1})$ are equal to $\tau = T/N$.

Now $t_0 = 0 < t_1 < t_2 \ldots < t_N = T$. To prove the existence of a measure $d W[0, T]$ lurking in the background we must find an upper bound for the total variations

$$\|d W_N\| = \sum |d W_N(x_0, x_1, x_2, \ldots, x_N)|$$

independent of N. Above the sum runs over all $x_0, x_1, \ldots, x_N = \pm 1$. The measure factorizes in the variables x_0 and $z_n, n = 1, 2, \ldots, N$. Hence the modulus may be put to every factor $w(\tau|z_n)$. We find $|w(\tau|z)| = \sqrt{(1 + z \cos \tau\omega)/2}$ so that summing over $z = \pm 1$

$$\begin{aligned} C(\tau) &= \sum |w(\tau|z)| = \sqrt{(1 + \cos \tau\omega)/2} + \sqrt{(1 - \cos \tau\omega)/2} \\ &= |\cos(\tau\omega/2)| + |\sin(\tau\omega/2)| = \sqrt{1 + |\sin \tau\omega|} \end{aligned}$$

and

$$\|d W_N\| = C(\tau)^N .$$

Lemma 1 The total variations of the cylinder measures $d W_N$, $N = 1, 2, \ldots$ satisfy the estimate

$$\|d W_N\| < \|d W[0, T]\| = \exp(+T\omega/2) .$$

Proof By a simple argument it is sufficient to show $[C(\tau)]^2 < C(2\tau)$ when we double N step by step. We check $C(\tau) > 1$, except for $\tau\omega$ equal to multiples of π when $C(\tau) = 1$ which is also seen by Pythagoras' theorem. So we obtain a monotone increasing sequence $\|d W_N\|$.

From $C^N \approx (1 + T\omega/N)^{N/2}$ for small $\tau = T/N$, we obtain the required upper bound. □

Since $\|d W[0, T]\|$ diverges to infinity when $T \to \infty$ we suspect that only for a finite time interval the measure is well defined.

5 Flip Number Sets

The expressions $d W_N$, for $N = 1, 2, 3, \ldots$, are complex-valued. We observe that $w(\tau|z) = u_{\pm}(-\tau)$, $z = \pm 1$ because of $U(t) = \exp(-\mathrm{i}\, tH)$, i. e. these complex numbers circulate in the opposite direction with time. Clearly, $w(\tau|z)$ are real (equal 0 or 1) only when $\tau\omega$ is equal to multiples of π. How looks like the polar decomposition and what about the phase? We obtain

$$\frac{d W_N}{d P_0} = w(\tau| - 1)^k \, w(\tau| + 1)^{N-k}$$

with

$$k = \frac{N - \sum z_n}{2}.$$

Indeed, the variables $k_n = (1 - z_n)/2$, $n = 1, 2, \ldots, N$, take values 0 or 1 and their sum $k = \sum k_n$ counts the number of "flips" $x_n = -x_{n-1}$ when $z_n = -1$. This leads to

Lemma 2 If $t_n = n\tau, n = 0, 1, 2, \ldots, N$ so that all time steps $(t_n - t_{n-1})$ are equal to $\tau = T/N$ the Feynman measure of a path $(x_0, x_1, \ldots, x_N)$ depends only on $k = \dfrac{N - \sum z_n}{2}$, where $z_n = x_{n-1}x_n$ and is given by the expression

$$d W_N = d P_0 \otimes \left[(\sin(\tau\omega/2))^k (\cos(\tau\omega/2))^{N-k}\right] \exp(\mathrm{i}\, \beta_k)$$

with angle

$$\beta_k = \frac{N\tau\omega - k\pi}{2}.$$

Proof

$$w(\tau| + 1) = \cos(\tau\omega/2) \exp(\mathrm{i}\, \tau\omega/2),$$
$$w(\tau| - 1) = -\mathrm{i}\, \sin(\tau\omega/2) \exp(\mathrm{i}\, \tau\omega/2).$$

These identities are visualized in Fig. 3a. We used $-\mathrm{i} = \exp(-\mathrm{i}\, \pi/2)$ and of course $T = N\tau$ is the total length of the time interval. For $0 \leq T\omega/2N \leq \pi/2$ both terms in the square brackets are nonnegative so that for N large enough the above formula gives the required decomposition into modulus and phase factor. □

We observe that the set $a \times S_N$ of trajectories $(a, x_1, \ldots, x_N)$ with fixed $x_0 = a = \pm 1$ may be devided into disjoint classes $\{a \times S_N(k)\}$, $k = 0, 1, 2, \ldots, N$ which contain precisely those elements with $(N - \sum z_n)/2 = k$.

Now we are ready to perform the limit $N \to \infty$ and simultaneously $\tau = T/N \to 0$.

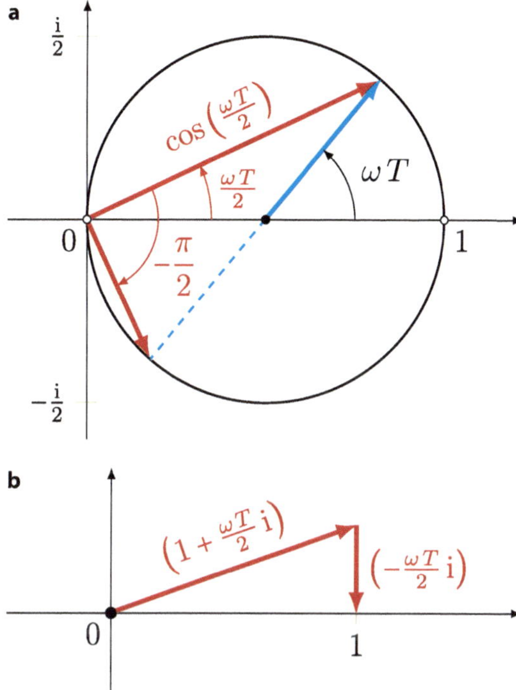

Fig. 3 Geometry of $w(\tau|\pm 1)$. **a** $V_0 = \exp(i\beta_0)$, **b** v_0, v_1 for T small

Theorem The Feynman integral measure of the set of trajectories $t \to x(t)$ starting at $x_0 = a$ and having $k = 0, 1, 2, \ldots$ "flips" within the time interval $(0, T]$ is given by

$$dP_0(a)\,\frac{(T\omega/2)^k}{k!}\,\exp(\mathrm{i}(T\omega - k\pi)/2)\,.$$

Proof With help of Stirling's formula we may approximate the number of elements in $\{a \times S_N(k)\}$ for large N by $N^k/k!$ so that

$$(\cos(T\omega/2N))^{N-k} \to 1 \quad \text{and} \quad [(\sin(T\omega/2N)N]^k \to T\omega/2\,.$$

Moreover

$$[\sin(T\omega/2N)]^k \to 0\,,$$

except $k = 0$. □

Hence when $N \to \infty$ only the single elements with $k = 0$ (i.e. no flips at all) survive with measure non zero which are exactly the trajectories $x(t) = -1$ for all $0 \le t \le T$, respectively $x(t) = +1$. The flip number sets remain measurable and we denote them by $\{a \times S(0, T](k)\}$, where $k = 0, 1, 2, \ldots$ with $a = \pm 1$.

Also subsets of trajectories and their measures $dW[T_1, T_2]$ for arbitrary intervals $[T_1, T_2]$ are easily defined because of time translation invariance, i.e. the expression depends only on $T = (T_2 - T_1)$. Applying the usual operations $\cap$, $\setminus$ for intersection and complement we obtain a rich algebra of subsets in $S[0, T]$ which contains also the standard cylinder sets as we will see just now.

In the following we concentrate on the flips and neglect the factor $dP_0(a) = 1/2$ remembering that every trajectory may start from two points with equal probability. Then we have a complex-valued Poisson distribution with parameter $(-\mathrm{i}\, T\omega/2)$ for the variable k

$$v_k = \exp\left(\frac{\mathrm{i}\, T\omega}{2}\right) \exp\left(\frac{-\mathrm{i}\, k\pi}{2}\right) c_k \,,$$

where

$$c_k = \frac{(T\omega/2)^k}{k!} = |v_k| \,, \quad k = 0, 1, 2, \ldots$$

It holds $c_0 = 1$. If $T > 0$ then $c_k > 0$, for $k = 1, 2, 3, \ldots$. The sum of all v_k is equal to the measure of all paths an hence equal to 1. But $|v_0| = 1$ and hence $\sum |v_k|^2 > 1$, except $T = 0$, the expressions $|v_k|^2$ do not define probabilities. The v_k are also defined by $v_0 = \exp(\mathrm{i}\, T\omega/2)$ and the recursion formula

$$v_k = \exp(-\mathrm{i}\, \pi/2) \frac{T\omega}{2k} v_{k-1} \,.$$

For small T compared with the period $T_0 = 2\pi/\omega$ of the Fermi oscillator we verify

$$v_0 = \exp\left(\frac{\mathrm{i}\, T\omega}{2}\right) \approx 1 + \frac{\mathrm{i}\, T\omega}{2} \qquad v_1 = -\mathrm{i}\frac{T\omega}{2} \exp\left(\frac{\mathrm{i}\, T\omega}{2}\right) \approx -\frac{\mathrm{i}\, T\omega}{2}$$

as drawn in Fig. 3b above for $T\omega \approx 1$. We observe that the phase vanishes exactly for $T = k(T_0/2)$ i. e. when there are in mean just two flips within one period. By a simple argument one can show that the modulus $(T\omega/2)^k/k!$ acquires its maximum $k_{\max}$ near $T\omega/2$, see Fig. 4.

We have $v_0 = \exp \mathrm{i}\, \beta_0$, $\beta_0 = T\omega/2$. This defines the line in the complex plane for all v_k with k even. In fact, the factors $\exp(-\mathrm{i}\, k\pi/2) = (-\mathrm{i})^k$ implement rotations by the angles $-k\pi/2$. Therefore, from the recursive formula in particular we obtain $v_1 = -\mathrm{i}(T\omega/2)\, v_0$, $v_2 = -(T\omega/2)^2/2\, v_0$, etc.

Every trajectory in $\{a \times S(0, T](k)\}$, for k even, connects $x_0 = a$ with $x(T) = a$. Because all those contribute to the QM transition their amplitudes should sum up to $w(T|z)$, $z = a\, a = 1$. Indeed, the sum of factors $(T\omega/2)^k/k!$ with k even is equal to $\cos(T\omega/2)$.

Similar for k odd, any trajectory in $\{a \times S(0, T](k)\}$ connects $x_0 = a$ with $x(T) = b = -a$. Hence we have reconstructed the cylindersets $\{a \times S(0, T) \times b\}$, $b = \pm a$, as

$$\bigcup \{a \times S(0, T](k)\} \,, \qquad \text{union over } k \text{ even/odd,}$$

and their Feynman path integrals. It seems worthwhile to compare all this with the formulae for the imaginary time stochastic process associated to our QM model.

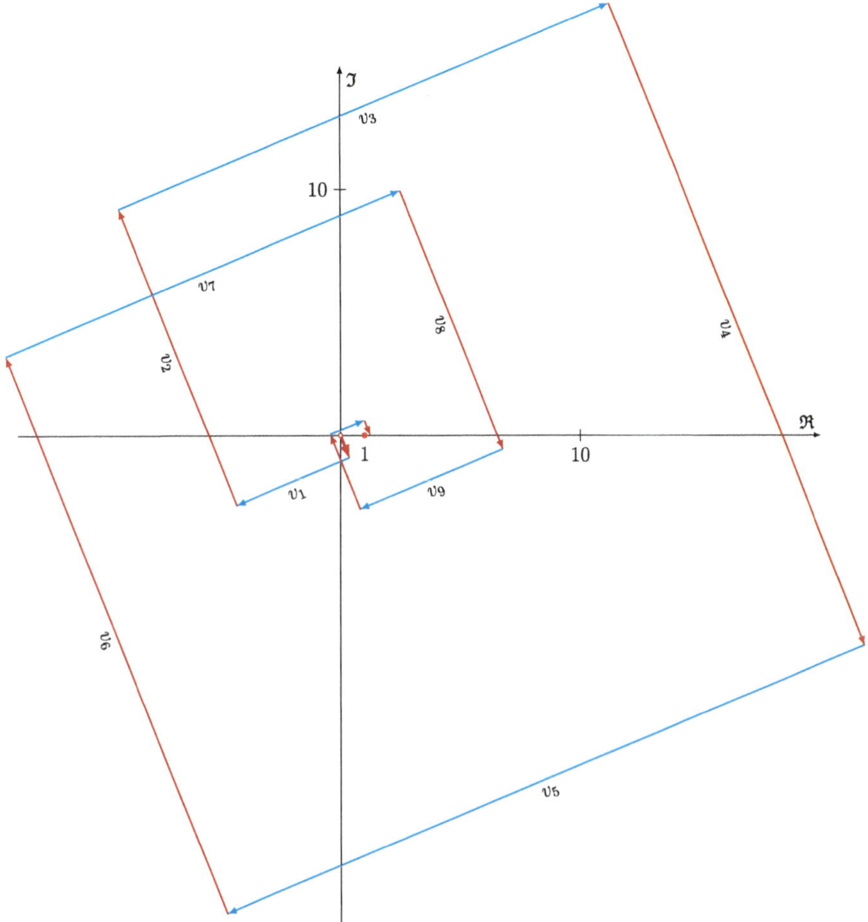

Fig. 4 Arrows v_k

6 Ferromagnetic Measure

Substituting formally $-\mathrm{i}\,\tau \to t \geq 0$ in the QM amplitudes $\big(1+\exp(\mathrm{i}\,\tau\omega)xy\big)/2$ we obtain

$$p(t|x,y) = \frac{1 \pm r}{2}, \quad y = \pm x$$

with $r = \exp(-t\omega)$. The expression $0 \leq p(t|x,y) \leq 1$ defines the transition function of a Markoff process $t \to x(t) = \pm 1$. We will denote the measure by dP. The identity (Chapman–Kolmogoroff equation)

$$\sum p(s|x,y)p(t|y,z) = p(s+t|x,z)\,,$$

summing over $y = 1$, means that $p(t|.,.)$, $t \geq 0$, build a semi-group. The pair $x(t), x(0)$ shows "ferromagnetic" correlation because $p(t|x, y) > 1/2$ for $xy = 1$. If we think about a spin chain it is more likely they are parallel directed. The correlation decays exponentially

$$p(t|x, y) \to 1/2 \quad \text{for } t \to \infty\,.$$

As for physical time let us start with discrete $t_n = n\tau$, $n = 0, 1, \ldots, N$, where $\tau = T/N$. The set $a \times S_N$ may be interpreted as family of linear chains $(x_0, x_1, x_2, \ldots, x_N)$ of $N + 1$ spins $x_n = \pm 1$ at "distances" τ. We may rewrite

$$p(\tau|x, y) = \frac{\exp(Jxy)}{\exp(J) + \exp(-J)}$$

and

$$dP_N = dP_0(x_0)\,\frac{\exp J \sum z_n}{(\exp J + \exp(-J))^N}$$

with $z_n = xy$, $x = x_{n-1}$ and $y = x_n$, $n = 1, 2, \ldots, N$. This is the well known expression used first by W. Lenz and E. Ising to describe ferromagnetism in $d = 1, 2$ dimensions. About 1973 a group of theoreticians around the world initiated the reconstruction of QFT models from stochastic processes with "ferromagnetic measures" [8].

The function $J = J(\tau)$ involves the Caley transformation

$$\exp(-2J) = \frac{1 - r}{1 + r}\,,$$

mapping the unit interval onto itself and is its own inverse. For small $\tau > 0$ spins nearby are strongly correlated i.e. we get big values of J and vice versa. What gives the analytic continuation $r \to \exp(\mathrm{i}\,t\omega)$? We find

$$\exp(-2J) \to -\mathrm{i}\,\tan(t\omega/2)\,,$$

which leads to the above formula for dW_N within the QM model.

In the variables $x_0, z_n, n = 1, 2, \ldots, N$, the measure dP_N factories into $dP_0(x_0)$ and the product of

$$dM(z_n) = \frac{1 + rz_n}{2}$$

Therefore x_0 and all $z_1, z_2, \ldots, z_N$ are mutually stochastic independent. The mean value of z is $\sum dM(z)z = r$. We may rewrite $x_n = x_0 z_1 z_2 \ldots z_n$ and similar for x_m. Hence

$$E[x_n] = \sum dP_0(x_0)x_0\Big[\sum dM(z)z\Big]^n = 0\,.$$

and

$$E[x_m\, x_n] = \sum dP_0(x_0)\Big[\sum dM(z)z\Big]^{n-m} = r^{n-m}\,.$$

Remark The dynamics of our system is described in terms of the flip variables z_n, whereas $x_0 = a$ is a reference variable which we can fix. This remembers the stochastic process called one-dimensional random walk

$$N \to x_N = a + (z_1 + z_2 + \ldots + z_N)$$

in terms of an initial position $x_0 = a$ and steps $z_n = \pm 1, n = 1, 2, \ldots, N$ forward and back with equal probability 1/2. Unfortunately, here no translation invariant regular measure exists when $N \to \infty$ because the x_N become unbounded. The counterpart of $dP_0(x_0)$ is the "flat" ergodic mean measure [9].

The measure dP_N of a chain $(x_0, x_1, \ldots, x_N)$ depends only on the initial value $x_0 = a$ and on the total number of flips $k = \sum k_n$, where again $k_n = (1 - z_n)/2$ with $n = 1, 2, \ldots, N$, see Fig. 5. For fixed $x_0 = a$ it holds

$$p_k = \frac{dP_N(a, x_1, x_2, \ldots, x_N)}{dP_0(a)} = q^k (1-q)^{N-k},$$

where

$$q = E[k_n] = \frac{1-r}{2} = \frac{1-\exp(-T\omega/N)}{2}.$$

Now $k_n = 0, 1$ and hence their sum k are true random variables. In fact, $(k_1, k_2, \ldots, k_N)$ is a Bernoulli chain [10] defined by the parameter q. As above multiplying the p_k by $N!/(k!(N-k)!)$ we obtain the probabilities for flip number sets $\{a \times S_N(k)\}$, $k = 0, 1, \ldots, N$. And finally, if we add these for all k even/odd by induction starting with $N = 1$ we recover the correct expressions $(1 \pm r^N)/2$ for the cylindersets $S_N(a \times b)$ i. e. the subset of paths in S_N with $x_0 = a$ and $x_N = b$.

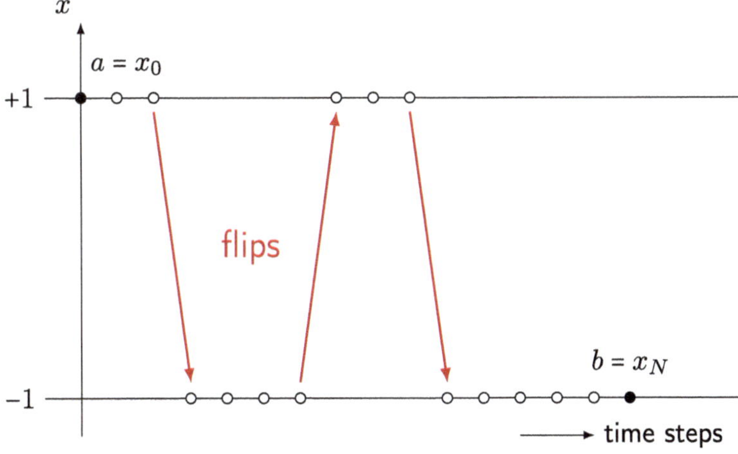

Fig. 5 Trajectory with 3 flips

For small $\tau = T/N$ the parameter q becomes close to $T\omega/2N$. As well known one may perform the limit $N \to \infty$ when keeping Nq finite. In result the random variable $k = (k_1 + k_2 + \ldots + k_n + \ldots)$ yields a Poisson distribution

$$\frac{\lambda^k}{k!}\exp(-\lambda)\,, \quad k = 0, 1, 2, \ldots$$

with

$$\lambda = \lim_{N\to\infty} \frac{N(1 - \exp(-T\omega/N))}{2} = \frac{T\omega}{2}\,.$$

7 Axioms

Within the 1960 years QFT models were reformulated in terms of vacuum expectation values called "Wightman functions" [11]. The possibility of their analytic continuation to imaginary time essentially results from the spectrum of the Hamiltonian.

The vacuum is the ground state $\Omega = 1$ with $H\Omega = 0$. In our QM model the fields are the time dependent position operators $Q(t)$. Using $U(t)\Omega = \Omega$, substituting $-\mathrm{i}t \to \tau \geq 0$ we obtain

$$\langle \Omega, Q(t)\Omega\rangle \to \sum dP(x)x = \frac{1}{2}\sum 1x = 0$$

Next

$$\begin{aligned}\langle \Omega, Q(0)Q(t)\Omega\rangle \to \langle \Omega, Q\exp(-\tau H)Q\Omega\rangle &= E[x(0)x(\tau)] \\ &= \sum dP(x, y)xy = \exp(-\tau\omega)\,, \quad \tau \geq 0\,.\end{aligned}$$

We observe that

$$D = Q(s)Q(t) = \exp(\mathrm{i}\,\tau\omega R), \quad \tau = (t - s)\,.$$

This follows from the explicit formulae for $Q(t)$ in terms of Q, P using $QQ = PP/\omega^2 = I$, $QP = \mathrm{i}\,\omega R$ and the identity $\cos(\tau\omega)I + \mathrm{i}\,\sin(\tau\omega)R \equiv \exp(\mathrm{i}\,\tau\omega R)$. Therefore time ordered products of pairs of "fields" $D_n = Q(t_{n-1})Q(t_n)$, $n = 2, 4, \ldots$, commute. This implies

$$\bigotimes D_n = \exp\Big(i\omega R \sum \tau_n\Big), \quad n = 2, 4, \ldots.$$

We conclude that $\otimes D_n$ applied to the vacuum Ω gives a phase factor leading to the well-known expression for the even N-point functions

$$< \Omega, Q(t_1)Q(t_2)\ldots Q(t_N)\Omega >= \exp\Big(i\omega \sum \tau_n\Big)\,.$$

Conversely, the moments of the measure dP called "Euclidean Green's functions"

$$E[x_1 x_2 \cdots x_N] = \sum dP x_1 x_2 \cdots x_N$$

satisfy axioms equivalent to those for physical time postulated by Osterwalder and Schrader [12]. Again, the reconstruction of the quantum dynamics crucially depends on a positivity condition. In order to formulate it we should symmetrize the probability space with respect to time reflection $\Theta: t \to -t$, glueing the interval $[0, T]$ together with $[-T, 0]$ to $[-T, T]$. Then the limit $T \to \infty$ may be performed and one obtains a stationary process $t \to x(t)$ with measure dP according to the general theory of Kolmogoroff.

We will identify QM vectors $f = a1 + bx$ in $\mathcal{H}_{\text{phys}}$ with functions F in $L^2(\mathbf{S}, dP)$ defining $F = f(x_0)$. For a pair f, g in $\mathcal{H}_{\text{phys}}$ respectively their images F, G in

$$E_0 L^2(\mathbf{S}, dP) = L^2(S_0, dP_0)$$

we have

$$\langle g, f \rangle = \sum dP G^* F = (G, F)\,.$$

We denote by $E[\cdot]$ the expectation with respect to the measure dP. Let $F = a1 + bx(t)$ be a function of one single "future time" $t \geq 0$ and $\Theta F(x(t)) = F(x(-t))$ be its mirror image.

Then it should hold

$$E[\Theta F^* F] = \sum dP \Theta F^* F \geq 0\,.$$

Explicitly

$$\begin{aligned}\sum dP(\Theta F^*)F &= \sum dP(z, x, y) F(z)^* F(y) \\ &= \left(\frac{1}{2}\right)^3 \sum (1 + rzx)(1 + rxy)(a1 + bz)^*(a1 + by) \\ &= \langle g, g \rangle\end{aligned}$$

where

$$g(x) = \frac{1}{2} \sum (1 + rxy) f(y) = a + rbx\,,$$

$r = \exp(-t\omega)$ and $z = x(-t)$, $x = x(0)$, $y = x(t)$ all summed over ± 1. Hence the expression in question becomes non-negative. With $V(t)$, $-\infty < t < +\infty$, the imaginary time translations acting as unitaries in $L^2(\mathbf{S}, dP)$ we may rewrite

$$g = \exp(-tH) f = E_0 V(t) F\,, \quad t \geq 0\,.$$

Conversely, according to the reconstruction of B. Szekefalvi-Nagy [13] the Hamiltonian semi-group admits a unique minimal dilation which are just the unitaries $V(t)$ in the big "Euclidean" Hilbert space. If dP is ergodic then $V(t)F = F$ implies $F = c1$ which implies $\exp(-tH)\Omega = \Omega$ [14].

E Nelson gave a rigorous derivation of the Feynman–Kac formula when the stochastic process $t \to x(t)$ governed by the measure dP satisfies the Markoff property and also the reflection property. The latter means that functions $F = F(x_0)$ in $L^2(S_0, dP_0)$ are Θ-invariant which is the case in our model.

Let $t_0 < t_1 < t_2 < \cdots < t_N$, rewrite $x_n = V(t_n)x_0V(t_n)^*$, for $n = 0, 1, 2, \ldots, N$ and define $F = x_1x_2 \cdots x_N$. Using the above mentioned properties of the measure dP and repeatetly the above definition of the Hamiltonian semi-group it follows

$$\begin{aligned} E[F] &= (1, E_0[F]) \\ &=< \Omega, \exp(-\tau_1 H)Q \exp(-\tau_2 H)Q \cdots \exp(-\tau_N H)Q\Omega > \\ &= \exp(-\sum \tau_n \omega)\,, \end{aligned}$$

with $\tau_n = (t_n - t_{n-1})$, for $n = 2, 4, \ldots, N$ even and otherwise zero. We observe that on the left side the function 1 representing the vacuum vector Ω may be substituted by any element in the time zero subspace $L^2(S_0, dP_0)$ which leads to the formula

$$E_0[F] = \exp(-\tau_1 H)Q \exp(-\tau_2 H) \cdots Q \exp(-\tau_N H)Q\Omega\,.$$

For N odd we obtain an additional factor $\exp(-\tau\omega)x_0$ so that the correlation functions must vanish because of $(1, x_0) = 0$.

8 Final Remarks

These notes are based on a lecture with title "Imaginary time path integrals with complex valued measure in QM models" given by the author October 12, 1994, in a seminar at the Dipartimento di Fisica a Pisa.

The main motivation for this work was to show "how natural are probabilistic methods in QFT", as wrote E Nelson. Certainly Gianni Morchio would have agreed. He very liked the functional integral point of view.

I suspect that several facts concerning the imaginary time formulation are hidden in the paper [8]. Nevertheless it was worthwhile to work out them myself and to present them in detail. Why? Unfortunately, for many interesting QM and QFT models the "physical" time Feynman path integral does not define a measure. We learned that for the Fermi oscillator the term $W[0, T]$ in question, for $T < \infty$, is well defined as limit of complex- valued cylinder measures.

Acknowledgements I am very greatful to my former colleagues Christian Zylka, Jürgen Schmidt and Frank Dittes from the University of Applied Sciences Erfurt who performed the translation of my text into LATEX and brought all figures into the present form while the author had to pass a chemotherapy in hospital.

References

1. Russold, F.: Die Inversion von N im NH_3 Molekül. Bachelor Thesis. Uni-Graz (2016)
2. Nelson, E.: Construction of quantum from Markoff fields. Journ. Func. Anal. **12**, 97 (1973). preprint 1972
3. Löffelholz, J.: On the Feynman-Kac formula for a QM system with memory. Ann. Phys. **46**(1), 55 (1989)
4. Löffelholz, J., Morchio, G., Strocchi, F.: The (QED) $0 + 1$ model and a possible dynamical solution of the strong CP problem. Ann. Phys. **250**(2), 367 (1995)
5. Löffelholz, J., Morchio, G., Strocchi, F.: Math structure of QED in temporal gauge. J. Math. Phys. **44**, 5095 (2003)
6. Uhlmann, A.: Die Grammatik der Quantenwelt. EAGLE, vol. 096. Edition am Gutenbergplatz, Leipzig (2017)
7. Daletsky, Y.L., Fomin, S.: Measures in infinte-dimensional spaces. Russ. Ed. NAUKA (1977)
8. Angelis, G.F., de Falco, D., Guerra, F.: Prob ideas in the theory of Fermi fields. Phys. Rev. D **23**(8), 1747 (1981)
9. Löffelholz, J., Morchio, G., Strocchi, F.: Spectral stochastic proecesses arising in QM models with a non-L^2 ground state. Lett. Math. Phys. **35**, 251 (1995)
10. Roepstorff, G.: Pfadintegrale in der Quantenphysik. Verlag Vieweg, Braunschweig (1991)
11. Streater, R.F., Wightman, A.: PCT – and all that. Benjamin, N.Y. (1964)
12. Osterwalder, K., Schrader, R.: Axioms for Euclidean Green's Functions. Commun. Math. Phys. **31**, 83 (1973)
13. Szögefalvi-Nagy, B., Fojas, C.: Harmonic analysis of operators in Hilbert spaces. Russ. Ed. MIR (1970)
14. Seiler, E.: Gauge Theories as a Problem of Constructive QFT and Statistical Mechanics. LNP, vol. 159. Springer (1982)

The Fröhlich–Morchio–Strocchi Mechanism: An Underestimated Legacy

Axel Maas

Contents

Abstract There is an odd tension in electroweak physics. Perturbation theory is extremely successful. At the same time, fundamental field theory gives manifold reasons why this should not be the case. This tension is resolved by the Fröhlich–Morchio–Strocchi mechanism. However, the legacy of this work goes far beyond the resolution of this tension, and may usher in a fundamentally and ontologically different perspective on elementary particles, and even quantum gravity.

1 Introduction

Non-Abelian gauge theories of Yang–Mills type [1–3], no matter the matter content, have a highly interesting feature. They are based on gauge (Lie-)groups, which do not form simple manifolds. This has far reaching consequences. Probably the most important one is that it is not possible to introduce global coordinate systems [4–6], an issue known as the Gribov–Singer ambiguity. This feature stems from the group structure, and is thus independent of the parameters of the theory, especially the value of any coupling constants. On top of this, non-Abelian gauge theories are

A. Maas (✉)
Institute of Physics, NAWI Graz, University of Graz, Graz, Austria
e-mail: axel.maas@uni-graz.at

A. Cintio, A. Michelangeli (Eds.), *Trails in Modern Theoretical and Mathematical Physics*, https://doi.org/10.1007/978-3-031-44988-8_11

affected, as any other quantum field theory, by Haag's theorem [7], which implies that a non-interacting theory and an interacting one are not unitarily equivalent. In a Yang–Mills theory, this is amplified by the fact that the free gauge theory would have a different character, namely to be reduced to N_g non-interacting Abelian gauge theories, where N_g is the number of generators of the gauge group.

This appears to imply that a conventional perturbative treatment [2] should not be possible at all. The elementary particles cannot act as asymptotic states due to Haag's theorem. And the required gauge-fixing for the employed saddle-point approximation in perturbation theory is not well-defined due to the Gribov–Singer ambiguity.

This appears to have "just" the consequence that genuine non-perturbative methods are required, and especially non-trivial asymptotic states are needed. The prime example is QCD. Here, asymptotic states are hadrons, and perturbation theory can at best be applied in special kinematics, where the involved field amplitudes are small enough that the group manifold is only probed within a single patch. Due to the strong coupling, however, this is generally agreed upon anyways [2, 3, 8, 9].

In the weak interactions, the situation is, on conceptual grounds, the same [3, 10–14]. Thus, it is not surprising that, e.g., there is no qualitative distinction between the strong-coupling case and the weak coupling case, as they are analytically connected states of the theory [10, 11]. But ignoring these fundamental questions and just applying perturbation theory turns out to be extremely successful in describing experimental results to high quantitative precision [2, 15]. This is attributed to the Brout–Englert–Higgs (BEH) effect [16–22]: It 'breaks' the gauge symmetry, effectively turning it into a non-gauge theory which does not need to take care of these issue. But, formally, a gauge symmetry cannot be broken by virtue of Elitzur's theorem [23], nor does this alleviates Haag's theorem.

It is precisely here, where Giovanni Morchio's legacy in form of the Fröhlich–Morchio–Strocchi (FMS) mechanism [13, 14] is the decisive puzzle piece. It explains how both aspects, the phenomenological success and the formal insights, can both be correct at the same time. How this happens will be discussed in Sect. 2. But while the original papers [13, 14] were mainly concerned with resolving this paradox, the legacy and implications of this work transcends in its importance the resolution of the paradox by far. In fact, it creates a framework, the FMS framework, to deal with a quite large class of theories effectively.

In the following, the FMS framework and the FMS mechanism, and some of their consequences, are presented, as there are:

- Experimental testability of the field theoretical underpinnings, Sect. 3.
- Consequences for non-Abelian Yang–Mills–Higgs theories beyond the standard model, Sect. 4.
- Applications beyond Yang–Mills–Higgs theories, Sect. 5.
- Ontological implications, Sect. 6.

In fact, the FMS mechanism, and the formal aspects on which it is build, have the potential to fundamentally transform our view of 'elementary' particle physics [24], and thus the way how we perceive reality. While the need to take gauge invariance seriously has been pointed out repeatedly before, and in fact on formal grounds [4, 5, 7, 10–12, 23, 25], it has been the work of Morchio and his collaborators [13, 14] to show how the subtleties work out in practice. They thereby paved the way for a more holistic picture of gauge symmetries, and how they are (not) relevant [24].

Given all these implications, it appears surprising that this has found so far no entry even in specialized textbook, much less has become the standard approach. Especially as the necessary additional effort is at best moderate, see Sect. 2. And the original papers [13, 14] are now more than 40 years old. While a full historical and sociological investigations is not (yet) available, superficial investigations [26] show that the insidious combination of the properties of the standard model and the success of the FMS mechanism itself appear to be the reason for that. Because in the particular case of the standard model, the FMS mechanism explains why only slight deviations can be expected, compared to a perturbative treatment. In fact, so slight, that they have not yet been observable in experiment, see Sect. 3. As a consequence, its additional layer of complexity has not been needed, as perturbation theory alone was sufficient. Thus, it got almost forgotten, and the (formally incorrect) idea of gauge symmetry breaking by the BEH effect has become accepted lore. Only within the philosophy of science community the challenged posed to our understanding by Elitzur's theorem and its contradiction to the BEH effect has remained a matter of importance [24, 27–33]. Especially, within the philosophy of gauge symmetry literature, even disbelief about the treatment of the issue by physicists was expressed.

Turning the whole story around, there is an important discovery awaiting. Either we are able to experimentally discover the correctness of the consequences of the FMS mechanism, or not. In the former case, this will make the FMS mechanism the accepted approach for treating the BEH effect, and will have far-reaching consequences for model building [26], see Sects. 4 and 5. Or, this will disprove our fundamental understanding of quantum gauge theories, as encoded in [4, 7, 10, 11, 13, 14, 23, 26], sending us back to the drawing board, and perhaps open entirely new avenues. As the effects are predictable and entirely fixed by the known parameters of the standard model, this decision can be performed. Even if it is a formidable, though manageable, task, see Sect. 3.

The only thing, which is not an option, is to ignore this tension. Because if the understanding of quantum gauge theories is correct, the consequences of ignoring the tension could easily be mistaken for signatures of physics beyond the standard model [26, 34–36]. Indeed, there is an off-chance that this may have already happened [37].

2 The FMS Mechanism

The starting point of the FMS framework is a simple statement. Given any expression $\mathcal{O}$, which transforms in a linear representation of a (continous) non-Abelian[1] gauge group G, an invariant group measure $\mathcal{D}\mu$, and an invariant action as weight factor $\exp(iS)$, it follows[2] that [13, 14]

$$\int \mathcal{D}\mu \mathcal{O} e^{iS} = 0, \tag{1}$$

because

$$\langle \mathcal{O} \rangle = \int \mathcal{D}\mu \mathcal{O} e^{iS} = \int \mathcal{D}\mu^{g^{-1}} \mathcal{O} e^{iS} = \int \mathcal{D}\mu \mathcal{O}^g e^{iS} = \langle \mathcal{O}^g \rangle,$$

where g is a gauge transformation. This can only be true for arbitrary g if $\langle \mathcal{O}^g \rangle = \langle \mathcal{O} \rangle = 0$. This statement is a generalization of Elitzur's theorem [23]. Thus, there can be no spontaneous breaking of a gauge symmetry by formation of a gauge-dependent condensate like in the BEH effect, which in turn would break the gauge symmetry. Thus, the gauge symmetry remains unbroken.

Additionally, this approach closes a loop hole in the original derivation, which assumed analyticity of the free energy in external sources, which is not necessarily the case [45]. In fact, this statement also applies to global groups [26, 46]. As a consequence, expectation values of gauge-dependent quantities necessarily vanish, if the gauge symmetry is unbroken. But this can then happen only by gauge-fixing [13, 14, 23].

Thus, without breaking gauge symmetry explicitly by gauge fixing, it remains necessarily unbroken. The BEH effect is therefore not a physical effect, but rather only a particular useful gauge choice implemented by, e.g., the 't Hooft-R_ξ gauges [47]. As a consequence, the Higgs vacuum expectation value is introduced by the gauge-fixing and thus gauge-dependent. Its actual value needs still to be determined from the gauge-fixed quantum effective potential, and whether it can be non-zero remains a dynamical, albeit gauge-dependent [48–51], question.

As a consequence, the Gribov–Singer ambiguity still applies, and thus the classification of physical states using BRST symmetry fails [52]. Rather, fully and manifestly gauge-invariant operators are needed to construct asymptotic states [12–14]. Fortuitously, this also elegantly satisfies Haag's theorem, as the asymptotic states are no longer necessarily non-interacting elementary particles.

---

[1] In fact, similar arguments do hold also in the Abelian case [38–40], and are then augmented by the usual subtleties of Abelian gauge theories [7]. This will not be detailed here, but follows along very similar lines, including the confirmation in lattice simulations [41, 42].

[2] The original work [13, 14] used a lattice regularization in Euclidean space-time to carefully bound expressions. While important on a formal level, this turns out to be transparent to the following, and therefore will be suppressed. Especially, behavior at sufficiently low energies compared to a (lattice) cutoff is likely independent on such details [43, 44].

While this is a field-theoretically satisfying prescription, this implies effectively to work with bound states. While non-perturbative methods exist to do so, they are much more demanding than perturbation theory. They work very well for theories like QCD [3, 9, 53, 54]. But the large hierarchy of the standard model, covering at least twelve orders of magnitude, make them practically not (yet) applicable. Moreover, the CP-breaking character of the weak interactions poses still conceptual challenges for some of them [26, 55].

It is here, where the second part of the FMS framework becomes central, the FMS mechanism [13, 14]. It can be argued that the Gribov–Singer ambiguity is quantitatively not important in the presence of a BEH effect [56], for which at least some circumstantial evidence exists [57, 58]. Similarly, the success of perturbation theory [2, 15] ignoring all of these issues requires understanding, but indicate that there exists some suppression mechanism.

The FMS mechanism now utilizes that any kind of perturbation theory is indeed also a small field-amplitude expansion [59]. Thus, if the dominating field configurations in the path integral are characterized by small-field amplitude fluctuations around some fixed field configuration, an expansion should be still quantitatively good. This could happen, e.g., due to a BEH effect, where the Higgs field develops after gauge fixing a vacuum expectation value as dominating field configuration. Hence, it should be possible to still expand accordingly, i.e. performing a saddle-point expansion around the Higgs vacuum expectation value. However, following the FMS framework, the expansion needs to be performed around the correct asymptotic states, which are manifestly gauge-invariant.

Consider as an example the simplest theory having all of these features, the Higgs sector of the standard model. Its Lagrangian is [2]

$$\begin{aligned}\mathcal{L} &= -\frac{1}{4}\mathrm{tr} W_{\mu\nu} W^{\mu\nu} + (D_\mu X)^\dagger D^\mu X + V(\det X)\\ W_{\mu\nu} &= \partial_\mu W_\nu - \partial_\nu W_\mu + ig\big[W_\mu, W_\nu\big]\\ D_\mu &= \partial_\mu + g W_\mu\end{aligned} \tag{2}$$

where g is the gauge coupling, $W_\mu = \tau^a W_\mu^a$ are algebra-valued gauge fields with the generators of the gauge group τ^a, in the standard model SU(2). X is the matrix-valued Higgs field derived from the complex doublet h [26]. This form makes explicit that the Higgs field is in the fundamental representation of the gauge group under left-multiplications, and also in a fundamental representation with respect to a right-multiplication of an additional global SU(2) group. It should be noted that X itself is not SU(2)-valued. The potential V is required to be invariant under both symmetries, which is ensured by construction.

The BEH effect is made possible by a suitable gauge-fixing, which explicitly breaks the gauge symmetry completely [2, 26]. After gauge-fixing, the Higgs field is then split conveniently as

$$X(x) = V + \eta(x) \tag{3}$$

where V is a constant. It is convenient, but not necessary, to choose $V = v\mathbf{1}$. If the quantum-effective action allows for $v \neq 0$, a BEH effect takes place[3].

The FMS framework demands to formulate matrix elements of physical observables in terms of gauge-invariant operators. To interpret them as particles in terms of asymptotic states requires them to be local. Local, manifestly gauge-invariant operators in a non-Abelian gauge theory are necessarily composite [6, 26]. Especially, this implies the split (3) is not applied at the level of the Lagrangian, like in perturbation theory. Rather, the FMS mechanism works by first writing down the desired matrix element in terms of local, composite operators and only then the split (3) is applied. Of course, the local composite operators can still carry global quantum numbers, especially spin or the those from the global SU(2) symmetry in (2).

The simplest case is the propagator of a scalar singlet. A suitable operator[4] would be $\det X$. The simplest, non-trivial, matrix element of this operator is the propagator. Taking only the connected part yields

$$\langle \det X(x) \det X(y) \rangle = v^2 \langle \mathrm{tr}\eta(x)\mathrm{tr}\eta(y) \rangle \tag{4}$$

$$+ v \langle \det \eta(x) \mathrm{tr}\eta(y) + \mathrm{tr}\eta(x) \det \eta(y) \rangle + \langle \det \eta(x) \det \eta(y) \rangle . \tag{5}$$

Since the left-hand side is gauge-invariant, so the sum on the right-hand side needs to be. However, the individual terms are not necessarily so. So far, this is an exact rewriting, and thus not a priori progress.

Because any kind of perturbative expansion is assuming that the quantities are analytic in the expansion parameter, a perturbative expansion cannot spoil the gauge-invariance of the right-hand side, if order-by-order the sum in powers of v is kept [60]. The first term (4) is the propagator of the elementary Higgs particle. Dropping the remaining terms (5) and expanding the elementary Higgs propagator in a perturbative series in the couplings yields that the bound-state propagator in this approximation is identical to the elementary Higgs propagator to all orders in perturbation theory. Especially, the poles coincide and thus the bound state mass is the same as the elementary one to all orders in perturbation theory[5]. Of course, at loop-level the elementary Higgs propagator is gauge-dependent [60, 62]. This was expected, as only the sum would be gauge-invariant. This gauge-invariance of the full sum has indeed been demonstrated explicitly at one-loop order [60, 62]. Furthermore, it was seen explicitly that only at energy scales much larger than v deviations from ordinary perturbative result arise. This explicitly shows how the perturbative success can be recovered by the FMS mechanism, at least in

[3] It is important to note that this is not [13, 14, 26] a background-field approach [2], as the splitting happens after gauge fixing and not before. Especially, there is no additional classical gauge symmetry of the split-off field. The previous gauge-fixing has already broken the gauge symmetry.

[4] At first sight, this appears to be the same as the fields appearing in unitary gauge [2]. However, unitary gauge introduces a non-trivial Faddeev–Popov operator due to gauge defects, and is thus different [26].

[5] At loop order, it is necessary to chose a pole scheme to fix the pole position, which is independent of the gauge parameter by virtue of the Nielsen identities [26, 61].

principle: The term (4) dominates over (5) in the experimentally probed regime. This is how the FMS mechanism resolves the paradox: The composite state has a FMS-dominant contribution, which coincides with one of the elementary particles.

Of course, this is only a self-consistency statement. While the agreement with experiment is certainly strong support, explicitly evaluation of both sides with non-perturbative methods provides another stringent tests. This has been done using lattice methods, confirming the FMS mechanism, see [26] for a review. Especially, no additional poles due to further bound states or resonances are observed below the inelastic threshold in this theory in all investigated channels, confirming the experimental findings.

Another important structure is visible when considering the triplet vector channel. Consider the following operator and its FMS expansion

$$\mathrm{tr}\tau^i X^\dagger D_\mu X = v^2 \delta^i_a W^a_\mu + \dots \tag{6}$$

Thus, this gauge-invariant vector operator, which is a triplet under the global SU(2) symmetry carried by the Higgs field, is mapped to the gauge boson field. With the same reasoning as before, this yields that the mass of the composite physical particle is the same as the one of the elementary particle. The W is hence the FMS-dominant contribution of the composite state. More importantly, there is a matrix $c^i_a = \delta^i_a$, carrying both global group i and local gauge group a indices, which yields a map of the global triplet to the gauge triplet. This is the mechanism by which the degeneracy from the gauge fields is transported to the gauge-invariant physical states. This mapping will be of special importance in Sect. 4. This structure was again confirmed at one-loop order [62] and on the lattice [26].

The whole construction can be repeated for the remainder of the standard model. This includes left-handed leptons [13, 14, 26, 37, 64] and hadrons [26, 37]. It turns out that only the very special structure of the standard model ensures the correct assignment of degeneracies and further quantum numbers like electric charge. Furthermore, always the physical, gauge-invariant states map to elementary states, if the latter exist, and to scattering states otherwise. Unfortunately, as noted before, the whole standard model cannot be yet treated non-perturbatively. Within simplified models of the lepton sector, however, also agreement is found [63], see Fig. 1. Hence, all possible theoretical tests of the FMS mechanism so far have confirmed it. Not to mention that it explains why perturbation theory and experiment agree so well. It should be noted in passing that the FMS mechanism implies at tree-level or in a pole scheme that most bound states have mass defects between 50% (for the Higgs) to 75% (for weak gauge bosons) to almost 100% (electrons and left-handed neutrinos). It is thus a highly relativistic effect, not accessible [26, 65] to quantum-mechanical models or heavy-particle effective field theories such as e.g. in [66].

All of this resolves the contradiction between the field-theoretical arguments and the success of perturbation theory, at least on the theoretical level.

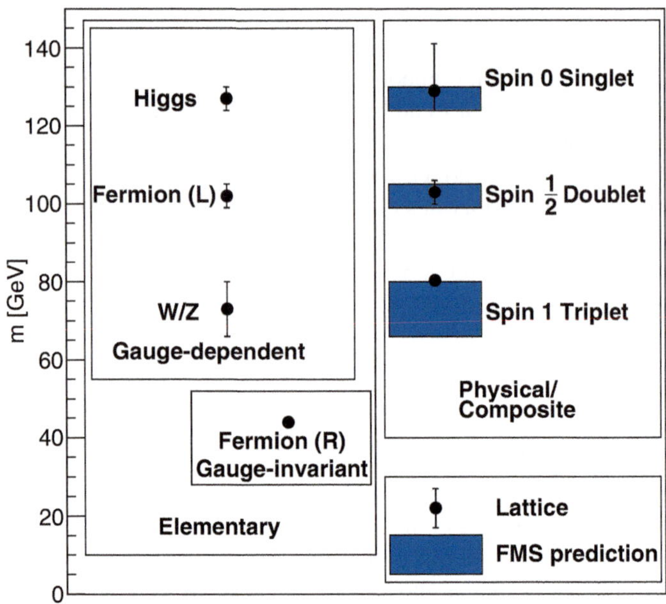

Fig. 1 A sample lattice spectrum for quenched vectorial fermions compared to the predictions from the FMS mechanism, alongside the elementary particle properties [63]. Note that a right-handed ungauged fermion is in addition necessary to construct in such a theory power-counting renormalizable Yukawa interaction with the Higgs [63]

3 Phenomenological Implications and Experimental Tests

Of course, it is not sufficient that analytical expressions and lattice results coincide. To ensure that this is actually the correct description of the standard model, experimental tests are needed. This needs to detect experimentally consequences of additional terms like (5). Since experiments usually involve scattering, it is required to address the scattering of composite states.

This is not a conceptual issue per se, as composite state scattering can be addressed within the LSZ formalism in the same way as the scattering of elementary states [2, 7, 67]. In particular, the choice of asymptotic operators does not matter.

But it furthermore needs a possibility to calculate such processes, as the appearing matrix elements and wave functions are intricate objects. Fortunately, the FMS mechanism can also be applied in such a case [37, 65, 68, 69]. Of course, there are now many more matrix elements on the right-hand side as in (4–5). Even for the simplest case at a lepton collider, i.e. two incoming, massless left-handed leptons described by composite operators L and likewise two outgoing composite fermion operators F [13, 14, 37, 63]

$$
\begin{aligned}
L &= X^{\dagger} l = vl + \eta^{\dagger} l \\
F &= X^{\dagger} f = vf + \eta^{\dagger} f,
\end{aligned} \tag{7}
$$

where l and f are the elementary left-hand fermion fields, this becomes highly complicated[6]. Symbolically, in the center-of-mass frame [37],

$$
\langle \overline{L}(-p)L(p)\overline{F}(-q)F(q)\rangle = v^4 \langle \overline{l}(-p)l(p)\overline{f}(-q)f(q)\rangle + \dots .
$$

The remaining terms have less powers in v, but always one or more composite operator of type $(\eta^{\dagger}a)(k)$, where a and k can be l and p or f and q, respectively. Taking only the leading term in the FMS mechanism recreates the usual perturbative results to all orders in the coupling constants [26, 37], again confirming the reason for perturbation theory to work quantitatively well.

Calculating the remaining terms requires to take the composite nature of the external fields into account. Such an augmented perturbation theory [68] works along the same lines as for matrix elements [60, 62]. In addition, it is necessary to supplement for the external states Bethe–Salpeter/Faddeev amplitudes [67, 68], instead of the non-interacting wave-functions of perturbation theory [2]. They also need to be calculated consistently using the FMS mechanism [68]. Furthermore, the FMS mechanism introduces in the amputated, connected matrix elements an additional vertex, a bound-state splitting vertex [60, 62, 68], which corresponds to a replacement of the composite states with elementary fields in the FMS mechanism [60, 62]. While these are only minor additions to the Feynman rules of perturbation theory to create an augmented perturbation theory, in practice this creates many more loop diagrams [60, 62, 70] for the additional 15 matrix elements involved. Thus, a full expression for any process is a formidable task, and remains still to be obtained.

However, there exist already a number of partial results [37, 71] as well as some lattice results [36, 72]. They indicate that the interactions of the electroweak composite bound states will work in a similar vain as the interactions of hadrons in (deep) (in)elastic scattering (DIS). This is depicted schematically in Fig. 2. There are three relevant energy regions, where energy here corresponds to some relevant energy scale $\sqrt{s}$. This could be, e.g., indeed the center-of-mass energy.

At very low energies, $m^2_{\text{bound state}} < s - N^2_{\text{constituent}} m^2_{\text{constituent}} \ll v^2$, the composite states are probed as a whole. This implies that their composite nature becomes readily apparent, e.g. the fact that they are not point-like [36, 72]. Similar to the case of hadron physics, the probed extension depends on the involved probe particles [9].

[6] Note that likewise to the vector triplet (6) this yields doublets of the global symmetry carried by the Higgs field X [13, 14, 37, 63]. Thus, left-handed electrons and electron neutrinos in the standard model are really distinguished by the same quantum number as the physical W and Z. The right-handed electron and electron neutrinos carry an independent right-handed flavor symmetry. The Yukawa couplings of the standard model eventually break both symmetries down to a common diagonal group [37, 63]. The same reasoning applies to quarks [26, 37].

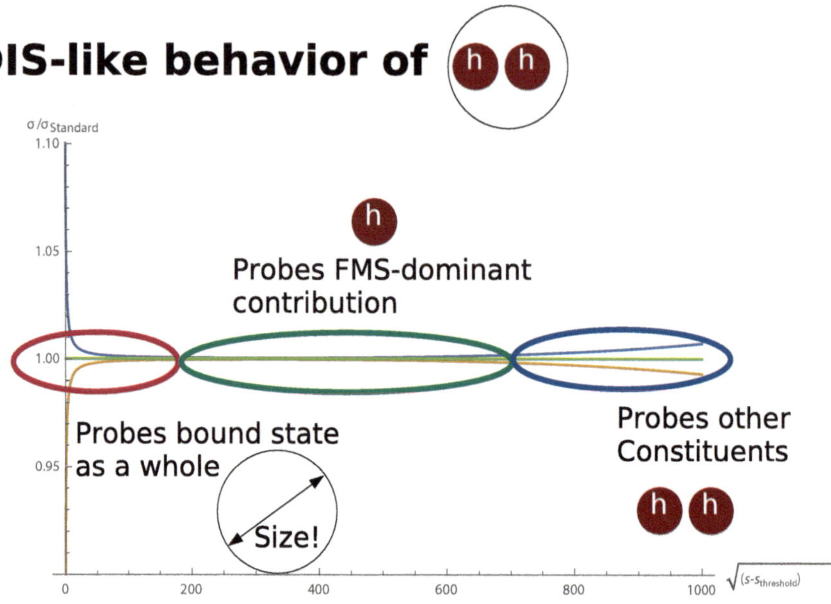

Fig. 2 A cartoon sketch of the behavior of interactions between gauge-invariant composite states, given as the ratio to perturbative results as a function of an arbitrary energy value above threshold. For a bound state like det $X = h^{\dagger}h$, at low energies the bound state as whole is probed. At intermediate energy, effectively only the first term in the expansion is relevant, giving the results closest to perturbation theory, as only the FMS-dominant contribution matters, here h. At very high energies the contribution of other valence particles and sea particles become probed

Since the extension is generated in the combination of weak and Higgs interactions, it likewise needs corresponding particles to probe it. While this information should be readily accessible from the Bethe–Salpeter/Faddeev amplitudes, these have not yet been calculated using augmented perturbation theory. However, lattice results indicate an effective size parameter of order a few inverse tens of GeV, both for the vector triplet and the scalar singlet [36, 72].

Such an extension provides also a decisive test for the composite nature of the observed particles, and thus of the FMS mechanism. Vector boson scattering (VBS) [75] appears as one suitable process to probe it [36]. Here, for extensions of a few $1/10\,\text{GeV}^{-1}$ as suggested by lattice simulations [36, 72], deviations within a few tens of GeV above the elastic threshold at $2m_V$ in the vector boson center of mass frame are expected, provided the vector bosons are on-shell, see Fig. 3.

This could be experimentally tested. Indeed, both the ATLAS experiment [73] and CMS experiment [74] at the LHC have shown that they can probe, in principle, the interesting regime. However, at the moment the contribution of VBS to the total yield is at most the same size as the uncertainties in the relevant kinematical regime [73, 74], and thus this awaits future improvements. At the same time, the theoretical

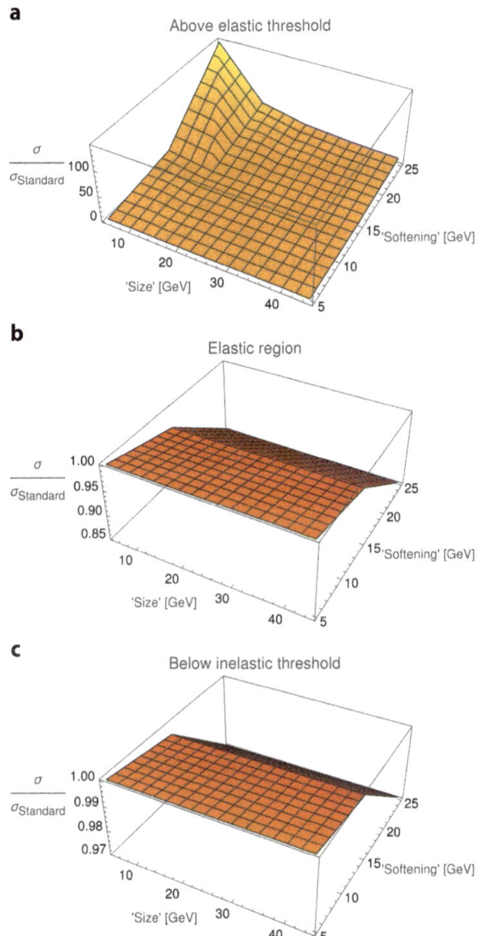

Fig. 3 The deviation of the cross section with respect to the (Born-level) cross section for on-shell vector boson scattering [36]. The three panels correspond to the integrated total cross section in three center of mass regions: Between 1 and 1.2 times the elastic threshold (**a**), between 1.2 and 1.5 times the elastic threshold (**b**), and the remainder up to the inelastic threshold at 2 (**c**). The choice of range corresponds at low energies to that of ATLAS in [73] and at intermediate energies to CMS in [74] for their control regions in the $ZZ \to 4l$ final states. Size and softening correspond to the scattering length and the suppression of bound states effect at the first non-vanishing term in the threshold expansion, respectively, see [36] for details.

estimates shown in figure 3 are from a truncated standard model, and will also need to be pushed to quantitative precision for a final evaluation.

In the intermediate momentum regime with $s \sim v^2$, the states appear to be dominated entirely by the component which leads to the two-point function, e.g. the W in (6) or the lepton l in (7). Thus, only a single valence particle dominates the state, the FMS-dominant constituent. Except for states like the scalar (4) this is the non-Higgs valence particle. The Higgs valence particles act as spectators to this FMS-dominant constituent [36, 60, 62, 72]. This is again very different from the QCD case, where all valence particles play qualitatively equal roles [8, 9]. It is here where the results resemble closest the non-augmented perturbative ones. In particular, if the particle is produced as a resonance in this regime, this implies little to no change close to the pole [26, 36, 60, 62].

At very high energies $s \gg v^2$, the other valence contributions [37, 71, 76], as well as sea contributions [77], start to become relevant. Depending on the colliding particles, this can have various impacts.

For colliding leptons, there will be additional contributions due to Higgs interactions [37]. Especially, perturbative violations [78] of the Bloch–Nordsieck theorem [79] will be canceled, as now full weak multiplets are present. This allows the treatment of collinear and soft radiation in the same way as in QCD [71]. This deviation from non-augmented perturbative results increases like a double Sudakov-logarithm, i.e. roughly like $\ln^2(s/m_W^2)$, at large energies. It can thus be a substantial effect at future lepton colliders [80].

At the current LHC, where hadrons collide, consequences are less obvious due to the strong interaction background. Still, at the very least, changes in the parton structure of the hadrons will be needed [71, 76] as most hadrons, and especially protons, need a Higgs valence contribution for gauge invariance [26, 37, 71]. These changes to the parton distribution functions may influence reactions involving particles coupling strongly to the Higgs, like top quarks and weak bosons [76]. But due to the need to include this information in parton distribution function fits, this is far from being really testable [71, 76] yet.

4 Applications Beyond the Standard Model

The decisive feature for perturbation theory to work so well in the standard model is the one-to-one mapping of the gauge multiplet structure to the global multiplet structure of the global Higgs symmetry, e.g. in (6) and (7). This similarly applies to all other standard model particles [13, 14, 26, 37]. In addition, there appear to be no additional bound states and resonances due to the weak bound state substructure, see [26]. However, this may not be generally true [37, 81].

It is natural to ask [35] whether the same applies to general Yang–Mills–Higgs theories, with or without further matter. In the original works [13, 14], it was already conjectured that adding further Higgs doublets to the standard model case would not alter the outcome qualitatively. Indeed, a detailed application of tree-level augmented perturbation theory to 2-Higgs doublet models [82] and the minimal supersymmetric standard model [83] confirm this conjecture. A check using non-perturbative methods has, however, not yet been performed. But it would be feasible to do so, e.g. using lattice methods [84–86]. Such models enlarge the global group, while leaving the gauge group fixed. However, because any phenomenological viable version of these theories supports global SU(2) multiplets [87, 88], the underlying FMS mechanism still works [82, 83].

The situation is potentially very different when enlarging the gauge group, while keeping or reducing the global group [26, 35]. This is the situation typical for grand unified theories (GUTs) [89, 90]. There are two separate aspects to be taken care of [91]. One is that such theories offer usually multiple breaking patterns in the BEH effect instead of the single one in the standard model. The other is that the accompanying change of the global group with respect to the standard model alters the possible multiplet structure of observable, gauge-invariant states.

Because the FMS mechanism is a map between the gauge-invariant operators and gauge-dependent operators, it needs a prescription to which gauge states it should map, a choice of gauge. If different breaking patterns can be realized by different gauge choices, this yields different possibilities for the map, and thus potentially different results of the FMS mechanism for the physical spectrum. This would jeopardize the usability of the FMS mechanism [91].

The simplest example for which this happens is an SU(3) gauge group with a single Higgs in the adjoint representation [90]. At tree-level, two equivalent breaking patterns occur, SU(2)×U(1) and U(1)×U(1), which differ, e.g. in the number of massless gauge bosons [50, 90, 91]. In such a case multiple possibilities exist how to set up the FMS mechanism [91], which will lead to different, and inequivalent, predictions for the physical spectrum. However, in general already one-loop corrections will lift the tree-level degeneracy [50, 51, 90]. Thus, there is again only one gauge choice in which the field fluctuations can be considered small, yielding again a unique setup for the FMS mechanism [50, 91, 92]. Thus, this appears to be not an issue, as long as no exception is found. However, this also implies that to apply augmented perturbation theory in such cases requires to first determine the corresponding allowed breaking pattern at the same order for the parameter set in question. This will be assumed to have happened in the following.

The physical spectrum can only form representations of the global symmetry groups. If these do not support the same multiplicities as the unbroken gauge groups, it cannot be expected that the one-to-one mapping of the standard model still works [35]. This is indeed the case.

The simplest example is a SU(3) Yang–Mills theory coupled to a Higgs field ϕ in the fundamental representation. The BEH effect creates the pattern SU(3)→SU(2), as the only possible little group in this case. Thus, this yields 3 massless gauge bosons in the adjoint of the unbroken SU(2), as well as 5 massive ones, of which four are degenerate and form two doublets under the unbroken SU(2), and one singlet under the unbroken SU(2) [2, 90, 91]. At the same time, the global symmetry is merely a U(1) acting as a global phase on the Higgs field [93]. Hence, in absence of accidental degeneracies, it can at most support two particles of the same mass. They form a particle and an antiparticle with respect to the U(1) group.

Augmented perturbation theory is in agreement with this analysis. The decisive ingredient is the matrix c from (6). The simplest operator to understand the difference creates a U(1)-neutral vector state. Using the FMS mechanism it follows that [93]

$$\phi_i D_\mu^{ij} \phi_j = v^2 c_a W_\mu^a + \dots,$$

where the mapping 'matrix' c now has the form $c_a = \delta_{a8}$, if the Higgs vacuum expectation value has been given the real 3-direction. In this way, the FMS mechanism maps the physical state to the most massive gauge boson, the singlet under the unbroken SU(2) gauge subgroup. This pattern continues to other channels [91]. Especially, as no symmetry exists to create a gauge-invariant triplet, the prediction from augmented perturbation theory is that the theory is gapped [91, 93, 94], in stark contrast to the gauge-dependent, ungapped spectrum. Again, augmented perturbation theory has been confirmed for this theory in lattice simulations [93–96].

Moreover, similar to QCD, it is not possible to construct a gauge-invariant state with only a single unit of U(1) charge, but at least 3 units are required. This is again in stark contrast to the perturbative case. Such states should perturbatively not appear asymptotically in a different way than as a scattering state. But non-perturbatively, because the U(1) charge is conserved, at least the lightest state with 3 units of U(1) (as well as its antistate) needs to be necessarily stable. Such a state is indeed observed in lattice simulations [94, 96]. However, because there is no charge-3 elementary state, its gauge-invariant operator cannot be mapped by the FMS mechanism onto an elementary state. Thus, the first non-zero matrix element in the FMS expression is not a propagator, but a higher n-point function [91]. At lowest order in a constituent-like model, this matrix element seems to provide at least the correct order of magnitude of the mass [91, 94, 96], but a detailed investigation is yet required. In total, the spectrum becomes indeed very involved and rich, once more channels are taken into account [95, 96], see Fig. 4.

Going beyond this simplest example, it becomes quickly apparent that one big difference is the number of physical degrees of freedom in contrast to the gauge degrees of freedom. Consider an SU(N) gauge theory coupled to a single fundamental Higgs. While the former is determined, up to internal excitations, by the possible multiplet structure of the global symmetry, which is fixed to U(1), the latter quickly increases with increasing dimension of the gauge group [91, 92]. Thus, the low-energy physics remains very similar, just as is the case with the glueball physics of large-N Yang–Mills theory. Also there the number of physical states remains fixed, and essentially unaltered, even when moving towards an infinite number of gauge degrees of freedom [97]. Thus, such a behavior is not without precedent.

Going beyond this case, many aspects of the spectrum become quickly dependent on the details of the representations of the Higgs fields and breaking patterns [91, 92]. But two more general features stand out.

The first is that any GUT [89, 90] needs to also include electromagnetism and thus the photon. The photon is exceptional, as it is a physically observable, massless vector state. This is well understood, but non-trivial, in QED [7, 98]. But in a GUT setting the requirement of gauge invariance requires that the photon is also a composite state, which is gauge-invariant with respect to the single unified gauge group. Hence, this requires that there exists a massless, uncharged composite vector bound state. This is indeed predicted with augmented perturbation theory [91, 92], but requires as minimal Higgs content a Higgs in the adjoint representation. While the calculation is involved, the appearance of such a state was confirmed in lattice gauge theory at an exploratory level [51, 99, 100]. This also provides an explicit example for the possibility to build a massless vector boson from massive constituents, without involving a Goldstone-type mechanism. In particular, given the explicit mass scale in the Lagrangian, even the usual argument of interpreting the photon as the Goldstone mode of broken dilatation symmetry [98] does not apply here.

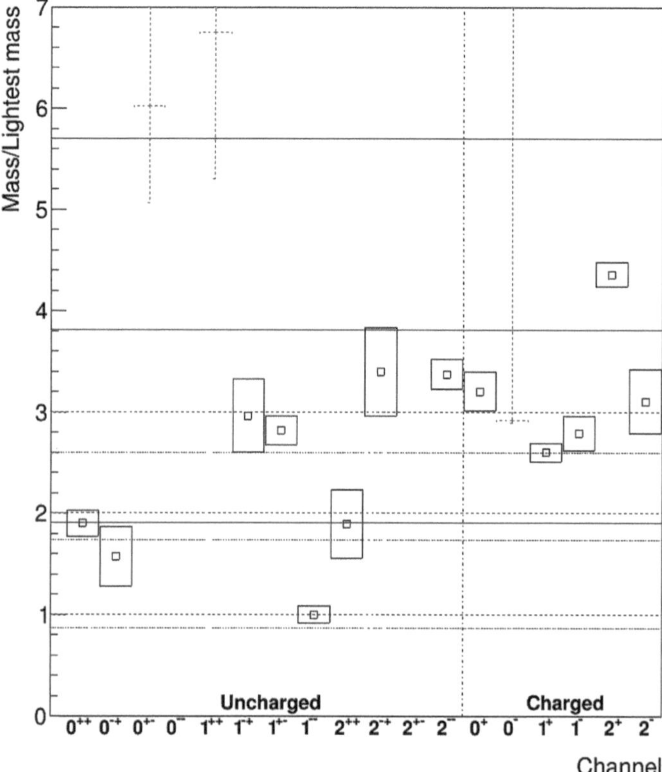

Fig. 4 The lowest level of SU(3) with a fundamental Higgs in the BEH Higgs region in various quantum number channels in units of the lightest state using lattice methods. The results are infinite-volume extrapolated and used a variational analysis. Simulations have been performed at $\beta = 6.85535$, $\kappa = 0.456074$, and $\lambda = 2.3416$, see [94–96] for further technical details. Results, partially preliminary, are from [95, 96]. Red dashed lines indicate upper limits and the horizontal blue lines correspond to the gauge-dependent mass scales, with the solid line the elementary Higgs mass, the dashed line the heavy gauge boson mass, and the dotted line the lighter gauge bosons' mass, as well as integer multiples. "(Un)Charged" refers to the U(1) charge and the label to the J^{PC} and J^{P} quantum numbers for the charged and uncharged states, respectively.

The second feature is the low-energy behavior. If the ideas of GUTs should work, it is required that the physical, gauge-invariant low-energy spectrum of the standard model is reproduced. While this is well reproduced in a perturbative approach [89], this is not true for the gauge-invariant spectrum [91, 92]. Here, the pattern of the simplest example repeats itself. Especially, many popular GUT candidates with minimal Higgs content are explicitly ruled out on a qualitative mismatch of the spectrum [92]. In fact, no qualitatively working candidate has yet been found [91, 92], not to mention obeying quantitative constraints like proton decay [2, 89]. Whether it is possible at all is a difficult question, and it could be even impossible in the conventional way [92]. At the very least it appears not possible without having

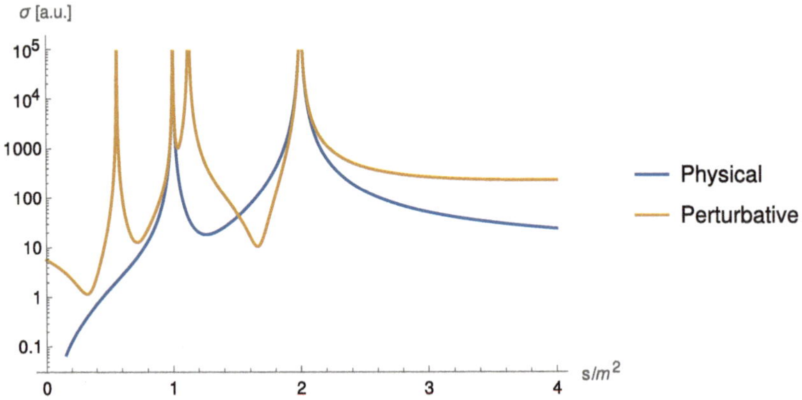

Fig. 5 The tree-level scattering cross-section in arbitrary units for Bhabba scattering at zero rapidity in units of the singlet 1^- mass in Fig. 4 for the theory described in the text [26]. Perturbative results are given in gold, and augmented perturbative results in blue

a Higgs content able to completely break the gauge group, while at the same time having a suitable global symmetry structure [92].

There are further far-reaching implications for searches for new physics because of the different spectrum. Consider again as the simplest example an SU(3) gauge theory with a Higgs in the fundamental representation and coupled in addition to a fermion in the fundamental representation coupled vectorially to the gauge interaction. This allows for an explicit mass, and does not require a Yukawa coupling to the Higgs. Performing Bhabba scattering it would usually be assumed that in the s-channel in the intermediate states all gauge bosons would show up, already at tree-level [2, 26]. However, treating the fermions correctly as bound states changes this qualitatively [26, 101]. The fermion bound state does no longer couple to all of the gauge bosons at tree-level, as the corresponding matrix element of bound states now contains a Higgs vacuum expectation value, which projects out all couplings, which are not mapped by the FMS mechanism to a vector bound state. Thus, only those vector states show up as resonances, which also appear in the physical spectrum. This is shown in Fig. 5. It is very satisfying that the resonances in the cross section are indeed only the physical ones. However, if experimental searches would not be based on augmented perturbation theory, but on perturbation theory, such a theory could be wrongly excluded. Especially if the energy reach is too small, where the gapped nature of the physical spectrum makes itself felt. Besides that, the fake resonances coming from unphysical degrees of freedom could easily misguide experimental searches. This does need to be taken properly into account. It should also be noted that close to the physical s-channel resonance (and the corresponding t-channel resonance) the results are nonetheless essentially indistinguishable. This had also been observed at next-to-leading order for two-point matrix elements [60, 62].

5 Applications Beyond Yang–Mills–Higgs Theory

While the FMS mechanism proper requires a BEH effect to work as described, the FMS framework can be extended far beyond. This potential has almost not been tapped at all yet.

5.1 *Theories Without a Higgs*

The FMS mechanism requires something to expand around. This is the reason, why it does not work, e.g., in QCD[7]. But the FMS framework still requires to take manifest gauge invariance into account, and always start from there.

To understand the implications, consider the idea that the Higgs is a composite state of new fermions [103–105], either in the context of technicolor or some other scenario. While the additional sector is strongly interacting, the weak interactions remains what it is. Thus, there are still weak gauge bosons, which need to be dressed to obtain the observable gauge-invariant (almost degenerate) vector state triplet (6), the physical version of the W-bosons and the Z boson [35].

If foregoing the possibility to construct a low-energy effective theory [105] and rather work with the ultraviolet theory, this immediately shows a problem. At least some of the additional fermions ψ need to be charged under both, the weak interactions and the new interactions. The former, because otherwise the new sector decouples. The latter, because otherwise the Higgs cannot be a bound state.

While the Higgs can be constructed as a straightforward singlet, the situation becomes involved for the vector bosons. The simplest possibility is to have two non-degenerate fermions, and to build an operator like

$$\tau_{ij}^a \bar{\psi}_{ru}^i \gamma^\mu D_\mu^{ru;sv} \bar{\psi}_{ru}^i \tag{8}$$

where aij are flavor indices, rs are new interaction indices, and uv are weak indices. The covariant derivative therefore contains both the weak gauge bosons and the new gauge bosons. This creates a state similar to the ρ meson of QCD, where the mass-splitting of the fermions can create the difference in the W and Z masses, and the SU(2) generator counts with a these states.

While this is formally working, this leads to two problems. On the one hand, phenomenology tends to require rather a large number of flavors [104, 105]. Their mass splitting must therefore be substantial to the lightest doublet, to avoid creating further vector states. At the same time, this is at odds with the foundational principle [103–105] of using a scaled-up version of QCD [106] to create masses by a condensate. In this scenario the pseudo-Goldstones would be absorbed as longitudinal degrees of freedom from the vector states, to avoid having light pseudoscalars around. With an operator like (8) this is neither possible, nor necessary. Also, such

[7] Though similar ideas can be pursued [102].

Goldstone bosons could not be defined in a gauge-invariant way with respect to the weak interactions, if they should play this role. Thus, such a scenario appears to be inconsistent with gauge symmetry. Moreover, quantitatively it would be required that a state created by (8) is the lightest (visible) state in the spectrum of the new sector. Whether this is possible, is unknown. At least, no working example has been found yet [26]. But essentially no dynamical investigations respecting manifest gauge symmetry with both sectors coupled have been conducted yet either.

Thus, also here the FMS framework shifts qualitatively the way how scenarios of this type needs to be addressed. Perhaps this will also open quite different possibilities, as this alleviates some of the problems due to light states in such theories.

5.2 Gravity and Supergravity

Gravity

The FMS mechanism relies on the single-field expansion (3). Therefore, only scalar fields can be used for the expansion, as long as global Poincaré invariance should be maintained. This is true in all quantum field theories. Hence, in many theories the FMS mechanism is not applicable. There is, however, a further exciting possibility beyond quantum field theory: Gravity.

General relativity can be considered a gauge theory of translations [107–109]. Local Lorentz symmetry in the tangent space can then be considered to be either dependent on the translation gauge symmetry, or can be considered as a second gauge interaction, leading to torsion [107]. Correspondingly, a quantum gravity theory will also be a quantum gauge theory. It is yet unclear if a, more or less extended, version of canonical gravity can be quantized using a path-integral, or similar, approach. But there is encouraging mounting circumstantial evidence that this is the case [110–119]. It will therefore be assumed here. The FMS framework then applies as well, requiring to start out with manifestly diffeomorphism-invariant (and local Lorentz-invariant) quantities as observables [110, 111, 120]

Observationally, at long distances, quantum gravity is indeed dominated by a special field configuration in our universe, a de Sitter metric[8]. This suggests that it should at least be possible to apply also the FMS mechanism in this case [120, 121]. Conceptually, this creates a BEH effect in quantum gravity. Similarly to the BEH effect in quantum field theory, the classical minimum will be the starting point. For the observational value of Newton's constant and the cosmological constant [15], this is indeed de Sitter. This is in agreement with the metric structure at long distances, supporting the possibility that the FMS mechanism should be possible.

There is one particularity which makes it different from the situation so far. In the previous cases, the field developing the vacuum expectation value and the

[8] On the question whether this should rather be a Friedmann–Lemaître–Robertson–Walker metric see [121].

gauge fields were not the same. Here, they are. While this does not introduce any conceptual obstacles, it makes it technically more involved. This is amplified by involvement of the gauge field, the metric, itself in all expressions.

Probably the most cumbersome feature is that a BEH effect in quantum gravity requires necessarily the use of a non-linear gauge condition. The FMS mechanism can only be applied after gauge-fixing and quantization is complete, and therefore does not alter it.

Consider as a minimal case the Einstein–Hilbert action

$$\begin{aligned} S &= \frac{1}{2\kappa}\int d^4x\sqrt{\det(-g)}(R+l) \\ \Gamma_{\mu\nu\rho} &= \frac{1}{2}\left(\partial_\rho g_{\mu\nu} + \partial_\nu g_{\mu\rho} - \partial_\mu r_{\nu\rho}\right) \\ R &= g^{\mu\nu}R_{\mu\nu} = g^{\mu\nu}g^{\rho\sigma}R_{\rho\mu\sigma\nu} \\ R_{\mu\nu\rho\sigma} &= \partial_\rho\Gamma_{\mu\nu\sigma} - \partial_\sigma\Gamma_{\mu\nu\rho} + g^{\alpha\beta}\Gamma_{\mu\alpha\rho}\Gamma_{\beta\nu\sigma} - g^{\alpha\beta}\Gamma_{\mu\alpha\sigma}\Gamma_{\beta\nu\rho} \end{aligned}$$

where $g_{\mu\nu}$ is the metric, R the curvature scalar, and κ and l suitably normalized versions of Newton's constant and the cosmological constant, respectively.

There is now a choice to be made, which was not previously present: Should the gauge condition be obeyed by the field configuration to be expanded around in the FMS mechanism? Before that, any choice here could be compensated by the gauge field. But now the vacuum expectation field and the gauge field is identical. It appears to be technically convenient [121] to choose a gauge, which is satisfied by the vacuum expectation value, in the present case the de Sitter metric $g_{\mu\nu}^{\mathrm{dS}}$. Due to reparametrization invariance, this can furthermore substantially alter the technical feasibility. One possible practical choice appears to be the Haywood gauge [121],

$$g^{\mu\nu}\partial_\mu g_{\mu\rho} = 0,$$

which is fulfilled by a flat metric, and the maximal symmetric de Sitter and anti-de Sitter metrics in standard Cartesian parametrization. This condition already shows the issue of non-linearity. In particular, because the inverse metric is not independent. This relation is in general highly non-linear.

The FMS mechanism constitutes now in splitting off the "vacuum expectation value" $g_{\mu\nu}^{\mathrm{dS}}$ [120]. Again, there is no unique way to do so. But if any approximations are good, the split-off part $\gamma_{\mu\nu}$ needs to be small, and thus at linear order many possibilities coincide [121], yielding

$$g_{\mu\nu} = g_{\mu\nu}^{\mathrm{dS}} + \gamma_{\mu\nu}.$$

The inverse of γ is not a metric, and determined by a Dyson-like relation

$$\gamma^{\mu\nu} = -(g^{\mathrm{dS}})^{\mu\sigma}\gamma_{\sigma\rho}g^{\rho\nu}. \tag{9}$$

Because $\gamma_{\mu\nu}$ is assumed small, the right-hand side can be expanded in a series in $\gamma_{\mu\nu}$. This creates an infinite series of tree-level vertices [121], but establishes a formulation in $\gamma_{\mu\nu}$ only. And a similar step is necessary for most observables as well.

While this is technically more cumbersome than in the quantum-field theory case, it is straightforward to use [120, 121]. At tree-level, it yields agreement with results from dynamical triangulation [110, 111, 121, 122] as well as a well-defined systematic limit to flat-space quantum-field theory as the lowest order in the FMS mechanism [120, 121].

The latter is probably best seen by considering how distances are measured in quantum gravity. Distance itself becomes in quantum gravity an expectation value [110, 117, 120]. A possible definition is given by

$$r(x,y) = \left\langle \min_{z(t)} \int_x^y dt g^{\mu\nu} \frac{dz_\mu(t)}{dt}\frac{dz_\nu(t)}{dt} \right\rangle$$
$$= \min_{z(t)} \int_x^y dt g_{\mu\nu}^{\mathrm{dS}} \frac{dz^\mu(t)}{dt}\frac{dz^\nu(t)}{dt} + \left\langle \min_{z(t)} \int_x^y dt \gamma_{\mu\nu} \frac{dz^\mu(t)}{dt}\frac{dz^\nu(t)}{dt} \right\rangle,$$

where x and y are points in the $\mathbb{R}^4$ underlying the manifold and the minimization requires to find the geodesic distance[9] between these points in the manifold configuration. This is averaged over the manifold configurations. The second line implements the FMS mechanism, which shows how the result splits between the contribution from the vacuum expectation value and the fluctuation field. Especially, if the fluctuations vanish, $\gamma \to 0$, this smoothly changes into the ordinary fixed curved-background quantum field theory distances.

Supergravity

Following the FMS framework through often leads to very surprising insights. Consider the concept of supersymmetry [88, 123]. In quantum field theory, supersymmetry appears to be essentially transparent for the FMS mechanism [83]. This is not surprising, as supersymmetry is a global symmetry, and thus should behave like, e.g., flavor symmetries.

However, despite all efforts and its inherently appealing nature [88, 123], no sign of supersymmetry in nature has been observed [15]. This leads to the claim that supersymmetry must be necessarily broken, at the expense of loosing some of its appeal [88, 123].

It is here were the FMS framework provides a possible way out. In our actual universe, it is not valid to consider supersymmetry as a stand-alone global symmetry, due to the existence of gravity. Because supersymmetry is part of the super Poincaré symmetry, this forces supersymmetry to become a local gauge symmetry,

[9] If singularities appear, or geodesics become incomplete, a suitable deformation has to be introduced.

supergravity [124, 125]. According to the FMS framework, physical observables cannot be gauge-dependent, and cannot change under the local supersymmetry transformations. Thus, the physical, observable spectrum is not and, in fact, cannot be supersymmetric. This alleviates the need to find a superpartner for, e.g., the electron, which is the usual argument for requiring supersymmetry to be broken [88]. Thus, it is possible to retain supersymmetry, and supergravity, as an intact symmetry of nature, without the need to observe a supersymmetric spectrum at experiments. Given the importance of supersymmetry to string theory [126], this can have far-reaching consequences.

In addition, similar to canonical quantum gravity, this implies the possibility to use the FMS mechanism on the same reasoning, this time introducing a BEH effect for the vierbein e_μ^a. Consider the simplest $\mathcal{N} = 1$ supergravity theory [123]

$$\begin{aligned} S &= \int dx \frac{\det e}{2\kappa} \left(e^{a\mu} e^{b\nu} R_{\mu\nu ab} - \bar{\Psi}_\mu \gamma^{\mu\nu\rho} D_\nu \Psi_\rho \right) \\ R_{\mu\nu ab} &= \partial_\mu \omega_{\nu ab} - \partial_\nu \omega_{\mu ab} + \omega_{\mu ac} \omega_\nu{}^c{}_b - \omega_{\nu ac} \omega_\mu{}^c{}_b \\ D_\nu &= \partial_\nu + \frac{1}{4} \omega_{\nu ab} \gamma^{ab} \\ \omega_{\nu ab} &= 2 e_{\mu[a} \partial_{[\nu} e_{b]}^{\mu]} - e_{\mu[a} e_{b]}^\sigma e_{\nu c} \partial^\mu e_\sigma^c \\ g_{\mu\nu} &= e_\mu^a g_{ab}^{\text{Flat}} e_\nu^b, \end{aligned}$$

with the Rarita–Schwinger graviton Ψ and $[i, j]$ implies antisymmetrization of i and j. Under a supersymmetry transformation δ_S

$$\begin{aligned} \delta_S e_\mu^a &= \frac{1}{2} \bar{\epsilon} \gamma^a \Psi_\mu \\ \delta_S \Psi_\mu &= D_\mu \epsilon, \end{aligned}$$

with local transformation function $\epsilon(x)$. As this is a gauge transformation, any physical observable needs to be invariant under it.

A possible example is the local composite operator[10]

$$e_a^\mu \gamma^a \Psi_\mu.$$

It is invariant under a supersymmetry transformation due to the compatibility of the tetrad and the Grassmann nature of the graviton. It therefore does not have a superpartner.

At the same time, applying the FMS mechanism in Haywood gauge with $e_\mu^a = \delta_\mu^a + \varepsilon_\mu^a$ for flat space and a small fluctuation field ε yields

$$e_a^\mu \gamma^a \Psi_\mu = \gamma^\mu \Psi_\mu + \varepsilon_\mu^a \gamma^a \Psi_\mu.$$

[10] Torsion will require a similar treatment for the γ matrices, probably using position-dependent γ matrices [127, 128] and another FMS mechanism for them [120].

Neglecting, as usual, the fluctuation part, the state describes a massless spin 1/2 particle. While promising to be more involved than the ordinary gravity case [121], it appears very appealing to follow-up on these exploratory heuristics.

6 Ontological Implications

The FMS framework is from the point of view of philosophy of physics quite remarkable from two perspectives [24, 27–33]. One is from the resolution of ambiguities in the BEH effect using the FMS mechanism. The other concerns the implications for the laws of nature by the FMS framework.

The FMS mechanism resolves the apparent paradox [29, 30, 33] of the gauge-dependence of the BEH effect and its apparent phenomenological success when treated as if it would be physical [26, 47–49]. It shows that the paradox is an artifact of the special structure of the standard model [26], which allows for a quantitatively effective possibility to ignore the issue of non-perturbative gauge invariance. Still, it was, and is, a source of some consternation in the philosophy of physics literature [24, 30, 31, 33], why this paradox has not been taken seriously, not realized, or even denied in large parts of the particle physics community. In fact, this dissonance even led to the odd situation that lattice approaches, which need by construction to take non-perturbative gauge invariance seriously and manifestly into account, denoted the composite states Higgs, Z, W, and so on [129–133], in obvious contradiction to their nature. Hence, despite having with the FMS mechanism a conceptually clean approach, the underestimation of the mechanism leaves still a kind of quagmire in notations in contemporary literature. Philosophically, of course, posing the question what is real, and what the role of gauge symmetry is, leads immediately to the necessity to find a resolution of the paradox.

This leads to the even more important perspective, this time with respect to the FMS framework. The question for the role of gauge symmetries is a very fundamental one. Since it appears possible to remove them from some theories explicitly [13, 14, 133–139], it is questionable whether they have any ontological relevance at all. This has already been formulated in terms of the Kretschmann objection [140], which in its generalized form states that any theory can be turned into a gauge theory, and in its inverse form that every gauge theory can be rewritten in terms of a (possibly non-local) non-gauge theory [24]. It is a most remarkable feature that such non-gauge theories seem at first sight to be again a theory of point-particles. However, due to the appearance of either an infinite series of polynomials in the Lagrangian and/or non-localities, it becomes quickly evident that this is just an artifact of tree-level perturbation theory [26]. In this context, it is important to note that the Aharanov–Bohm effect [141], often cited as supporting that gauge fields are physically real, can indeed also be described entirely without resorting to gauge degrees of freedom [24, 139].

It thus appears that gauge symmetries are merely redundant degrees of freedom[11], which are however technically indispensable. However, this is not a very precise phrasing, see [24] for a more detailed discussion.

A possible stance is that only measurable, and thereby at least gauge-invariant, entities should be ontological, i.e. possible candidates for being part of reality. If one accepts this premise, the FMS framework fundamentally reshuffles the building blocks of nature. Aside from hypothetical right-handed neutrinos, all observed particles are necessarily extended, and described by composite, gauge-invariant operators. This is a fundamental paradigmatic shift compared to the idea of fundamental point particles. It was also the latter idea, which gave rise to string theory [146], due to the problems entailed by the point-like nature of elementary particles. Having as fundamental entities composite ones would change this premise at least partly.

Furthermore, allowing the fundamental laws of nature to be build from extended objects would possibly open up alternatives to the idea of ever smaller structures, or higher energies. Especially, as the concept of energy itself becomes in quantum gravity ontologically doubtful, as energy is no longer gauge-invariant. Such a recast of the approach to the fundamental laws of nature would be nothing but transformative, and would even affect school textbooks and popular science fundamentally.

7 Summary

The most obvious consequence of the FMS framework [13, 14], and with this one aspect of Giovanni Morchio's legacy, is to reconcile the foundations of field theory with the phenomenological success of the perturbative treatment of the BEH effect. With the FMS mechanism, this delivered a tool to turn the very fundamental considerations of the FMS framework into phenomenological applications and even paved the way to experimental tests. In fact, it will allow a guaranteed discovery. Either, experimental tests will confirm the FMS framework, and will show that elementary particles like the Higgs are actually composite, extended objects even within the framework of the standard model of particle physics. Or, this will show that the current formulation of the standard model of particles as a quantum gauge field theory is insufficient, either on formal grounds or because the model is incomplete. Either way, the decision will change drastically our view of the world.

While even this aspect has been drastically underestimated, the far-reaching consequences of the FMS framework are even more so. The insistence on forcing a manifestly and non-perturbative gauge-invariant approach even at arbitrarily weak coupling and a convenient hiding of the gauge symmetry by the BEH effect shows that it was possibly to take field theory seriously without loosing the technical

[11] It has been claimed, see e.g. [142] for an introduction, that semi-classical considerations of black holes make gauge symmetries physical. It could not yet been substantiated whether this holds true in full quantum gravity [143], and gauge-invariant formulations of canonical quantum gravity appear to disfavor such a possibility [110, 111, 120, 144, 145].

ability to be predictive. In fact, in view of the Gribov–Singer ambiguity and the theorems of Haag and Elitzur, it provides a much better understanding of why (and when) perturbation theory can be a quantitatively viable approach.

At the same time, this reasoning is a role model. The FMS framework showed how further quantum gauge theories beyond the standard model should be approached: From the question of physical observables, and maintaining formal consistency. Approximations need to maintain consistency to a much better degree as standard perturbation theory does, which even in non-gauge theories runs afoul of Haag's theorem. Applications beyond the standard model showed explicitly that results based on the FMS framework are in much better agreement with full, non-perturbative results, even at very weak coupling, than those which break formal consistency like perturbation theory.

Moreover, the FMS framework shows that what is usually called confinement is not a distinct phenomena, but could really be viewed as an aspect of manifest gauge invariance [6]. This unifies the way how physical observable particles in the standard model should be treated, and removes the necessity to separate between the strong interaction and the electroweak one in conceptual terms [26]. This generalizes then to arbitrary other theories, up to and including quantum gravity ones. Especially, it implies that any gauge theory needs to be considered ontologically to be a theory of extended objects, rather than point-like elementary particles. Confirming this in the standard-model case experimentally would indeed change disruptively the way how we think about the laws of nature. Thus, Giovanni Morchio's legacy could very well become a crucial stepping stone in particle physics and the search for the most fundamental laws of nature and what reality is.

Acknowledgements I am grateful to P. Berghofer and S. Plätzer for a critical reading of the manuscript. Part of this work has been supported by the the Austrian Science Fund FWF, grant P32760 and doctoral school grant W1203-N16. Part of the results have been obtained using the HPC clusters at the University of Graz and the Vienna Scientific Cluster (VSC).

References

1. Yang, C.N., Mills, R.L.: Phys. Rev. **96**, 191 (1954). https://doi.org/10.1103/PhysRev.96.191
2. Böhm, M., Denner, A., Joos, H.: Gauge theories of the strong and electroweak interaction. Teubner, Stuttgart, p. 784 (2001)
3. Montvay, I., Münster, G.: Quantum fields on a lattice. Cambridge monographs on mathematical physics. Cambridge University Press, Cambridge, p. 491 (1994)
4. Singer, I.M.: Commun. Math. Phys. **60**, 7 (1978). https://doi.org/10.1007/BF01609471
5. Gribov, V.N.: Nucl. Phys. **B139**, 1 (1978). https://doi.org/10.1016/0550-3213(78)90175-X
6. Lavelle, M., McMullan, D.: Phys. Rept. **279**, 1 (1997). https://doi.org/10.1016/S0370-1573(96)00019-1
7. Haag, R.: Local quantum physics: Fields, particles, algebras. Texts and monographs in physics. Springer, Berlin, p. 356 (1992)
8. Dissertori, G., Knowles, I.G., Schmelling, M.: Quantum Chromodynamics: High energy experiments and theory. Oxford University Press, Oxford, p. 538 (2003)

9. Yndurain, F.J.: The Theory of Quark and Gluon Interactions. Theoretical and Mathematical Physics. Springer, Berlin (2006). https://doi.org/10.1007/3-540-33210-3
10. Osterwalder, K., Seiler, E.: Ann. Phys. **110**, 440 (1978). https://doi.org/10.1016/0003-4916(78)90039-8
11. Fradkin, E.H., Shenker, S.H.: Phys. Rev. **D19**, 3682 (1979). https://doi.org/10.1103/PhysRevD.19.3682
12. Banks, T., Rabinovici, E.: Nucl. Phys. **B160**, 349 (1979). https://doi.org/10.1016/0550-3213(79)90064-6
13. Fröhlich, J., Morchio, G., Strocchi, F.: Phys. Lett. **B97**, 249 (1980). https://doi.org/10.1016/0370-2693(80)90594-8
14. Fröhlich, J., Morchio, G., Strocchi, F.: Nucl. Phys. **B190**, 553 (1981). https://doi.org/10.1016/0550-3213(81)90448-X
15. Workman, R.L., et al.: PTEP **2022**, 083C01 (2022). https://doi.org/10.1093/ptep/ptac097
16. Englert, F., Brout, R.: Phys. Rev. Lett. **13**, 321 (1964). https://doi.org/10.1103/PhysRevLett.13.321
17. Higgs, P.W.: Phys. Rev. Lett. **13**, 508 (1964). https://doi.org/10.1103/PhysRevLett.13.508
18. Higgs, P.W.: Phys. Lett. **12**, 132 (1964). https://doi.org/10.1016/0031-9163(64)91136-9
19. Higgs, P.W.: Phys. Rev. **145**, 1156 (1966). https://doi.org/10.1103/PhysRev.145.1156
20. Guralnik, G.S., Hagen, C.R., Kibble, T.W.B.: Phys. Rev. Lett. **13**, 585 (1964). https://doi.org/10.1103/PhysRevLett.13.585
21. Kibble, T.W.B.: Phys. Rev. **155**, 1554 (1967). https://doi.org/10.1103/PhysRev.155.1554
22. Englert, F., Brout, R., Thiry, M.F.: Nuovo Cim. **A43**(2), 244 (1966). https://doi.org/10.1007/BF02752859
23. Elitzur, S.: Phys. Rev. **D12**, 3978 (1975). https://doi.org/10.1103/PhysRevD.12.3978
24. Berghofer, P., François, J., Friederich, S., Gomes, H., Hetzroni, G., Maas, A., Sondenheimer, R. Gauge Symmetries, Symmetry Breaking, and Gauge-Invariant Approaches. Elements in the foundations of contemporary physics. Cambridge University Press, Cambridge (2023). https://doi.org/10.1017/9781009197236
25. 't Hooft, G.: Nato Adv. Study Inst. Ser. B Phys. **59**, 101 (1980)
26. Maas, A.: Prog. Part. Nucl. Phys. **106**, 132 (2019). https://doi.org/10.1016/j.ppnp.2019.02.003
27. Friederich, S.: Eur. J. Philos. Sci. **3**(2), 157 (2013)
28. Friederich, S.: J. Gen. Philos. Sci **45**, 335 (2014)
29. Lyre, H.: Lokale Symmetrien und Wirklichkeit: Eine naturphilosophische Studie über Eichtheorien und Strukturenrealismus. Mentis, Paderborn (2004)
30. Lyre, H.: Int. Stud. Phil. Sci. **22**, 119 (2008). https://doi.org/10.1080/02698590802496664
31. François, J.: Philos. of Sci. **86**(3), 472 (2019)
32. Smeenk, C.: Philos. of Sci. **73**(5), 487 (2006)
33. Struyve, W.: Stud. Hist. Philos. Mod. Phys. **42**, 226 (2011). Higgs effect https://doi.org/10.1016/j.shpsb.2011.06.003
34. Maas, A., Mufti, T.: JHEP **1404**, 6 (2014). https://doi.org/10.1007/JHEP04(2014)006
35. Maas, A.: Mod. Phys. Lett. **A30**(29), 1550135 (2015). https://doi.org/10.1142/S0217732315501357
36. Jenny, P., Maas, A., Riederer, B.: Phys. Rev. D **105**, 114513 (2022). https://doi.org/10.1103/PhysRevD.105.114513
37. Egger, L., Maas, A., Sondenheimer, R.: Mod. Phys. Lett. **A32**, 1750212 (2017). https://doi.org/10.1142/S0217732317502121
38. Lenz, F., Naus, H., Ohta, K., Thies, M.: Ann. Phys. **233**, 17 (1994). Broken global symmetry Higgs https://doi.org/10.1006/aphy.1994.1059
39. Dudal, D., Peruzzo, G., Sorella, S.P.: JHEP **10**, 39 (2021). https://doi.org/10.1007/JHEP10(2021)039
40. Dudal, D., van Egmond, D.M., Guimaraes, M.S., Holanda, O., Palhares, L.F., Peruzzo, G., Sorella, S.P.: JHEP **02**, 188 (2020). https://doi.org/10.1007/JHEP02(2020)188

41. Woloshyn, R.M.: Phys. Rev. D **95**(5), 54507 (2017). https://doi.org/10.1103/PhysRevD.95.054507
42. Lewis, R., Woloshyn, R.: Phys. Rev. D **98**(3), 34502 (2018). https://doi.org/10.1103/PhysRevD.98.034502
43. Seiler, E.: Gauge Theories as a Problem of Constructive Quantum Field Theory and Statistical Mechanics. Lect. Notes Phys.. (1982)
44. Hasenfratz, A., Hasenfratz, P.: Phys. Rev. D **34**, 3160 (1986). https://doi.org/10.1103/PhysRevD.34.3160
45. Maas, A., Zwanziger, D.: Phys. Rev. D **89**, 34011 (2014). https://doi.org/10.1103/PhysRevD.89.034011
46. Sartori, G.: Riv. Nuovo Cim. **14N11**, 1 (1991). https://doi.org/10.1007/BF02810048
47. Lee, B., Zinn-Justin, J.: Phys. Rev. D **5**, 3137 (1972). https://doi.org/10.1103/PhysRevD.5.3137
48. Caudy, W., Greensite, J.: Phys. Rev. D **78**, 25018 (2008). https://doi.org/10.1103/PhysRevD.78.025018
49. Maas, A.: Mod. Phys. Lett. A **27**(38), 1250222 (2012). https://doi.org/10.1142/S0217732312502227
50. Dobson, E., Maas, A., Riederer, B.: Multiple breaking patterns in the Brout–Englert–Higgs effect beyond perturbation theory. Ann. Phys. (2022). https://doi.org/10.1016/j.aop.2023.169404
51. Kajantie, K., Laine, M., Rajantie, A., Rummukainen, K., Tsypin, M.: JHEP **11**, 11 (1998). Adjoint https://doi.org/10.1088/1126-6708/1998/11/011
52. Fujikawa, K.: Nucl. Phys. B **223**, 218 (1983). https://doi.org/10.1016/0550-3213(83)90102-5
53. Brambilla, N., et al.: Eur. Phys. J. C **74**(10), 2981 (2014). https://doi.org/10.1140/epjc/s10052-014-2981-5
54. Gross, F., et al.: 50 years of quantum chromodynamics (2022)
55. Hasenfratz, P., von Allmen, R.: JHEP **02**, 79 (2008). Parity violation lattice https://doi.org/10.1088/1126-6708/2008/02/079
56. Lenz, F., Negele, J.W., O'Raifeartaigh, L., Thies, M.: Ann. Phys. **285**, 25 (2000). https://doi.org/10.1006/aphy.2000.6072
57. Maas, A.: Eur. Phys. J. C **71**, 1548 (2011). https://doi.org/10.1140/epjc/s10052-011-1548-y
58. Capri, M., Dudal, D., Guimaraes, M., Justo, I., Sorella, S.: Ann. Phys. **343**, 72 (2013). https://doi.org/10.1016/j.aop.2014.01.014
59. Rivers, R.J.: Path integral methods in quantum field theory. Cambridge monographs on mathematical physics). Cambridge University Press, Cambridg, p. 339 (1987)
60. Maas, A., Sondenheimer, R.: Phys. Rev. D **102**, 113001 (2020). https://doi.org/10.1103/PhysRevD.102.113001
61. Nielsen, N.K.: Nucl. Phys. B. **101**, 173 (1975). https://doi.org/10.1016/0550-3213(75)90301-6
62. Dudal, D., van Egmond, D., Guimaraes, M., Palhares, L., Peruzzo, G., Sorella, S.: Eur. Phys. J. C **81**(3), 222 (2020). https://doi.org/10.1140/epjc/s10052-021-09008-9
63. Afferrante, V., Maas, A., Sondenheimer, R., Törek, P.: SciPost Phys. **10**, 62 (2021). https://doi.org/10.21468/SciPostPhys.10.3.062
64. Shrock, R.E.: Nucl. Phys. B **278**, 380 (1986). https://doi.org/10.1016/0550-3213(86)90219-1
65. Maas, A.: Mod. Phys. Lett. A **28**, 1350103 (2013). https://doi.org/10.1142/S0217732313501034
66. Grifols, J.: Phys. Lett. B **264**, 149 (1991). Higgs bound state Schroedinger equation https://doi.org/10.1016/0370-2693(91)90719-7
67. Meißner, U.G., Rusetsky, A.: Effective Field Theories. Cambridge University Press (2022)
68. Maas, A., Plätzer, S., Sondenheimer, R.: Cross sections in augmented perturbation theory (2024). unpublished

69. Maas, A., Egger, L.: PoS **EPS-HEP2017**, p. 450 (2017). http://inspirehep.net/record/1628431/files/arXiv:1710.01182.pdf
70. Maas, A., Plätzer, S., Veider, F.: Lepton collisions in augmented perturbation theory (2024). unpublished
71. Maas, A., Reiner, F.: Restoring the Bloch–Nordsieck theorem in the electroweak sector of the standard model. Phys. Rev. D (2022). https://doi.org/10.1103/PhysRevD.108.013001
72. Maas, A., Raubitzek, S., Törek, P.: Phys. Rev. D **99**(7), 74509 (2019). https://doi.org/10.1103/PhysRevD.99.074509
73. Aad, G., et al.: Evidence of off-shell Higgs boson production from ZZ leptonic decay channels and constraints on its total width with the ATLAS detector. Phys. Lett. B (2023) https://doi.org/10.1016/j.physletb.2023.138223
74. Tumasyan, A., et al.: Nature Phys. **18**(11), 1329 (2022). https://doi.org/10.1038/s41567-022-01682-0
75. Buarque, D., et al.: Vector boson scattering processes: Status and prospects (2021). https://doi.org/10.1016/j.revip.2022
76. Fernbach, S., Lechner, L., Maas, A., Plätzer, S., Schöfbeck, R.: Phys. Rev. D **101**(11), 114018 (2020). https://doi.org/10.1103/PhysRevD.101.114018
77. Bauer, C.W., Provasoli, D., Webber, B.R.: JHEP **11**, 30 (2018). https://doi.org/10.1007/JHEP11(2018)030
78. Ciafaloni, M., Ciafaloni, P., Comelli, D.: Nucl. Phys. B. **589**, 359 (2000). https://doi.org/10.1016/S0550-3213(00)00508-3
79. Bloch, F., Nordsieck, A.: Phys. Rev. **52**, 54 (1937). https://doi.org/10.1103/PhysRev.52.54
80. Ellis, R.K., et al.: Physics briefing book: Input for the European strategy for particle physics update 2020 (2019)
81. Greensite, J.: Excitations of elementary fermions in gauge Higgs theories. Phys. Rev. D (2020) https://doi.org/10.1103/PhysRevD.102.054504
82. Maas, A., Pedro, L.: Phys. Rev. D **93**(5), 56005 (2016). https://doi.org/10.1103/PhysRevD.93.056005
83. Maas, A., Schreiner, P.: The manifestly gauge-invariant spectrum of the minimal supersymmetric standard model (2023)
84. Wurtz, M., Lewis, R., Steele, T.: Phys. Rev. D **79**, 74501 (2009). Multiple Higgs https://doi.org/10.1103/PhysRevD.79.074501
85. Lewis, R., Woloshyn, R.: Phys. Rev. D **82**, 34513 (2010). https://doi.org/10.1103/PhysRevD.82.034513
86. Sarkar, M., Catumba, G., Hiraguchi, A., Hou, G.W.S., Jansen, K., Kao, Y.J., Lin, C.J.D., Ramos Martinez, A.: PoS **LATTICE2022**, p. 388 (2023). https://doi.org/10.22323/1.430.0388
87. Branco, G., Ferreira, P., Lavoura, L., Rebelo, M., Sher, M.: Phys. Rept. **516**, 1 (2012). https://doi.org/10.1016/j.physrep.2012.02.002
88. Aitchison, I.: Supersymmetry in Particle Physics. An Elementary Introduction. Cambridge University Press (2007). Supersymmetry
89. Langacker, P.: Phys. Rept. **72**, 185 (1981). https://doi.org/10.1016/0370-1573(81)90059-4
90. O'Raifeartaigh, L.: Group structure of gauge theories. Cambridge Monographs On Mathematical Physics. Cambridge University Press, Cambridge, p. 172 (1986)
91. Maas, A., Sondenheimer, R., Törek, P.: Ann. Phys. **402**, 18 (2019). https://doi.org/10.1016/j.aop.2019.01.010
92. Sondenheimer, R.: Phys. Rev. D **101**(5), 56006 (2020). https://doi.org/10.1103/PhysRevD.101.056006
93. Maas, A., Törek, P.: Phys. Rev. D **95**(1), 14501 (2017). https://doi.org/10.1103/PhysRevD.95.014501
94. Maas, A., Törek, P.: Ann. Phys. **397**, 303 (2018). https://doi.org/10.1016/j.aop.2018.08.018
95. Dobson, E., Maas, A., Riederer, B.: The spectrum of SU(3) Yang-Mills theory with a fundamental Higgs (2024). unpublished

96. Dobson, E., Maas, A., Riederer, B.: PoS LATTICE2022, p. 210 (2022). https://doi.org/10.22323/1.430.0210
97. Lucini, B., Panero, M.: Phys. Rept. **526**, 93 (2013). https://doi.org/10.1016/j.physrep.2013.01.001
98. Lenz, F., Naus, H., Ohta, K., Thies, M.: Ann. Phys. **233**, 51 (1994). https://doi.org/10.1006/aphy.1994.1060
99. Lee, I.H., Shigemitsu, J.: Nucl. Phys. B **263**, 280 (1986). https://doi.org/10.1016/0550-3213(86)90117-3
100. Afferrante, V., Maas, A., Törek, P.: Phys. Rev. D **101**(11), 114506 (2020). https://doi.org/10.1103/PhysRevD.101.114506
101. Törek, P., Maas, A.: PoS **ALPS2018**, p. 27 (2018)
102. Hoyer, P.: Few Body Syst. **58**(3), 128 (2017). https://doi.org/10.1007/s00601-017-1291-5
103. Lane, K.: Two lectures on technicolor (2002)
104. Sannino, F.: Acta Phys. Polon. B **40**, 3533 (2009). Technicolor
105. Andersen, J., Antipin, O., Azuelos, G., Del Debbio, L., Del Nobile, E.: Eur. Phys. J. Plus **126**, 81 (2011). Technicolor https://doi.org/10.1140/epjp/i2011-11081-1
106. Quigg, C., Shrock, R.: Phys. Rev. D **79**, 96002 (2009). https://doi.org/10.1103/PhysRevD.79.096002
107. Hehl, F.W., Von Der Heyde, P., Kerlick, G.D., Nester, J.M.: Rev. Mod. Phys. **48**, 393 (1976). https://doi.org/10.1103/RevModPhys.48.393
108. Kibble, T.W.B.: J. Math. Phys. **2**, 212 (1961). 168(1961) https://doi.org/10.1063/1.1703702
109. Sciama, D.W.: Recent Developments in General Relativity. Pergamon Press, p. 415 (1962). chap. On the analogy between charge and spin in general relativity
110. Ambjorn, J., Goerlich, A., Jurkiewicz, J., Loll, R.: Phys. Rept. **519**, 127 (2012). https://doi.org/10.1016/j.physrep.2012.03.007
111. Loll, R.: Class. Quant. Grav. **37**(1), 13002 (2020). https://doi.org/10.1088/1361-6382/ab57c7
112. Niedermaier, M., Reuter, M.: Living Rev. Rel. **9**, 5 (2006). https://doi.org/10.12942/lrr-2006-5
113. Reuter, M., Saueressig, F.: New J. Phys. **14**, 55022 (2012). https://doi.org/10.1088/1367-2630/14/5/055022
114. Reuter, M., Saueressig, F.: Quantum Gravity and the Functional Renormalization Group. Cambridge University Press (2019)
115. Christiansen, N., Knorr, B., Meibohm, J., Pawlowski, J.M., Reichert, M.: Phys. Rev. D **92**(12), 121501 (2015). https://doi.org/10.1103/PhysRevD.92.121501
116. Bosma, L., Knorr, B., Saueressig, F.: Phys. Rev. Lett. **123**(10), 101301 (2019). https://doi.org/10.1103/PhysRevLett.123.101301
117. Schaden, M.: Causal space-times on a null lattice (2015)
118. Asaduzzaman, M., Catterall, S., Unmuth-Yockey, J.: Tensor network formulation of two dimensional gravity. Phys. Rev. D (2019). https://doi.org/10.1103/PhysRevD.102.054510
119. Catterall, S., Ferrante, D., Nicholson, A.: Eur. Phys. J. Plus **127**, 101 (2012). https://doi.org/10.1140/epjp/i2012-12101-4
120. Maas, A.: Scipost Phys. **8**(4), 51 (2020). https://doi.org/10.21468/SciPostPhys.8.4.051
121. Maas, A., Markl, M., Müller, M.: Phys. Rev. D **107**(2), 25013 (2023). https://doi.org/10.1103/PhysRevD.107.025013
122. Dai, M., Laiho, J., Schiffer, M., Unmuth-Yockey, J.: Phys. Rev. D **103**(11), 114511 (2021). https://doi.org/10.1103/PhysRevD.103.114511
123. Weinberg, S.: The quantum theory of fields. Vol. 3: Supersymmetry. Univ. Pr., Cambridge, UK (2000)
124. Van Nieuwenhuizen, P.: Phys. Rept. **68**, 189 (1981). https://doi.org/10.1016/0370-1573(81)90157-5
125. Freedman, D.Z., Van Proeyen, A.: Supergravity. Cambridge University Press, Cambridge (2012). http://www.cambridge.org/mw/academic/subjects/physics/theoretical-physics-and-mathematical-physics/supergravity?format=AR

126. Polchinski, J.: String theory. Vol. 2: Superstring theory and beyond. Cambridge Monographs on Mathematical Physics. Cambridge University Press (2007). https://doi.org/10.1017/CBO9780511618123
127. Gies, H., Lippoldt, S.: Phys. Rev. D **89**(6), 64040 (2014). https://doi.org/10.1103/PhysRevD.89.064040
128. Gies, H., Lippoldt, S.: Phys. Lett. **B743**, 415 (2015). https://doi.org/10.1016/j.physletb.2015.03.014
129. Lang, C.B.: Private communication (2017)
130. Wittig, H.: Private communication (2018)
131. Evertz, H., Jansen, K., Jersak, J., Lang, C., Neuhaus, T.: Nucl. Phys. B **285**, 590 (1987). Scalar QED https://doi.org/10.1016/0550-3213(87)90356-7
132. Langguth, W., Montvay, I., Weisz, P.: Nucl. Phys. B **277**, 11 (1986). https://doi.org/10.1016/0550-3213(86)90430-X
133. Philipsen, O., Teper, M., Wittig, H.: Nucl. Phys. B **469**, 445 (1996). https://doi.org/10.1016/0550-3213(96)00156-3
134. Langguth, W., Montvay, I.: Phys. Lett. B **165**, 135 (1985). https://doi.org/10.1016/0370-2693(85)90707-5
135. Masson, T., Wallet, J.C.: A remark on the spontaneous symmetry breaking mechanism in the standard model (2010)
136. Attard, J., François, J., Lazzarini, S., Masson, T.: Foundations of Mathematics and Physics one Century After Hilbert, New Perspectives. Springer (2018). chap. The dressing field method of gauge symmetry reduction, a review with examples
137. Kondo, K.I.: Eur. Phys. J. C **78**(7), 577 (2018). https://doi.org/10.1140/epjc/s10052-018-6051-2
138. Gambini, R., Pullin, J.: Loops, knots, gauge theories and quantum gravity. Cambridge Monographs on Mathematical Physics. Cambridge University Press, Cambridge (2000). http://www.cambridge.org/mw/academic/subjects/physics/theoretical-physics-and-mathematical-physics/loops-knots-gauge-theories-and-quantum-gravity?format=PB
139. Strocchi, F., Wightman, A.: Charge gauge invariance Maxwell theory Aharonov Bohm effect. J. Math. Phys. **15**, 2198 (1974). https://doi.org/10.1063/1.1666601
140. Pooley, O.: Towards a Theory of Spacetime Theories. Einstein Studies, vol. 13. Birkhauser, New York, NY, pp. 105–143 (2017). hap. Background Independence, Diffeomorphism Invariance and the Meaning of Coordinates
141. Aharonov, Y., Bohm, D.: Phys. Rev. **115**, 485 (1959). https://doi.org/10.1103/PhysRev.115.485
142. Reece, M.: TASI lectures: (No) global symmetries to axion physics (2023)
143. Perry, M.: Private communication (2018)
144. Donnelly, W., Giddings, S.B.: Phys. Rev. D **93**(2), 24030 (2016). [Erratum: Phys. Rev.D94,no.2,029903(2016) https://doi.org/10.1103/PhysRevD.93.024030] https://doi.org/10.1103/PhysRevD.94.029903
145. Giddings, S., Weinberg, S.: Gauge-invariant observables in gravity and electromagnetism: Black hole backgrounds and null dressings. Phys. Rev. D (2019). https://doi.org/10.1103/PhysRevD.102.026010
146. Polchinski, J.: String theory. Vol. 1: An introduction to the bosonic string. Cambridge Monographs on Mathematical Physics. Cambridge University Press (2007). https://doi.org/10.1017/CBO9780511816079

Curvature of the Determinant Line Bundle for Elliptic Boundary Problems over an Interval

Simon Scott

Contents

Abstract Details are given on the zeta function metric and connection on the determinant line bundle over the Grassmannian associated to boundary value problems over an interval.

1 Introduction

In the collaboration [2] of Giovanni Morchio with coauthors Bernhelm Booss-Bavnbek and Krzysztof Wojciechowski, in which the author here was kindly included, aspects of the global analysis of elliptic boundary value problems were analysed. The topic presented in this note are some details of one particular evolution of those ideas towards geometric index theory but restricted to the toy-model case of a 1-dimensional manifold. Specifically, though these results appeared elsewhere in greater generality on higher dimensional manifolds, what is given here are the computations for the case of the finite dimensional Grassmannian of global boundary problems for a compact 1-manifold, but which were not submitted to the arXiv or elsewhere.

S. Scott (✉)
Department of Mathematics, King's College, London, UK
e-mail: simon.scott@kcl.ac.uk

A. Cintio, A. Michelangeli (Eds.), *Trails in Modern Theoretical and Mathematical Physics*, https://doi.org/10.1007/978-3-031-44988-8_12

1.1 Relative Zeta-Determinants

Let A be an operator on a Hilbert space $\mathcal{H}$. If A is bounded of the form $I+\alpha$ with α of trace class, then it has a Fredholm determinant $\det_F A = \sum \mathrm{Tr}\,(\wedge^k \alpha)$, equal by Lidskii's Theorem to the product of its eigenvalues, and satisfying the characteristic properties of the determinant in finite-dimensions. If A is an unbounded operator, then to make sense of its determinant a choice of regularization procedure is needed. To define the zeta-determinant regularization we assume that A has discrete spectrum with principal angle θ, meaning there is a neighbourhood of the ray $R_\theta = \{re^{i\theta} \mid r \geq 0\}$ disjoint from spec(A), and that the operator norm of $(A-\lambda)^{-1}$ decays like $1/|\lambda|$ as $\lambda \to \infty$ along R_θ. For Re$(s) > 0$ one then has the convergent integral

$$A_\theta^{-s} = \frac{i}{2\pi}\int_{\Gamma_\theta} \lambda_\theta^{-s}(A-\lambda I)^{-1}\, d\lambda,$$

where $\lambda_\theta^{-s} = |\lambda|^{-s}e^{-is\arg(\lambda)}$, $\theta \leq \arg\lambda \leq \theta + 2\pi$, is the branch of λ^{-s} defined by R_θ, and Γ_θ is the contour beginning at ∞, traversing R_θ to a small circle around the origin, and then back along R_θ to ∞.

Assume $(A-\lambda)^{-m}$ is trace class for $m > -\alpha_0 > 0$ and that as $\lambda \to \infty$ along R_θ there is an asymptotic expansion of the form

$$\mathrm{Tr}\,(A-\lambda)^{-m} \sim \sum_{i=0}^{\infty}\sum_{k=0}^{k_i} a_{ik}\lambda^{-m-\alpha_i}\log^k\lambda, \tag{1}$$

where $-\infty < \alpha_0 < \alpha_1 < \ldots$ and $\alpha_i \to \infty$. This means that the spectral zeta function

$$\mathbf{z}_\theta(s, A) = \mathrm{Tr}\, A_\theta^{-s}$$

defined in the standard way for Re$(s) > -\alpha_0$ extends meromorphically to all of $\mathbb{C}$. If furthermore for $\alpha_J = 1$ one has $a_{J,k} = 0$ for $k > 0$, A is said to be *admissible*. $\mathbf{z}_\theta(s, A)$ is then holomorphic near $s = 0$, and the $\mathbf{z}$-determinant is defined by

$$\det{}_{\mathbf{z},\theta} A := \exp\left(-\frac{d}{ds}_{|s=0}\mathbf{z}_\theta(s, A)\right). \tag{2}$$

Seeley [9] showed that any elliptic (pseudo-)differential operator D of positive order over a closed manifold is admissible, and $\mathbf{z}_\theta(0, D)$ is then a local invariant depending only on the leading symbol of D which can be read off from (1). In contrast $\log\det_{\mathbf{z},\theta} D = -\mathbf{z}_\theta^{'}(0, D)$ is highly non-local (its first variation is also non-local) and this makes it a hard invariant to compute, except on certain symmetric spaces where an exact identification of the eigenvalues is possible.

An instructive example is the Laplacian $\Delta_a = D_a^* D_a$ on the interval $[0, 2\pi]$ with $D_a = i d/dx + a$ and $0 < a \leq 1$ and subject to period boundary conditions $\phi(0) = \phi(2\pi)$; note that this is a *global* boundary condition – it is not specified pointwise but rather depends on the entire boundary manifold (just two points in this case). Δ_a has eigenvalues $\{(n+a)^2 \mid n \in \mathbb{Z}\}$ and so in terms of the classical Riemann–Hurwitz zeta function $\mathbf{z}(s,a) = \sum_{n=0}^{\infty} 1/(n+a)^s$, one has $\mathbf{z}_\pi(\Delta, s) = \mathbf{z}(2s,a) + \mathbf{z}(2s, 1-a)$ for $\mathrm{Re}(s) > 1/2$. The meromorphically continued value $\mathbf{z}'(0,a) = \log(\Gamma(a)/\sqrt{2\pi})$ then yields $\det_{\mathbf{z}}\Delta_a = 4\sin^2 \pi a$. On the other hand, one can completely ignore zeta functions and formally compute

$$\det \Delta_a = \prod_{n\in\mathbb{Z}} (n+a)^2 = \left(\prod_{n\in\mathbb{Z}\backslash 0} n^2\right).a^2. \prod_{n\in\mathbb{Z}\backslash 0} (1 - \frac{a^2}{n^2})^2$$

$$= \left(\prod_{n\in\mathbb{Z}\backslash 0} n^2\right) \frac{\sin^2 \pi a}{\pi^2},$$

so that for $a_1, a_2 \in (0,1)$

$$\frac{\det \Delta_{a_1}}{\det \Delta_{a_2}} = \frac{\sin^2 \pi a_1}{\sin^2 \pi a_2} = \frac{\det_{\mathbf{z}}\Delta_{a_1}}{\det_{\mathbf{z}}\Delta_{a_2}}, \tag{3}$$

identifying the ratio of the ad-hoc but canonical determinant $\det \Delta_a$ with the ratio of the rigorous $\mathbf{z}$-function regularized determinant for Δ_a.

Equation (3) portrays a certain relativity principle for determinants of admissible operators, which asserts that for comparable operators A_1, A_2 the relative $\mathbf{z}$-determinant $\det_{\mathbf{z},\theta}(A_1, A_2)$ can be computed as a Fredholm determinant of an operator canonically determined by the relative resolvent. Here, comparable means that A_1, A_2 have the same principal angle θ and that the relative resolvent $(A_1 - \lambda)^{-1} - (A_2 - \lambda)^{-1}$ is trace class with

$$\mathrm{Tr}\left((A_1-\lambda)^{-1} - (A_2-\lambda)^{-1}\right) = -\partial_\lambda \log \det\nolimits_F S_\lambda, \tag{4}$$

where the *scattering matrix* $S_\lambda = S_\lambda(A_1, A_2)$ is an operator on a certain Hilbert space with a Fredholm determinant $\det_F S_\lambda$. For large enough real s one has the relative zeta function

$$\mathbf{z}(s, A_1, A_2) = \frac{i}{2\pi} \int_\Gamma \lambda^{-s} \mathrm{Tr}\left((A_1-\lambda)^{-1} - (A_2-\lambda)^{-1}\right) d\lambda.$$

If we further assume that the relative resolvent trace has an expansion as $\lambda \to \infty$

$$\mathrm{Tr}\left((A_1-\lambda)^{-1} - (A_2-\lambda)^{-1}\right) \sim \sum_{j=1}^{\infty} \sum_{k=0}^{1} b_{j,k} (-\lambda)^{-\alpha_j} \log^k(-\lambda), \tag{5}$$

where $\alpha_j \nearrow +\infty$, then this defines the meromorphic continuation of $\mathbf{z}(s, A_1, A_2)$ to all of $\mathbb{C}$.

For a function with an asymptotic expansion

$$f(\lambda) \sim \sum_{j=0}^{\infty} \sum_{k=0}^{1} c_{jk}(-\lambda)^{-\beta_j} \log^k(-\lambda) + c_0 \log(-\lambda) + c_1$$

as $\lambda \to \infty$ in $\Lambda_{\theta,\varepsilon}$, where $\beta_j \nearrow +\infty$ and $\beta_j \neq 0$, its regularized limit is defined to be the constant term in the expansion: $\mathrm{LIM}^{\theta}_{\lambda\to\infty} f(\lambda) = c_1$. Then (with $S := S_0$) there is the following precise form of the relativity principle for determinants:

Theorem [7] *For operators A_1, A_2 which are $\mathbf{z}$-comparable and $\mathbf{z}$-admissible, one has*

$$\frac{\det_{\mathbf{z},\theta} A_1}{\det_{\mathbf{z},\theta} A_2} = \det{}_F S \, . \, e^{-\mathrm{LIM}^{\theta}_{\lambda\to\infty} \log \det_F S_\lambda} \, . \tag{6}$$

Here, we show that for a family of boundary problems over an interval this principle is also central in understanding the Hermitian structure on the determinant line bundle defined by zeta-function.

1.2 Families of Global Boundary Problems

Let $X = [0, \beta]$ where β is a positive real number and let E be a complex Hermitian vector bundle over X of rank n. Relative to a choice of trivialization of E, a first-order elliptic differential operator $\mathcal{D}$ acting on $C^\infty(X; E)$ has the form $A(x)d/dx + B(x)$, where A, B are complex $n \times n$ matrices and $\det A(x) \neq 0$. The space $\mathrm{Ell}_{1,n}$ of all first-order elliptic operators on E is therefore identified with $C^\infty(X, \mathrm{Gl}_n(\mathbb{C})) \times C^\infty(X, \mathrm{End}(\mathbb{C}^n))$. The operator $\mathcal{D}$ extends to a continuous map $\mathcal{D}: H^1(X; E) \to L^2(X; E)$ on the first Sobolev completion. The bundle over the boundary for the Cauchy data is the $2n$-dimensional graded vector space $C^0(\partial X, (E_0 \oplus E_\beta)) \cong E_0 \oplus E_\beta$, and we have the restriction map to the boundary

$$\gamma: H^1(X; E) \longrightarrow E_0 \oplus E_\beta, \; \gamma(\psi) = (\psi(0), \psi(\beta)).$$

A *global boundary condition* for $\mathcal{D} \in Ell_{1,n}$ is specified by a projection P on $E_0 \oplus E_\beta$, where by projection we mean self-adjoint indempotent. The pair $(\mathcal{D}, P)$ combine to define the elliptic boundary value problem:

$$\mathcal{D}_P = \mathcal{D}: \mathrm{dom}(\mathcal{D}_P) \longrightarrow L^2(X; E),$$

where

$$\mathrm{dom}(\mathcal{D}_P) = \{\psi \in H^1(X; E) \mid P\gamma\psi = 0\} \, .$$

The parameter space of global boundary conditions for $\mathcal{D}$ is therefore the complex Grassmannian $Gr(E_0 \oplus E_\beta)$, consisting of one component

$$\mathrm{Gr}_k(E_0 \oplus E_\beta) = \{P \in \mathrm{End}(E_0 \oplus E_\beta) \mid P^2 = P, \; P^* = P, \; \mathrm{tr}(P) = k\},$$

for each integer $k = 0, \ldots, 2n$. A point P of the $k(2n-k)$-dimensional complex manifold $\mathrm{Gr}_k(E_0 \oplus E_\beta)$ is equivalently specified by $W = \mathrm{range}\,(P) \subset E_0 \oplus E_\beta$. In particular, the *Calderon projection* $P(\mathcal{D})$ onto the Cauchy data subspace $K(\mathcal{D}) = \{v \in E_0 \oplus E_\beta \mid \exists\, \psi \in C^\infty(X; E),\ \mathcal{D}\psi = 0, \gamma\psi = v\}$ defines a distinguished element of $Gr_n(E_0 \oplus E_\beta)$. More precisely, any element of $\mathrm{Ker}(\mathcal{D})$ is of the form $h(x)v$ for some $v \in E_0$, where $h(x) \in \mathrm{End}(E_0, E_x)$ is the fundamental solution matrix uniquely solving $\mathcal{D}h(x) = 0$ subject to $h(0) = I$. Hence there is a canonical isomorphism $\gamma\colon \mathrm{Ker}(\mathcal{D}) \to K(\mathcal{D})$ and

$$K(\mathcal{D}) = \mathrm{graph}(h := h(\beta)\colon E_0 \to E_\beta) \subset E_0 \oplus E_\beta. \tag{7}$$

For any two $P_1, P_2 \in Gr(E_0 \oplus E_\beta)$ we have a finite-rank operator

$$(P_1, P_2) := P_2 \circ P_1\colon W_1 \to W_2,$$

where $W_i = \mathrm{range}\,(P_i)$, and $\mathrm{ind}\,(P_1, P_2) = \dim W_1 - \dim W_2$. The pivotal fact is that the unbounded operator $\mathcal{D}_P$ is modeled by the finite-rank operator on boundary data

$$S(P) := (P(\mathcal{D}), P)\colon K(\mathcal{D}) \to W\ .$$

The operator $\mathcal{D}_P$ is a Fredholm operator with kernel and cokernel consisting of smooth sections, and $Gr_k(E_0 \oplus E_\beta)$ parameterizes EBVPs of index

$$\mathrm{ind}\,\mathcal{D}_P = \mathrm{ind}\,S(P) = n - k. \tag{8}$$

This implies the relative index formula $\mathrm{ind}\,\mathcal{D}_{P_1} - \mathrm{ind}\,\mathcal{D}_{P_2} = \mathrm{ind}\,(P_2, P_1)$.

Definition 1 By a smooth family of first-order elliptic differential operators over $[0, \beta]$ parameterized a manifold B we shall mean an element $\mathbb{D} \in C^\infty(B, \mathrm{Ell}_{1,n})$. A Grassmann section means an element $\mathbb{P} \in C^\infty(B, \mathrm{Gr}_k(E_0 \oplus E_\beta)) := Gr(B, E, k)$. A smooth family of elliptic boundary value problems means a pair $(\mathbb{D}, \mathbb{P})$.

For each $b \in B$, $(\mathbb{D}, \mathbb{P})$ parameterizes $\mathcal{D}_b \in \mathrm{Ell}_{1,n}$ and $P_b \in \mathrm{Gr}_k(E_0 \oplus E_\beta)$ and hence the EBVP $\mathcal{D}_{P_b} = \mathcal{D}_b\colon \mathrm{dom}(\mathcal{D}_{P_b}) \longrightarrow L^2(X; E)$. Equivalently one may think of $(\mathbb{D}, \mathbb{P})$ as a bundle homomorphism

$$(\mathbb{D}, \mathbb{P})\colon \mathcal{H}_{\mathbb{P}} \longrightarrow \mathcal{H},$$

where $\mathcal{H}$ is the trivial bundle $\mathcal{H} = B \times C^\infty(X; \mathbb{C}^n)$ and $\mathcal{H}_{\mathbb{P}}$ is the weak vector bundle with fibre $\mathrm{dom}_\infty(\mathcal{D}_{P_b}) = \{\psi \in C^\infty(X; E) \mid P_b\gamma\psi = 0\}$.

Proposition 1 A Grassmann section $\mathbb{P} \in Gr(B, E, k)$ is equivalent to a smooth rank k complex bundle $\mathcal{W} \longrightarrow B$ with fibre $W_b := \mathrm{range}\,(P_b)$. The bundle $\mathcal{W}$ is endowed with a natural Hermitian metric $g^{\mathcal{W}}$ and compatible connection $\nabla^{\mathcal{W}} = \mathbb{P}\cdot d\cdot\mathbb{P}$ with curvature 2-form $\mathbf{R}^{\mathcal{W}} = \mathbb{P}\,d\,\mathbb{P}\,d\,\mathbb{P} \in \Omega^2(B; \mathrm{End}(\mathcal{W}))$. The induced connection on the complex line bundle $\mathrm{Det}\,(\mathcal{W})$ has curvature $\mathrm{tr}(\mathbf{R}^{\mathcal{W}}) = \mathrm{tr}(\mathbb{P}\,d\,\mathbb{P}\,d\,\mathbb{P}) \in \Omega^2(B)$.

We omit the proof.

To each pair of Grassmann sections $\mathbb{P}^0, \mathbb{P}^1$ there is thus the smooth finite-rank family

$$(\mathbb{P}^0, \mathbb{P}^1) \in C^\infty(B; \mathrm{Hom}(\mathcal{W}^0, \mathcal{W}^1)), \quad (\mathbb{P}^0, \mathbb{P}^1)_b \equiv P_b^1 P_b^0 \colon W_{0,b} \longrightarrow W_{1,b},$$

where $\mathcal{W}^i$ are the bundles of Proposition 1. In particular, associated to $\mathbb{D} \in C^\infty(B, \mathrm{Ell}_{1,n})$ is a preferred *Calderon section* $P(\mathbb{D}) \in Gr(B, E, n)$, defined by $b \mapsto P(\mathcal{D}_b)$ and constructed from global data. The bundle $\mathcal{K}(\mathbb{D}) \to B$ associated to $P(\mathbb{D})$ is canonically trivial by (7).

Abstractly, the determinant associated to a smooth family of Fredholm operators $\mathcal{A} = \{A_b \colon H_b^1 \longrightarrow \mathcal{H}_b^2 \mid b \in B\}$ arises as a canonical section $b \mapsto \det(A_b)$ of the determinant line bundle $\mathrm{DET}\,\mathcal{A} = \cup_{b\in B} \mathrm{Det}\, A_b$. the fibre $\mathrm{Det}\, A_b$ of the complex line bundle $\mathrm{DET}\,\mathcal{A}$ is canonically isomorphic to $\mathrm{Det}\,\mathrm{Ker}(A_b)^* \otimes \mathrm{Det}\,\mathrm{Coker}(A_b)$, where $\mathrm{Det}\, V := \wedge^{\max} V$. See [3, 6] for details.

For each smooth family of EBVPs $(\mathbb{D}, \mathbb{P})$ we have a determinant line bundle $\mathrm{DET}\,(\mathbb{D}, \mathbb{P})$ equipped with its determinant section $b \mapsto \det(\mathcal{D}_{P_b})$, and for $(\mathbb{P}^1, \mathbb{P}^2) \colon \mathcal{W}_1 \to \mathcal{W}_2$, we have a determinant bundle $\mathrm{DET}\,(\mathbb{P}^0, \mathbb{P}^1)$. In particular, associated to $(\mathbb{D}, \mathbb{P})$ is the finite rank family $\mathbb{S}(\mathbb{P}) = (P(\mathbb{D}), \mathbb{P})$ with determinant bundle $\mathrm{DET}\,(\mathbb{S}(\mathbb{P}))$ and canonical section $b \mapsto \det(S(P_b))$. There is a canonical line bundle isomorphism

$$\mathrm{DET}\,(\mathbb{D}, \mathbb{P}) \cong \mathrm{DET}\,(\mathbb{S}(\mathbb{P})) = \mathrm{Det}\;\mathcal{K}(\mathbb{D})^* \otimes \mathrm{Det}\,\mathcal{W}, \tag{9}$$

preserving the determinant sections $\det(\mathcal{D}_{P_b}) \longleftrightarrow \det(S(P_b))$. $\mathrm{DET}\,(\mathbb{D}, \mathbb{P})$ is therefore classified by the isomorphism class of the complex line bundle $\mathrm{Det}\,\mathcal{W}$: in terms of Chern classes, $c_1(\mathrm{DET}\,(\mathbb{D}, \mathbb{P})) = c_1(\mathrm{Det}\,\mathcal{W})$.

1.3 Statement of Results

The identification (9) means that by pull-back $\mathrm{DET}\,(\mathbb{D}, \mathbb{P})$ inherits a metric $\|\,.\,\|_C$ and compatible connection ∇^C from $\mathrm{DET}\,(\mathbb{S}(\mathbb{P}))$. Over the open subset U of B where the operators $\mathcal{D}_{P_b}$ are invertible the *canonical metric* on $\mathrm{DET}\,(\mathbb{D}, \mathbb{P})$ is defined by

$$\|\det \mathcal{D}_{P_b}\|_C^2 = \det{}_C(\Delta_{P_b})$$

where $\Delta_P = (\mathcal{D}_P)^* \mathcal{D}_P$, and the canonical regularization of $\det(\Delta_P)$ is the finite-rank determinant on $K(\mathcal{D})$, $\det_C(\Delta_P) := \det(S(P)^* S(P))$. On the other hand, $\mathrm{DET}\,(\mathbb{D}, \mathbb{P})$ has a Quillen metric defined over U by $\|\det \mathcal{D}_P\|_{\mathbf{z}}^2 = \det_{\mathbf{z}}(\Delta_P)$.

Theorem 1 *Let $\mathbb{P}^1$, $\mathbb{P}^2$ be Grassmann sections for $\mathbb{D}$ and let $\mathcal{D}_{P_1} \in (\mathbb{D}, \mathbb{P}^1)$, $\mathcal{D}_{P_2} \in (\mathbb{D}, \mathbb{P}^2)$ be invertible at $b \in B$. Then*

$$\frac{\|\det(\mathcal{D}_{P_1})\|_{\mathbf{z}}}{\|\det(\mathcal{D}_{P_2})\|_{\mathbf{z}}} = \frac{\|\det(\mathcal{D}_{P_1})\|_C}{\|\det(\mathcal{D}_{P_2})\|_C}. \tag{10}$$

That is,

$$\frac{\det_{\mathbf{z}}(\Delta_{P_1})}{\det_{\mathbf{z}}(\Delta_{P_2})} = \frac{\det_C(\Delta_{P_1})}{\det_C(\Delta_{P_2})}. \tag{11}$$

Equivalently, since $S(P(\mathcal{D})) = Id$,

$$\det_{\mathbf{z}}(\Delta_P) = \det_{\mathbf{z}}(\Delta_{P(\mathcal{D})})\det_C(\Delta_P). \tag{12}$$

The canonical metric is the natural metric on DET $(\mathbb{D}, \mathbb{P})$ induced from the Hermitian metrics on the bundles $\mathcal{K}(\mathbb{D})$ and $\mathcal{W}$. We therefore obtain by functoriality a canonical connection on DET $(\mathbb{D}, \mathbb{P})$ compatible with $\|.\|_C$, defined over U by

$$\nabla^{C,\mathbb{P}} \det(\mathcal{D}_P) = \operatorname{Tr}_C(\mathcal{D}_P^{-1}\nabla\mathcal{D}_P)\det(\mathcal{D}_P),$$

where[1] $\operatorname{Tr}_C(\mathcal{D}_P^{-1}\nabla\mathcal{D}_P) := \operatorname{tr}_K\big(S(P)^{-1}\nabla^{K,\mathcal{W}}S(P)\big)$. Here $\nabla^{\mathcal{K},\mathcal{W}}$ is the induced connection on $\operatorname{Hom}(\mathcal{K}(\mathbb{D}), \mathcal{W})$,

$$\nabla^{\mathcal{K},\mathcal{W}}(B)(\xi) = \nabla^{\mathcal{W}}(B(\xi)) - B\nabla^{\mathcal{K}}\xi, \tag{13}$$

for $B \in \operatorname{Hom}(\mathcal{K}(\mathbb{D}), \mathcal{W})$, where $\nabla^{\mathcal{K}}, \nabla^{\mathcal{W}}$ are the connections of Proposition 1.

On the other hand, we can use $\mathbb{P}$ to define a modified Bismut connection $\widetilde{\nabla}^{\mathbb{P}}$ on $\operatorname{Hom}(\mathcal{H}, \mathcal{H}_{\mathbb{P}})$. A **z**-function connection on DET $(\mathbb{D}, \mathbb{P})$ can then be defined over U analogously to [3, 6] by setting

$$\frac{\nabla^{\mathbf{z},\mathbb{P}} \det \mathcal{D}_P}{\det \mathcal{D}_P} = \operatorname{Tr}_{\mathbf{z}}(\mathcal{D}_P^{-1}\nabla\mathcal{D}_P) := \frac{d}{ds}_{|s=0}(s\theta_{\mathbb{P}}(s)) \tag{14}$$

where, with $\widetilde{\Delta}_P = \mathcal{D}_P\mathcal{D}_{P^*}^*$,

$$\theta_{\mathbb{P}}(s) = -\operatorname{Tr}(\widetilde{\Delta}_P^{-s}\mathcal{D}\widetilde{\nabla}^{\mathbb{P}}\mathcal{D}_P^{-1}),$$

is defined around zero by analytic continuation[2].

Theorem 2 *Let* $\mathbb{P}^1$, $\mathbb{P}^2$ *be choices of Grassmann sections. Let* $\Omega_C^{\mathbb{P}^1}, \Omega_{\mathbf{z}}^{\mathbb{P}^1}$ *be the curvature 2-forms of the canonical and zeta connection on DET* $(\mathbb{D}, \mathbb{P}^1)$, *and let* $\Omega_C^{\mathbb{P}^2}, \Omega_{\mathbf{z}}^{\mathbb{P}^2}$ *be curvature forms on DET* $(\mathbb{D}, \mathbb{P}^2)$. *Then one has*

$$\Omega_{\mathbf{z}}^{\mathbb{P}^1} - \Omega_{\mathbf{z}}^{\mathbb{P}^2} = \Omega_C^{\mathbb{P}^1} - \Omega_C^{\mathbb{P}^2}. \tag{15}$$

Equivalently,

$$\Omega_{\mathbf{z}}^{\mathbb{P}} = \Omega_{\mathbf{z}}^{P(\mathbb{D})} + \Omega_C^{\mathbb{P}} \tag{16}$$

$$= \Omega_{\mathbf{z}}^{P(\mathbb{D})} + \operatorname{tr}(\mathbf{R}^{\mathcal{W}}) - \operatorname{tr}(\mathbf{R}^{\mathcal{K}(\mathbb{D})}). \tag{17}$$

---

[1] Throughout, $\operatorname{tr} = \operatorname{tr}_V$ denotes the trace on a finite-dimensional vector space V, Tr an operator trace, and Tr_C, $\operatorname{Tr}_{\mathbf{z}}$ the canonical and zeta regularized traces.

[2] We differ from [3] by a sign since we use the form on the dual bundle.

The second identity (16), which says that $\Omega_{\mathbf{z}}^{\mathbb{P}}$ consists of an interior part plus a boundary correction term, follows from (15) because $\nabla^{C,P(\mathbb{D})}$ is the trivial connection. (17) then follows from Proposition 1 and the definition of $\nabla^{C,\mathbb{P}}$.

As an example, consider the case of the 'universal' family of EBVPs

$$(\mathbb{D},\mathbb{P}) = \{\mathcal{D}_P : P \in Gr(E_0 \oplus E_\beta)\}$$

relative to a fixed operator $\mathcal{D}$. Let $\Omega_{\mathbf{z}}$ be the $\mathbf{z}$ curvature of the corresponding determinant bundle. Then the first and third terms in (17) vanish, and we obtain:

Corollary 1

$$\Omega_{\mathbf{z}} = i\omega_{Gr},$$

where ω_{Gr} is the Kahler form on the Grassmannian. □

2 Relative Zeta-Function Metric: Proof of Theorem 1

For smooth sections ψ, ϕ of E one has the Green's form

$$< \mathcal{D}\psi, \phi >_X - < \psi, \mathcal{D}^*\phi >_X = < \sigma\gamma\psi, \gamma\phi >, \tag{18}$$

where, if $A(x)$ is the leading coefficient of $\mathcal{D}$, $\sigma = -A(0) \oplus A(\beta) \in \mathrm{Gl}(E_0 \oplus E_\beta)$. By definition, (18) vanishes for all $\psi \in \mathrm{dom}\,(\mathcal{D}_P)$ if and only if $\phi \in \mathrm{dom}\,(\mathcal{D}^*_{P^*})$, where P^* denotes the adjoint boundary problem. If $A(x)$ is unitary then

$$P^* = \sigma(I - P)\sigma^{-1} \tag{19}$$

(cf. [4]). In order to simplify some of the formulas, we will assume this to be the case, so that (19) holds, but this assumption is easily removed.

To study the Laplacian boundary problem

$$\begin{cases} \Delta_P = \mathcal{D}^*\mathcal{D} : \mathrm{dom}\,(\Delta_P) \longrightarrow L^2(X;E) \\ \mathrm{dom}(\Delta_P) = \{\psi \in H^2(X;E) \mid P^*\gamma\mathcal{D}\psi = 0,\ P\gamma\psi = 0\}\,, \end{cases}$$

observe that $\mathrm{dom}(\Delta_P)$ is a subspace of the domain of the first-order EBVP

$$\widehat{\Delta}_P = \widehat{\Delta} : \mathrm{dom}(\widehat{\Delta}_P) \to L^2(X; E \oplus E)\,.$$

Here

$$\widehat{\Delta} := \begin{pmatrix} \mathcal{D} & -I \\ 0 & \mathcal{D}^* \end{pmatrix} : H^1(X; E \oplus E) \longrightarrow L^2(X; E \oplus E),$$

with $\mathrm{dom}(\widehat{\Delta}_P) = \{(\psi, \phi) \in H^1(X; E \oplus E) \mid \widehat{P}\widehat{\gamma}(\psi, \phi) = 0\}$, where $\widehat{P} := P \oplus P^*$ and $\widehat{\gamma}(\psi, \phi) := (\gamma\psi, \gamma\phi)$. The map $\psi \longmapsto \widehat{\psi} = (\psi, \mathcal{D}\psi)$ defines a canonical embedding $H^2(X; E) \longrightarrow H^1(X; E \oplus (X; E)$ and we have

$$\widehat{\Delta}\widehat{\psi} = \begin{pmatrix} 0 \\ \Delta\psi \end{pmatrix}, \tag{20}$$

identifying the solution spaces of the operators Δ and $\widehat{\Delta}$: if $\{\psi_1, \ldots, \psi_k\}$ is a basis for $\mathrm{Ker}(\Delta)$ then $\{\widehat{\psi}_1, \ldots, \widehat{\psi}_k\}$ is a basis for $\mathrm{Ker}(\widehat{\Delta})$. Moreover, there is a preferred such basis formed by the columns of the fundamental solution matrix $\widehat{h}(x)\colon E_0 \oplus E_0 \to E_x \oplus E_x$ for $\widehat{\Delta}$, solving uniquely $\widehat{\Delta}\widehat{h}(x) = 0,\ \widehat{h}(0) = I$. We define

$$S(\widehat{P}) := \widehat{P} \circ P(\widehat{\Delta})\colon K(\widehat{\Delta}) \longrightarrow \widehat{W} = \mathrm{range}\,(\widehat{P})\,, \tag{21}$$

where $K(\widehat{\Delta}) = \mathrm{graph}(\widehat{h} = \widehat{h}(\beta)\colon E_0 \oplus E_0 \to E_\beta \oplus E_\beta)$ is the Cauchy space for $\widehat{\Delta}$.

For a linear operator $A\colon E_0 \oplus E_1 \to F_0 \oplus F_1$ considered as a block 2×2 matrix relative to the direct sums, $[A]_{(1,2)}\colon E_1 \to F_0$ refers to the top-right entry in the $(1,2)$ position. From (20)

$$(\Delta_P - \lambda)^{-1} = \left[\widehat{\Delta}_{P,\lambda}^{-1}\right]_{(1,2)}, \tag{22}$$

where $\widehat{\Delta}_{P,\lambda} = \widehat{\Delta}_\lambda = \begin{pmatrix} \mathcal{D} & -I \\ -\lambda & \mathcal{D}^* \end{pmatrix}$, with domain $\mathrm{dom}\,(\widehat{\Delta}_P)$. Indeed, we compute

$$\widehat{\Delta}_{P,\lambda}^{-1} = \begin{pmatrix} \mathcal{D}_P^*(\tilde{\Delta}_P - \lambda)^{-1} & (\Delta_P - \lambda)^{-1} \\ \lambda(\tilde{\Delta}_P - \lambda)^{-1} & \mathcal{D}_P(\Delta_P - \lambda)^{-1} \end{pmatrix}, \tag{23}$$

where $\tilde{\Delta} = \mathcal{D}^*\mathcal{D}$, $\tilde{\Delta}_P = \mathcal{D}_P\mathcal{D}_P^*$.

The Poisson operator of Δ is the operator

$$\begin{aligned} &\widehat{\mathcal{K}}\colon (E_0 \oplus E_0) \oplus (E_\beta \oplus E_\beta) \to C^\infty(X, E), \\ &\widehat{\mathcal{K}}(v)(x) = \widehat{h}(x)p_0 P(\widehat{\Delta})v, \end{aligned} \tag{24}$$

where p_0 is the projection map $(E_0 \oplus E_0) \oplus (E_\beta \oplus E_\beta) \to (E_0 \oplus E_0)$. The restriction of $\widehat{\mathcal{K}}$ to $K(\widehat{\Delta})$ is an isomorphism $\widehat{\mathcal{K}}\colon K(\widehat{\Delta}) \longrightarrow \mathrm{Ker}\widehat{\Delta} \cong \mathrm{Ker}\Delta$ while

$$\widehat{\gamma} \circ \widehat{\mathcal{K}} = P(\widehat{\Delta}) \tag{25}$$

as operators on $(E_0 \oplus E_0) \oplus (E_\beta \oplus E_\beta)$.

The invertibility of the operators $\mathcal{D}_P, \Delta_P, \widehat{\Delta}_P, S(\widehat{P})$ are equivalent statements and in this case we can define the *Poisson operator* of Δ_P by

$$\widehat{\mathcal{K}}(\widehat{P}) = \widehat{\mathcal{K}}S(\widehat{P})^{-1}\widehat{P}\colon (E_0 \oplus E_0) \oplus (E_\beta \oplus E_\beta) \to C^\infty(X, E). \tag{26}$$

The restriction $\widehat{\mathcal{K}}(\widehat{P})\colon \text{range}\,(\widehat{P}) \to \text{Ker}(\widehat{\Delta})$ is an isomorphism with left-inverse $\widehat{P}\gamma_{|Ker(\widehat{\Delta})}$, for

$$\widehat{P}\gamma\widehat{\mathcal{K}}(\widehat{P}) = \widehat{P}\gamma\widehat{\mathcal{K}}S(\widehat{P})^{-1}\widehat{P} = P(\widehat{\Delta})S(\widehat{P})^{-1}\widehat{P} = \widehat{P}\,. \tag{27}$$

The following relative inverse formula holds:

Proposition 2 If $\Delta_{P_1}, \Delta_{P_2}$ are invertible, then

$$\Delta_{P_1}^{-1} = \Delta_{P_2}^{-1} - [\widehat{\mathcal{K}}(\widehat{P}_1)\widetilde{\gamma}\widehat{\Delta}_{P_2}^{-1}]_{(1,2)}. \tag{28}$$

In particular, $\Delta_{P_1}^{-1} - \Delta_{P_2}^{-1}$ is a smoothing operator.

Proof We have

$$\widehat{\Delta}_P^{-1}\widehat{\Delta} = I - \widehat{\mathcal{K}}(\widehat{P})\gamma \tag{29}$$

and hence

$$\begin{aligned}\Delta_{P_1}^{-1} &= [\widehat{\Delta}_{P_1}^{-1}]_{(1,2)} = [\widehat{\Delta}_{P_1}^{-1}\widehat{\Delta}\widehat{\Delta}_{P_2}^{-1}]_{(1,2)}\\ &= [(I - \widehat{\mathcal{K}}(\widehat{P}_1)\gamma)\widehat{\Delta}_{P_2}^{-1}]_{(1,2)} = \Delta_{P_2}^{-1} - [\widehat{\mathcal{K}}(\widehat{P}_1)\widetilde{\gamma}\widehat{\Delta}_{P_2}^{-1}]_{(1,2)}\,.\end{aligned}$$

To see (29), one can either check it directly using (41), or invariantly as in [8]. □

For later use, note that there is a similar relative inverse for the EBVP $\mathcal{D}_P$. $\mathcal{D}$ has Poisson operator

$$\mathcal{K}\colon E_0 \oplus E_\beta \longrightarrow C^\infty(X, E), \quad \mathcal{K}(u)(x) = h(x)p_0 P(\mathcal{D})u,$$

with p_0 the projection map $E_0 \oplus E_\beta \to E_0$, which restricts to an isomorphism $\mathcal{K}\colon K(\mathcal{D}) \to \text{Ker}(\mathcal{D})$. If $\mathcal{D}_P$ is invertible

$$\mathcal{D}_P^{-1}\mathcal{D} = I - \mathcal{K}(\widehat{P})\gamma\,, \tag{30}$$

where $\mathcal{K}(P) = \mathcal{K}S(P)^{-1}P\colon E_0 \oplus E_\beta \longrightarrow C^\infty(X, E)$, and by a similar argument to Proposition 2

$$\mathcal{D}_{P_1}^{-1} = \mathcal{D}_{P_2}^{-1} - \mathcal{K}(P_1)\gamma\mathcal{D}_{P_2}^{-1}\,. \tag{31}$$

2.1 Stiefel Coordinates

An element of $\mathrm{Hom}(E_0 \oplus E_\beta, E_0)$ can be written $M = \begin{bmatrix} M_0 & M_\beta \end{bmatrix}$ where $M_0 \in \mathrm{Hom}(E_0, E_0)$ and $M_\beta \in \mathrm{Hom}(E_\beta, E_0)$. The complex Stiefel manifold St_k parameterizes elements of $\mathrm{Hom}(E_0 \oplus E_\beta, E_0)$ of rank k (at least one invertible $k \times k$ minor), and the projection map

$$\begin{aligned} \pi : St_k &\longrightarrow \mathrm{Gr}_k(E_0 \oplus E_\beta), \\ M &\longmapsto P_{[M_0, M_\beta]} = \begin{pmatrix} M_0^* M_{0,\beta}^{-1} M_0 & M_0^* M_{0,\beta}^{-1} M_\beta \\ M_\beta^* M_{0,\beta}^{-1} M_0 & M_\beta^* M_{0,\beta}^{-1} M_\beta \end{pmatrix}, \end{aligned} \tag{32}$$

where $M_{0,\beta} := M M^* = M_0 M_0^* + M_\beta M_\beta^*$, defines St_k as a principal $Gl(\mathbb{C}^k)$ bundle over $\mathrm{Gr}_k(E_0 \oplus E_\beta)$, the Stiefel frame bundle. Over the index zero component of the Grassmannian $\mathrm{dom}(\mathcal{D}_P)$ has the following description in Stiefel coordinates $[M_0, M_\beta]$:

Lemma 1 For $P = P_{[M_0, M_\beta]} \in \mathrm{Gr}_n(E_0 \oplus E_\beta)$ one has

$$\mathrm{dom}(\mathcal{D}_P) = \{\psi \in H^1(X; E) \mid M_0 \psi(0) + M_\beta \psi(\beta) = 0\}. \tag{33}$$

Proof The lemma states that

$$P \begin{pmatrix} \psi(0) \\ \psi(\beta) \end{pmatrix} = 0 \quad \text{and} \quad M_0 \psi(0) + M_\beta \psi(\beta) = 0 \tag{34}$$

are the same statement. But from (32) the first equality gives

$$\begin{cases} M_0^* M_{0,\beta}^{-1} (M_0 \psi(0) + M_\beta \psi(\beta)) = 0 \\ M_\beta^* M_{0,\beta}^{-1} (M_0 \psi(0) + M_\beta \psi(\beta)) = 0, \end{cases}$$

while multiplying these equations respectively by M_0 and M_β and summing them is the second equation in (34). The reverse implication is obvious. □

Note that (33) has a $\mathrm{Gl}(\mathbb{C}^n)$'s worth of ambiguity in describing $\mathrm{dom}(\mathcal{D}_P)$ corresponding to a choice of generator $[M_0 \;\; M_\beta]$ in the fibre of St_n over P.

We have the following Stiefel coordinate formula for the canonical metric:

Proposition 3 Let $P = P_{[M_0, M_\beta]} \in \mathrm{Gr}_n(E_0 \oplus E_\beta)$ and let $\mathcal{M} = M_0 + M_\beta h \in \mathrm{End}(E_0)$. Then

$$\mathrm{det}_C \Delta_P = \det Q_h^{-1} \det M_{0,\beta}^{-1} \, |\det \mathcal{M}|^2. \tag{35}$$

Proof We have $S(P)^*S(P) = P(\mathcal{D})PP(\mathcal{D})\colon K(\mathcal{D}) \to K(\mathcal{D})$ and $K(\mathcal{D}) = \{(\xi, h\xi)\colon \xi \in E_0\} \subset E_0 \oplus E_\beta$. $\mathrm{End}(E_0)$ acts on $K(\mathcal{D})$ by $q.(\xi, h\xi) = (q\xi, hq\xi)$. So, using (32),

$$P(\mathcal{D})PP(\mathcal{D})\begin{pmatrix}\xi \\ h\xi\end{pmatrix} = \begin{pmatrix} Q_h^{-1} & Q_h^{-1}h^* \\ hQ_h^{-1} & hQ_h^{-1}h^* \end{pmatrix}\begin{pmatrix} M_0^* M_{0,\beta}^{-1}\,\mathcal{M}\,\xi \\ M_\beta^* M_{0,\beta}^{-1}\,\mathcal{M}\,\xi \end{pmatrix}$$
$$= Q_h^{-1}\mathcal{M}^* M_{0,\beta}^{-1}\,\mathcal{M}\begin{pmatrix}\xi \\ h\xi\end{pmatrix}.$$

Hence $\det_{\mathbb{C}}\Delta = \det(Q_h^{-1}M_{0,\beta}^{-1}\mathcal{M}^*\mathcal{M})$, and we reach the conclusion. □

We also need a Stiefel coordinate formula for $\widehat{\Delta}_P^{-1}$. First:

Lemma 2 Let $P = P_{[M_0,M_\beta]}$. Then

$$\mathrm{dom}\,(\Delta_P) = \left\{\psi \in H^2(X;E) \mid \widehat{M}_0\begin{pmatrix}\psi(0) \\ \mathcal{D}\psi(0)\end{pmatrix} + \widehat{M}_\beta\begin{pmatrix}\psi(\beta) \\ \mathcal{D}\psi(\beta)\end{pmatrix} = 0\right\}, \tag{36}$$

where

$$\widehat{M}_0 = \begin{pmatrix} M_0^* M_{0,\beta}^{-1} M_0 & M_0^* M_{0,\beta}^{-1} M_0 A_0^{-1} - A_0^{-1} \\ M_\beta^* M_{0,\beta}^{-1} M_0 & M_\beta^* M_{0,\beta}^{-1} M_0 A_0^{-1} \end{pmatrix},$$
$$\widehat{M}_\beta = \begin{pmatrix} M_0^* M_{0,\beta}^{-1} M_\beta & -M_0^* M_{0,\beta}^{-1} M_\beta A_\beta^{-1} \\ M_\beta^* M_{0,\beta}^{-1} M_\beta & A_\beta^{-1} - M_\beta^* M_{0,\beta}^{-1} M_\beta A_\beta^{-1} \end{pmatrix}, \tag{37}$$

are canonically defined by $\widehat{P}$. Here $A_0 := A(0)$, $A_\beta := A(\beta)$. With respect to the decomposition $(E_0 \oplus E_0) \oplus (E_\beta \oplus E_\beta)$ of the space of boundary data, one has

$$\widehat{P} = P_{[\widehat{M}_0,\widehat{M}_\beta]}. \tag{38}$$

Proof From (19) we have $P\gamma\psi = 0,\ P^*\gamma\mathcal{D}\psi = 0$ is equivalent to

$$P\begin{pmatrix}\psi(0) \\ \psi(\beta)\end{pmatrix} + \sigma^{-1}P^*\begin{pmatrix}\mathcal{D}\psi(0) \\ \mathcal{D}\psi(\beta)\end{pmatrix} = 0. \tag{39}$$

But $\sigma = -A_0 \oplus A_\beta$, and from (32) we obtain (37) by substituting in (39). The identity (38) follows as in Lemma 1. □

Let $\widehat{h}_\lambda(x)\colon E_0 \oplus E_0 \to E_x \oplus E_x$ be the fundamental solution matrix for $\widehat{\Delta}_\lambda$,

$$\widehat{\Delta}_\lambda \widehat{h}_\lambda(x) = 0, \quad \widehat{h}_\lambda(0) = I. \tag{40}$$

Then we have:

Proposition 4 Let $P = P_{[M_0,M_\beta]}$. Then $\Delta_{P,\lambda}$ is invertible if and only if

$$\widehat{\mathcal{M}}_\lambda = \widehat{M}_0 + \widehat{M}_\beta \widehat{h}_\lambda,$$

is invertible, and in that case $\Delta_{P,\lambda}^{-1}$ has kernel

$$k_{P,\lambda}(x,y) = \begin{cases} -\Big[\widehat{h}_\lambda(x)(\widehat{\mathcal{M}}_\lambda^{-1}\widehat{M}_\beta\widehat{h}_\lambda)\widehat{h}_\lambda(y)^{-1}\widehat{A}(y)^{-1}\Big]_{(1,2)} & x < y\,, \\ \Big[\widehat{h}_\lambda(x)(I - \widehat{\mathcal{M}}_\lambda^{-1}\widehat{M}_\beta\widehat{h}_\lambda)\widehat{h}_\lambda(y)^{-1}\widehat{A}(y)^{-1}\Big]_{(1,2)} & x > y\,, \end{cases} \tag{41}$$

where $\widehat{A}(x) = A(x) \oplus -A^*(x)$. In particular, if $P_1 = P_{[M_0,M_\beta]}$, $P_2 = P_{[N_0,N_\beta]}$ then $\Delta_{P_1,\lambda}^{-1} - \Delta_{P_2,\lambda}^{-1}$ has smooth kernel

$$-\Big[\widehat{h}_\lambda(x)(\widehat{\mathcal{M}}_\lambda^{-1}\widehat{M}_\beta - \widehat{\mathcal{N}}_\lambda^{-1}\widehat{N}_\beta)\widehat{h}_\lambda\widehat{h}_\lambda(y)^{-1}\widehat{A}(y)^{-1}\Big]_{(1,2)}.$$

Proof For each fixed y, $\widehat{k}_P(x,y)$ must satisfy $\widehat{P}\begin{pmatrix}\widehat{k}_P(0,y)v \\ \widehat{k}_P(\beta,y)v\end{pmatrix} = 0$, for all $v \in E_y$. By Lemma 2 this is equivalent to $\widehat{M}_0\widehat{k}_P(0,y) + \widehat{M}_\beta\widehat{k}_P(\beta,y) = 0$. On the other hand, from (40) we have $\widehat{\Delta}_{P,\lambda}^{-1} = \widehat{h}_\lambda(x)\big(\frac{d}{dx}\big)_{P'}^{-1}\widehat{h}_\lambda(x)^{-1}\widehat{A}(x)^{-1}$, where P' is the gauge transformed boundary condition with respect to $\widehat{h}_\lambda^{-1}$. Since the derivative of the Heaviside function is the Dirac delta distribution (41) now follows.

For the first statement, note that $\Delta_{P,\lambda}$ is invertible if and only if $S_\lambda(\widehat{P}) = \widehat{P} \circ P(\widehat{\Delta}_\lambda)$ is invertible, while by a direct computation

$$S_\lambda(\widehat{P})^{-1} = \begin{pmatrix} \widehat{\mathcal{M}}_\lambda^{-1}\widehat{M}_0 & \widehat{\mathcal{M}}_\lambda^{-1}\widehat{M}_\beta \\ \widehat{h}\widehat{\mathcal{M}}_\lambda^{-1}\widehat{M}_0 & \widehat{h}\widehat{\mathcal{M}}_\lambda^{-1}\widehat{M}_\beta \end{pmatrix}. \tag{42}$$

□

Note that the Stiefel coordinate formula (41) also follows from (28) by a direct substitution using (42).

2.2 *Proof of Theorem 1*

We know that $(\Delta_P - \lambda)^{-1}$ is trace class. In fact, there is the following precise formula:

Proposition 5

$$\mathrm{Tr}\,(\Delta_P - \lambda)^{-1} = -\frac{\partial}{\partial\lambda}\log\det\widehat{\mathcal{M}}_\lambda, \tag{43}$$

Proof The proof follows Prop. 3.1 of Lesch and Tolksdorf's article [5].

For $C(x)\colon \mathbb{C}^n \oplus \mathbb{C}^n \to \mathbb{C}^n \oplus \mathbb{C}^n$ one has

$$\mathrm{tr}[C(x)]_{1,2} = -\mathrm{tr}(\frac{\partial}{\partial\lambda}(\widehat{\Delta}_\lambda)C(x)),$$

while from (40)

$$\frac{\partial}{\partial\lambda}\widehat{\Delta}_\lambda \cdot \widehat{h}_\lambda(x) = -\widehat{\Delta}_\lambda \frac{\partial}{\partial\lambda}\widehat{h}_\lambda(x).$$

Therefore

$$\begin{aligned}
&\mathrm{Tr}\,(\Delta_P - \lambda)^{-1}\\
&= \int_0^\beta \mathrm{tr}\{k_{P,\lambda}(x,x)\}\,dx\\
&= \int_0^\beta \mathrm{tr}\left\{\left[\widehat{h}_\lambda(x)\widehat{\mathcal{M}}_\lambda^{-1}\widehat{M}_\beta \widehat{h}_\lambda \widehat{h}_\lambda(x)^{-1}\widehat{A}(x)^{-1}\right]_{(1,2)}\right\} dx\\
&= -\int_0^\beta \mathrm{tr}\left\{\frac{\partial}{\partial\lambda}\widehat{\Delta}_\lambda \widehat{h}_\lambda(x)\widehat{\mathcal{M}}_\lambda^{-1}\widehat{M}_\beta \widehat{h}_\lambda \widehat{h}_\lambda(x)^{-1}\widehat{A}(x)^{-1}\right\} dx\\
&= -\int_0^\beta \mathrm{tr}\left\{\widehat{\Delta}_\lambda \frac{\partial}{\partial\lambda}(\widehat{h}_\lambda(x))\widehat{\mathcal{M}}_\lambda^{-1}\widehat{M}_\beta \widehat{h}_\lambda \widehat{h}_\lambda(x)^{-1}\widehat{A}(x)^{-1}\right\} dx\\
&= -\int_0^\beta \mathrm{tr}\left\{(\widehat{h}_\lambda(x)^{-1}\widehat{A}(x)^{-1}\widehat{\Delta}_\lambda \widehat{h}_\lambda(x)).\widehat{h}_\lambda(x)^{-1}\frac{\partial}{\partial\lambda}(\widehat{h}_\lambda(x))\widehat{\mathcal{M}}_\lambda^{-1}\widehat{M}_\beta \widehat{h}_\lambda\right\} dx\\
&= -\int_0^\beta \mathrm{tr}\left\{\frac{d}{dx}\left(\widehat{h}_\lambda(x)^{-1}\frac{\partial}{\partial\lambda}(\widehat{h}_\lambda(x))\widehat{\mathcal{M}}_\lambda^{-1}\widehat{M}_\beta \widehat{h}_\lambda\right)\right\} dx\\
&= -\left[\mathrm{tr}\left\{\widehat{h}_\lambda(x)^{-1}\frac{\partial}{\partial\lambda}(\widehat{h}_\lambda(x))\widehat{\mathcal{M}}_\lambda^{-1}\widehat{M}_\beta \widehat{h}_\lambda\right\}\right]_{x=0}^{\beta}\\
&= -\mathrm{tr}\left\{\widehat{h}_\lambda^{-1}\frac{\partial}{\partial\lambda}(\widehat{h}_\lambda)\widehat{\mathcal{M}}_\lambda^{-1}\widehat{M}_\beta \widehat{h}_\lambda\right\} \quad \text{since } \frac{\partial}{\partial\lambda}\widehat{h}_\lambda(0) = 0\\
&= -\frac{\partial}{\partial\lambda}\log\det\widehat{\mathcal{M}}_\lambda. \quad \square
\end{aligned}$$

If $\Delta_{P_1}, \Delta_{P_2}$ are invertible, with $P_1 = P_{[M_0,M_\beta]}$, $P_2 = P_{[N_0,N_\beta]}$, then from (43) we have

$$\mathrm{Tr}\left((\Delta_{P_1} - \lambda)^{-1} - (\Delta_{P_2} - \lambda)^{-1}\right) = -\frac{\partial}{\partial\lambda}\log\frac{\det\widehat{\mathcal{M}}_\lambda}{\det\widehat{\mathcal{N}}_\lambda}, \tag{44}$$

so the scattering matrix is $S_\lambda = \widehat{\mathcal{N}}_\lambda^{-1}\widehat{\mathcal{M}}_\lambda$. Because the boundary problems Δ_{P_i} are elliptic in the classical sense of Seeley [9], they have asymptotic expansions as $\lambda \to \infty$ in $\Lambda_{\pi,\varepsilon}$

$$\operatorname{Tr}(\Delta_{P_i} - \lambda)^{-1} \sim c_{-1}^{(i)}(-\lambda)^{1/2} + \sum_{k\geq 1} c_k^{(i)}(-\lambda)^{-k/2}. \tag{45}$$

Hence we find that $\Delta_{P_1}, \Delta_{P_2}$ are $\mathbf{z}$-comparable and $\mathbf{z}$-admissible, and so by (6) (or by [5]) we have

$$\frac{\det_{\mathbf{z}}(\Delta_{P_1})}{\det_{\mathbf{z}}(\Delta_{P_2})} = \frac{\det\widehat{\mathcal{M}}}{\det\widehat{\mathcal{N}}}.\exp\left[-\mathrm{LIM}_{\lambda\to\infty}\log\det\frac{\det\widehat{\mathcal{M}}_{-\lambda}}{\det\widehat{\mathcal{N}}_{-\lambda}}\right]. \tag{46}$$

Proposition 6

$$\det\widehat{\mathcal{M}} = \det(A_0)^{-1}\det(h^*)^{-1}\det M_{0,\beta}^{-1}|\det\mathcal{M}|^2 \tag{47}$$

Proof We compute that

$$\widehat{h}(x) = \begin{pmatrix} h(x) & J(x) \\ 0 & A(x)(h(x)^*)^{-1}A_0^{-1} \end{pmatrix},$$

where $h(x)$ is the parallel transport for $\mathcal{D}$, and $J(x)$ is the unique solution to

$$\mathcal{D}J(x) = A(x)(h(x)^*)^{-1}A_0^{-1}, \quad J(0) = 0.$$

Hence since $\widehat{h} = \widehat{h}(\beta)$, and setting $J := J(\beta)$, we have from Lemma 2

$$\begin{aligned}\widehat{\mathcal{M}} &= \begin{pmatrix} M_0^*M_{0,\beta}^{-1}\mathcal{M} & M_0^*M_{0,\beta}^{-1}M_\beta J + M_0^*M_{0,\beta}^{-1}(M_0 - M_\beta(h^*)^{-1})A_0^{-1} - A_0^{-1} \\ M_\beta^*M_{0,\beta}^{-1}\mathcal{M} & M_0^*M_{0,\beta}^{-1}M_\beta J + M_\beta^*M_{0,\beta}^{-1}(M_0 - M_\beta(h^*)^{-1}) + (h^*)^{-1}A_0^{-1} \end{pmatrix} \\ &= \begin{pmatrix} 0 & I \\ (h^*)^{-1}\mathcal{M}^* & -(h^*)^{-1} \end{pmatrix} \\ &\quad\times \begin{pmatrix} I & M_{0,\beta}^{-1}(M_\beta J + M_0 - M_\beta(h^*)^{-1}) \\ M_0^* & M_0^*M_{0,\beta}^{-1}(M_\beta J + M_0 - M_\beta(h^*)^{-1}) - I \end{pmatrix}\begin{pmatrix} M_{0,\beta}^{-1}\mathcal{M} & 0 \\ 0 & A_0^{-1} \end{pmatrix}. \end{aligned} \tag{48}$$

Since a 2×2 block matrix $\left(\begin{smallmatrix} A & B \\ C & D \end{smallmatrix}\right): E_0 \oplus E_1 \longrightarrow E_0 \oplus E_1$ where $A: E_0 \to E_0$, $B: E_1 \to E_0$ etc, has determinant $\det(A).\det(D - CA^{-1}B)$ if A is invertible, and determinant $\det(D).\det(A - BD^{-1}C)$ if D is invertible, we find that the determinant of the second matrix in (48) reduces to $\det(-I)$, in particular the J term disappears, and term by term we obtain

$$\det\widehat{\mathcal{M}} = \det(-(h^*)^{-1}\mathcal{M}^*).\det(-I).\Big(\det(M_{0,\beta}^{-1}\mathcal{M})\det(A_0)^{-1}\Big),$$

and this is (47). $\square$

In view of (35) and (47), we can rewrite (46) as

$$\frac{\det_{\mathbf{z}}(\Delta_{P_1})}{\det_{\mathbf{z}}(\Delta_{P_2})} = \frac{\det_C(\Delta_{P_1})}{\det_C(\Delta_{P_2})} \exp\left(-\mathrm{LIM}_{\lambda\to\infty} \log\det \frac{\det \widehat{\mathcal{M}}_{-\lambda}}{\det \widehat{\mathcal{N}}_{-\lambda}}\right). \tag{49}$$

It remains to show that the LIM term vanishes. Consider first the case where $\mathcal{D}$ is self-adjoint. Then $h(x) \in U(n)$ and we can gauge transform Δ_P to $\Delta^0_{U^{-1}PU}$, where Δ^0 is a flat Laplacian and $U = \gamma h(x) = I \oplus h$. By continuity it is enough to work over the dense open subset $U_{Gl} \subset Gr_{2n}(E_0 \oplus E_\beta)$, parameterizing graphs of invertible $T\colon E_0 \to E_\beta$, defining Stiefel coordinates $M_0 = I$ and $M_\beta := T^*$. Set $P_1 := P_T$. Then $\det \widehat{M}_\beta = \det(Q_T^{-1}T^*)$ and so $\widehat{M}_\beta$ is invertible, and we have

$$\log\det \widehat{\mathcal{M}}_\lambda = \log\det \widehat{M}_\beta + \log\det(\widehat{M}_\beta^{-1}\widehat{M}_0 + \widehat{h}_\lambda).$$

A similar computation to [5], Prop. 3.4, yields for $\lambda \longrightarrow \infty$

$$\log\det(\widehat{M}_\beta^{-1}\widehat{M}_0 + \widehat{h}_\lambda) = n\beta\sqrt{\lambda} + \log 2^{-n} - \log\det \widehat{M}_\beta + O(\frac{1}{\sqrt{\lambda}}),$$

and hence that $\log\det \widehat{\mathcal{M}}_\lambda = n\beta\sqrt{\lambda} + \log 2^{-n} + O(1/\sqrt{\lambda})$. Repeating the argument for $\widehat{\mathcal{N}}_\lambda$ we therefore have

$$\log\det \frac{\det \widehat{\mathcal{M}}_{-\lambda}}{\det \widehat{\mathcal{N}}_{-\lambda}} = O(\frac{1}{\sqrt{\lambda}}),$$

as $\lambda \longrightarrow \infty$, and hence (49) reduces to (11).

We extend this to general $\mathcal{D}$ through a variational method:

Proposition 7 Let $\mathcal{D}^r$ be a one parameter family of Dirac operators. Let $P_1, P_2 \in Gr_n(E_0 \oplus E_\beta)$ with $\mathcal{D}^r_{P_i}$ invertible. Then the $\Delta_{P_i} = \Delta^r_{P_i}$ are invertible and

$$\frac{d}{dr}\log\frac{\det_{\mathbf{z}}(\Delta_{P_1})}{\det_{\mathbf{z}}(\Delta_{P_2})} = \frac{d}{dr}\log\frac{\det_C(\Delta_{P_1})}{\det_C(\Delta_{P_2})}. \tag{50}$$

Proof Let $h_r, \widehat{h}_r$ denote the respective paths of parallel transport operators of the first-order elliptic operators $\mathcal{D}^r, \widehat{\Delta}^r$. Let $\mathcal{M}_r = M_0 + M_\beta h_r$, $\mathcal{N}_r = N_0 + N_\beta h_r$, and let $\widehat{\mathcal{M}}_r = \widehat{M}_0 + \widehat{M}_\beta \widehat{h}_r$, $\widehat{\mathcal{N}}_r = \widehat{N}_0 + \widehat{N}_\beta \widehat{h}_r$. Then, from (47),

$$\frac{d}{dr}\log\frac{\det_C(\Delta_{P_1})}{\det_C(\Delta_{P_2})} = \frac{d}{dr}\log\frac{\det(\mathcal{M}_r^*\mathcal{M}_r)}{\det(\mathcal{N}_r^*\mathcal{N}_r)}. \tag{51}$$

On the other hand, a straightforward application of Duhamel's Principle yields

$$\frac{d}{dr}\log\frac{\det_{\mathbf{z}}(\Delta_{P_1})}{\det_{\mathbf{z}}(\Delta_{P_2})} = \mathrm{Tr}\left\{\dot{\mathcal{D}}^*\left((\mathcal{D}^*_{P_1^*})^{-1} - (\mathcal{D}^*_{P_2^*})^{-1}\right)\right\} + \mathrm{Tr}\left\{\dot{\mathcal{D}}\{\mathcal{D}_{P_1}^{-1} - \mathcal{D}_{P_2}^{-1})\right\},$$

while from (23) and $\dot{\widehat{\Delta}} = \dot{\mathcal{D}} \oplus \dot{\mathcal{D}}^*$ we find

$$\dot{\widehat{\Delta}}\left(\widehat{\Delta}_{P_1}^{-1} - \widehat{\Delta}_{P_2}^{-1}\right) = \begin{pmatrix} \dot{\mathcal{D}}\left(\mathcal{D}_{P_1}^{-1} - \mathcal{D}_{P_2}^{-1}\right) & \Delta_{P_1}^{-1} - \Delta_{P_2}^{-1} \\ 0 & \dot{\mathcal{D}}^*\left((\mathcal{D}^*_{P_1^*})^{-1} - (\mathcal{D}^*_{P_2^*})^{-1}\right) \end{pmatrix},$$

and hence that

$$\frac{d}{dr} \log \frac{\det_{\mathbf{z}}(\Delta_{P_1})}{\det_{\mathbf{z}}(\Delta_{P_2})} = \mathrm{Tr}\left\{\dot{\widehat{\Delta}}\left(\widehat{\Delta}_{P_1}^{-1} - \widehat{\Delta}_{P_2}^{-1}\right)\right\}. \tag{52}$$

We have $\widehat{\Delta}_{P_1}^{-1} = \widehat{\Delta}_{P_2}^{-1} - \widehat{\mathcal{K}}_r(\widehat{P}_1)\widehat{\gamma}\widehat{\Delta}_{P_2}^{-1}$, where $\widehat{\mathcal{K}}_r(\widehat{P}_1)\widehat{\gamma} = \widehat{H}_r p_0 P(\widehat{\Delta}) S_r(\widehat{P}_1)^{-1}\widehat{P}_1\gamma$, and $(\widehat{H}_r v)(x) = \widehat{h}_r(x)v$, and since $\widehat{\Delta}^r\widehat{H}_r = 0$, then $\dot{\widehat{\Delta}}\widehat{H}_r = -\widehat{\Delta}\dot{\widehat{H}}_r$. Therefore using (29)

$$\begin{aligned}
&\mathrm{Tr}\left\{\dot{\widehat{\Delta}}\left(\widehat{\Delta}_{P_1}^{-1} - \widehat{\Delta}_{P_2}^{-1}\right)\right\} \\
&= -\mathrm{Tr}\left\{\dot{\widehat{\Delta}}\widehat{H}_r p_0 S_r(\widehat{P}_1)^{-1}\widehat{P}_1\gamma\widehat{\Delta}_{P_2}^{-1}\right\} \\
&= \mathrm{Tr}\left\{\widehat{\Delta}\dot{\widehat{H}}_r p_0 S_r(\widehat{P}_1)^{-1}\widehat{P}_1\gamma\widehat{\Delta}_{P_2}^{-1}\right\} \\
&= \mathrm{Tr}\left\{\widehat{P}_1\gamma\widehat{\Delta}_{P_2}^{-1}\widehat{\Delta}\dot{\widehat{H}}_r p_0 S_r(\widehat{P}_1)^{-1}\widehat{P}_1\right\} \\
&= \mathrm{Tr}\left\{\widehat{P}_1\gamma\left(I - \widehat{\mathcal{K}}_r(\widehat{P}_1)\widehat{\gamma}\right)\dot{\widehat{H}}_r p_0 S_r(\widehat{P}_1)^{-1}\widehat{P}_1\right\} \\
&= \mathrm{Tr}\left\{\widehat{P}_1\gamma\dot{\widehat{H}}_r p_0 S_r(\widehat{P}_1)^{-1}\widehat{P}_1\right\} - \mathrm{Tr}\left\{\widehat{P}_1\gamma\widehat{\mathcal{K}}_r(\widehat{P}_1)\widehat{\gamma}\dot{\widehat{H}}_r p_0 S_r(\widehat{P}_1)^{-1}\widehat{P}_1\right\} \\
&= \mathrm{Tr}\left\{S_r(\widehat{P}_1)^{-1}\widehat{P}_1\frac{d}{dr}P(\widehat{\Delta})\right\} - \mathrm{Tr}\left\{S_r(\widehat{P}_1)S_r(\widehat{P}_2)^{-1}\widehat{P}_2\frac{d}{dr}P(\widehat{\Delta})S_r(\widehat{P}_1)^{-1}\right\} \\
&= \mathrm{Tr}\left\{S_r(\widehat{P}_1)^{-1}\frac{d}{dr}S_r(\widehat{P}_1)\right\} - \mathrm{Tr}\left\{S_r(\widehat{P}_2)^{-1}\frac{d}{dr}S_r(\widehat{P}_2)\right\},
\end{aligned}$$

using the symmetry of the trace, and $\frac{d}{dr}S(\widehat{P}_i) = \widehat{P}_i\frac{d}{dr}P(\widehat{\Delta})$ and (25).

Now choose Stiefel coordinates for $P = P_{[M_0,M_\beta]}$. The finite-rank operator $S_r(\widehat{P})\colon K(\widehat{\Delta}^r) \to W \oplus W^*$, where $W = \mathrm{range}\,(P)$, $W^* = \mathrm{range}\,(P^*)$, has derivative

$$\frac{d}{dr}\left\{S_r(\widehat{P})\right\}v_r = \frac{d}{dr}\left\{S_r(\widehat{P})v_r\right\} - S_r(\widehat{P})\frac{d}{dr}v_r.$$

An element $v_r \in K(\widehat{\Delta}^r)$ has the form $(\xi, \widehat{h}_r\xi)$, $\xi \in E_0 \oplus E_0$, and so using Stiefel coordinate representation for $\widehat{P} = \widehat{P}_{[\widehat{M}_0,\widehat{M}_\beta]}$, we have

$$\begin{aligned}
\frac{d}{dr}\left\{S_r(\widehat{P})\right\}v_r &= \frac{d}{dr}\begin{pmatrix}\widehat{M}_0^*\widehat{M}_{0,\beta}^{-1}\widehat{\mathcal{M}}_r\xi \\ \widehat{M}_\beta^*\widehat{M}_{0,\beta}^{-1}\widehat{\mathcal{M}}_r\xi\end{pmatrix} - S_r(\widehat{P})\begin{pmatrix}0 \\ \dot{\widehat{h}}_r\xi\end{pmatrix} \\
&= \begin{pmatrix}\widehat{M}_0^*\widehat{M}_{0,\beta}^{-1}\left(\widehat{M}_\beta - \widehat{\mathcal{M}}_r Q_{\widehat{h}}^{-1}\widehat{h}^*\right)\dot{\widehat{h}}_r\xi \\ \widehat{M}_\beta^*\widehat{M}_{0,\beta}^{-1}\left(\widehat{M}_\beta - \widehat{\mathcal{M}}_r Q_{\widehat{h}}^{-1}\widehat{h}^*\right)\dot{\widehat{h}}_r\xi\end{pmatrix},
\end{aligned}$$

and so (42) implies $S_r(\widehat{P})^{-1}\frac{d}{dr}S_r(\widehat{P})v_r = \left\{\widehat{\mathcal{M}}_r^{-1}\left(\widehat{M}_\beta - \widehat{\mathcal{M}}_r Q_{\widehat{h}}^{-1}\widehat{h}^*\right)\dot{\widehat{h}}_r\right\}v_r$. Therefore

$$\begin{aligned}\mathrm{Tr}\left\{S_r(\widehat{P})^{-1}\frac{d}{dr}S_r(\widehat{P})\right\} &= \mathrm{tr}\left\{\widehat{\mathcal{M}}_r^{-1}\widehat{M}_\beta\dot{\widehat{h}}_r\right\} - \mathrm{tr}\left\{Q_{\widehat{h}}^{-1}\widehat{h}^*\dot{\widehat{h}}_r\right\} \\ &= \frac{d}{dr}\log\det\,\widehat{\mathcal{M}}_r - \mathrm{tr}\left\{Q_{\widehat{h}}^{-1}\widehat{h}^*\dot{\widehat{h}}_r\right\} \\ &= \frac{d}{dr}\log\det\,(\mathcal{M}_r^*\mathcal{M}_r) + \alpha(\widehat{\Delta}^r),\end{aligned}$$

where $\alpha(\widehat{\Delta}^r) = -\mathrm{tr}\{(h_r^* A_0^r)^{-1}\frac{d}{dr}(h_r^* A_0^r\} - \mathrm{tr}\{Q_{\widehat{h}}^{-1}\widehat{h}^*\dot{\widehat{h}}_r\}$, and we use (47). The term $\alpha(\widehat{\Delta}^r)$ depends only on the operator $\widehat{\Delta}^r$, not on $\widehat{P}$, and therefore

$$\mathrm{Tr}\left\{S_r(\widehat{P}_1)^{-1}\frac{d}{dr}S_r(\widehat{P}_1)\right\} - \mathrm{Tr}\left\{S_r(\widehat{P}_2)^{-1}\frac{d}{dr}S_r(\widehat{P}_2)\right\} = \frac{d}{dr}\log\frac{\det(\mathcal{M}_r^*\mathcal{M}_r)}{\det(\mathcal{N}_r^*\mathcal{N}_r)},$$

which completes the proof. □

Next, let $\mathcal{D}^r$ be a path of operators in $Ell_{1,n}$ connecting $\mathcal{D} := \mathcal{D}^1$ with a self-adjoint first-order elliptic operator $\mathcal{D}^0$ with $\mathcal{D}^0_{P_i}$ invertible. The path can always be chosen such that $\mathcal{D}^r_{P_1}$ is invertible for each r; equivalently, such that $\mathcal{M}_r$ is invertible for each r. If it occurs that an $\mathcal{N}_r$ is not invertible at some point along the path, then the path can be perturbed slightly to remove the singularity without affecting the invertibility of $\mathcal{M}_r$, since that is an open condition. Hence we can integrate (50) along this path to obtain

$$\frac{\det_{\mathbf{z}}(\Delta_{P_1})}{\det_{\mathbf{z}}(\Delta_{P_2})}\left(\frac{\det_{\mathbf{z}}(\Delta^0_{P_1})}{\det_{\mathbf{z}}(\Delta^0_{P_2})}\right)^{-1} = \frac{\det_{\mathcal{C}}(\Delta_{P_1})}{\det_{\mathcal{C}}(\Delta_{P_2})}\left(\frac{\det_{\mathcal{C}}(\Delta^0_{P_1})}{\det_{\mathcal{C}}(\Delta^0_{P_2})}\right)^{-1},$$

where $\Delta^0 = (\mathcal{D}^0)^*\mathcal{D}^0$. Since $\mathcal{D}^0$ is self-adjoint and we know that (10) holds for such operators, this completes the proof of Theorem 1.

3 Relative Zeta Function Curvature: Proof of Theorem 2

We define a **z**-function connection on DET $(\mathbb{D}, \mathbb{P})$ following a modified version of the prescription of Quillen–Bismut–Freed. The **z**-function connection form (14) is defined over $U \subset B$ by

$$\omega_{\mathbf{z}} := \frac{d}{ds}_{|s=0}(s\theta_{\mathbb{P}}(s)), \tag{53}$$

where $\theta_{\mathbb{P}}(s) = -\mathrm{Tr}\,(\widetilde{\Delta}_P^{-s}\mathcal{D}\widetilde{\nabla}^{\mathbb{P}}\mathcal{D}_P^{-1})$ is defined around zero by analytic continuation. Here $\widetilde{\nabla}^{\mathbb{P}}$ is a connection on the infinite-dimensional smooth bundle $\mathrm{Hom}(\mathcal{H}, \mathcal{H}_{\mathbb{P}})$

induced (13) from connections on $\mathcal{H}$ and $\mathcal{H}_{\mathbb{P}}$. Since $\mathcal{H}$ is the trivial bundle we can choose the trivial de-Rham connection 'd'. The bundle $\mathcal{H}_{\mathbb{P}}$, however, with fibre dom $_\infty(\mathcal{D}_{P_b})$ is non-trivial whenever the finite-rank bundle $\mathcal{W}$ defined by the Grassmann section $\mathbb{P}$ is non-trivial. Indeed, a section of $\mathcal{H}$ is the same thing as a C^∞ section of the trivial finite-rank vertical bundle $E^\nu \to B \times X$ equal to E along the fibres of the trivial fibration $B \times X \to B$. A section of $\mathcal{H}_{\mathbb{P}}$, on the other hand, is a C^∞ section of a non-trivial finite-rank vertical bundle $E^\nu_{\mathbb{P}} \to B \times X$; a section of $E^\nu_{\mathbb{P}}$ is required to satisfy $P_b\begin{pmatrix} s(b,0) \\ s(b,\beta)\end{pmatrix}$ at each $b \in B$. Consequently, the trivial connection d on $\mathcal{H}$ does not descend to a connection on the subbundle $\mathcal{H}_{\mathbb{P}}$ due to the variation of P_b. Hence a modified connection $\nabla^{\mathbb{P}}$ is needed which takes sections of $\mathcal{H}_{\mathbb{P}}$ to sections of $\mathcal{H}_{\mathbb{P}}$. Defining $\nabla^{\mathbb{P}}$ is the same thing as defining an 'honest' connection on the finite-rank bundle $E^\nu_{\mathbb{P}}$ and one can work entirely in that framework. Here we shall work directly with the bundle $\mathcal{H}_{\mathbb{P}}$ and define $\nabla^{\mathbb{P}}$ as follows.

First, we define the bundle restriction map

$$\gamma : \mathcal{H} \longrightarrow \mathbf{C}^{2n}, \quad \gamma s_b = \begin{pmatrix} s_b(0) \\ s_b(\beta)\end{pmatrix} \tag{54}$$

for $s_b \in \mathcal{H}_b = C^\infty(X, E)$, where $\mathbf{C}^{2n}$ is the trivial complex bundle over B of rank $2n$ (with fibre $\cong E_0 \oplus E_\beta$). Next, fix a smooth non-decreasing function $\phi : X = [0, \beta] \to [0, 1]$ with

$$\phi(x) = 0 \quad \text{in } [0, \beta/4], \quad \phi(x) = 1 \quad \text{in } [3\beta/4, \beta],$$

and define the extension operator $m_\phi : \mathbb{C}^{2n} \to C^\infty(X, \mathbb{C}^n)$ by $(\mp v)(x) = \phi(x)v$. Let p_0, p_β be the projection maps $\mathbb{C}^{2n} \cong E_0 \oplus E_\beta$ to $E_0 \cong \mathbb{C}^n$, $E_\beta \cong \mathbb{C}^n$, respectively. Then we define $\mathcal{M}_\phi : \mathbf{C}^{2n} \to \mathcal{H}$, $v \longmapsto \mathcal{M}_\phi v$ by $(\mathcal{M}_\phi v)(x) = m_{1-\phi} p_0 v + \mp p_\beta v$. We then have

$$\gamma(\mathcal{M}_\phi v) = \begin{pmatrix} (1-\phi(0))p_0 v + \phi(0)p_\beta v \\ (1-\phi(\beta))p_0 v + \phi(\beta)p_\beta v \end{pmatrix} = \begin{pmatrix} p_0 v \\ p_\beta v \end{pmatrix} = v\ . \tag{55}$$

The bundle maps $\gamma, \mathcal{M}_\phi$ induce the corresponding maps between the spaces $C^\infty(B; \mathcal{H})$ and $C^\infty(B; \mathbf{C}^{2n})$, and we also denote these by γ and $\mathcal{M}_\phi$.

We now define a connection on $\mathcal{H}_{\mathbb{P}}$ by

$$\nabla^{\mathbb{P}} = d + \mathcal{M}_\phi \mathbb{P} d \mathbb{P} \gamma, \tag{56}$$

or pointwise on B, $\nabla^{\mathbb{P}} = d + \mathcal{M}_\phi P_b d P_b \gamma$. We may drop the b subscript in the following.

Proposition 8 $\nabla^{\mathbb{P}}$ defines a connection on the bundle $\mathcal{H}^{\mathbb{P}}$. Let $\mathbb{P}^1, \mathbb{P}^2$ be Grassmann sections, then

$$\nabla^{\mathbb{P}^1} - \nabla^{\mathbb{P}^2} = \mathcal{M}_\phi(P^1 dP^1 - P^2 dP^2)\gamma \,. \tag{57}$$

Proof We have $\nabla^{\mathbb{P}}: \Omega^0(B, \mathcal{H}_{\mathbb{P}}) \longrightarrow \Omega^1(B, \mathcal{H})$, and one easily checks that $\nabla^{\mathbb{P}}$ satisfies the Leibnitz rule $\nabla^{\mathbb{P}}(fs) = df.s + f\nabla^{\mathbb{P}}s$, for $f \in C^\infty(B), s \in \Omega^0(B, \mathcal{H}_{\mathbb{P}})$, noting that $f: B \to \mathbb{C}$ is not affected by the restriction map γ. We need to see that $\nabla^{\mathbb{P}}$ has range in $\Omega^1(B, \mathcal{H}_{\mathbb{P}})$. But if $s \in \Omega^0(B, \mathcal{H}_{\mathbb{P}})$, then $P_b\gamma s(b) = 0$ and hence $dP_b.\gamma s(b) = -P_b\gamma ds(b)$, so that

$$P_b dP_b.\gamma s(b) = -P_b\gamma ds(b) \,. \tag{58}$$

Therefore using (55) and (58)

$$P_b\gamma\nabla^{\mathbb{P}}s = P_b\gamma ds + P_b\gamma\mathcal{M}_\phi P_b dP_b\gamma s = -P_b dP_b\gamma s + P_b dP_b\gamma s = 0,$$

and hence $\nabla^{\mathbb{P}}s \in \Omega^1(B, \mathcal{H}_{\mathbb{P}})$. Finally, the identity (57) is immediate from the definition of $\nabla^{\mathbb{P}}$. $\square$

With the connections $\nabla^{\mathbb{P}}, \nabla^{\text{triv}} = d$ on $\mathcal{H}_{\mathbb{P}}, \mathcal{H}$ at hand we have an induced connection $\widetilde{\nabla}^{\mathbb{P}}$ (pointwise $\widetilde{\nabla}^P := \widetilde{\nabla}^{P_b}$) on $\text{Hom}(\mathcal{H}, \mathcal{H}_{\mathbb{P}})$. For large $\text{Re}(s) > 0$, $\widetilde{\Delta}_P^{-s}\mathcal{D}_P\widetilde{\nabla}^P\mathcal{D}_P^{-1} = \widetilde{\Delta}_P^{-s}\mathcal{D}\widetilde{\nabla}^P\mathcal{D}_P^{-1}$ is trace class, and for small t

$$\phi_t(s) := \text{Tr}\,((I + t\mathcal{D}\widetilde{\nabla}_X^P\mathcal{D}_P^{-1})\widetilde{\Delta}_P^{-s}),$$

where $X \in \text{Vect}(B)$, has a meromorphic continuation to $\mathbb{C}$ which is regular at $s = 0$. Hence (by [1], Prop 2.9)

$$\frac{d}{dt}_{t=0}\phi_t(s) := -s\text{Tr}\,(\widetilde{\Delta}_P^{-(s+1)}\mathcal{D}\widetilde{\nabla}_X^P\mathcal{D}_P^{-1}\widetilde{\Delta}) = s\theta_P(s)(X)$$

has a meromorphic continuation to $\mathbb{C}$ with a simple pole at $s = 0$.

Proposition 9 Let $\mathbb{P}^1, \mathbb{P}^2$ be Grassmann sections and for $i = 1, 2$ let

$$\theta_i(s) = \text{Tr}\,(\widetilde{\Delta}_{P^i}^{-s}\mathcal{D}\widetilde{\nabla}^i\mathcal{D}_{P^i}^{-1}),$$

where $\widetilde{\nabla}^i := \widetilde{\nabla}^{P^i}$. Then

$$\begin{aligned}\theta_1(s) - \theta_2(s) = &\ \text{Tr}\,\{\widetilde{\Delta}_{P^1}^{-s}\mathcal{D}\widetilde{\nabla}^1(\mathcal{K}(P^1)\gamma\mathcal{D}_{P^2}^{-1})\} \\ &- \text{Tr}\,\{\widetilde{\Delta}_{P^1}^{-s}\mathcal{D}\mathcal{M}_\phi(P^1 dP^1 - P^2 dP^2)\gamma\mathcal{D}_{P^2}^{-1}\} \\ &- \text{Tr}\,\{(\widetilde{\Delta}_{P^1}^{-s} - \widetilde{\Delta}_{P^2}^{-s})\mathcal{D}\widetilde{\nabla}^2\mathcal{D}_{P^2}^{-1})\}\,.\end{aligned} \tag{59}$$

Proof The relative connection form is the 1-form

$$\theta_1(s) - \theta_2(s) = \mathrm{Tr}\,(\widetilde{\Delta}_{P1}^{-s}\mathcal{D}\widetilde{\nabla}^1\mathcal{D}_{P1}^{-1}) - \mathrm{Tr}\,(\widetilde{\Delta}_{P2}^{-s}\mathcal{D}\widetilde{\nabla}^2\mathcal{D}_{P2}^{-1})\,.$$

We have from (31) that

$$\begin{aligned}\mathcal{D}_{P1}^{-1} - \mathcal{D}_{P2}^{-1} &= -\mathcal{K}(P)\gamma\mathcal{D}_{P2}^{-1},\\ \widetilde{\Delta}_{P1}^{-s} - \widetilde{\Delta}_{P2}^{-s} &= \frac{i}{2\pi}\int_{\Gamma_\pi}\lambda^{-s}[\widetilde{\mathcal{K}}_\lambda(\widetilde{P}^1)\gamma\widetilde{\Delta}_{P2,\lambda}^{-1}]_{(1,2)}\,d\lambda\end{aligned}\tag{60}$$

are smoothing operators, where the terms $\widetilde{\mathcal{K}}_\lambda$ and so on, are the operators for $\widetilde{\Delta}$ corresponding to those in (28) for Δ.

For a section A of $\mathrm{Hom}(\mathcal{H},\mathcal{H}_{\mathbb{P}})$ one has $\widetilde{\nabla}^P(A)(s) = \nabla^P(A(s)) - A ds$, $s \in C^\infty(B,\mathcal{H})$, and we can extend this to $\mathrm{Hom}(\mathcal{H},\mathcal{H})$ by the same formula. Then

$$\begin{aligned}\widetilde{\nabla}^1(\mathcal{D}_{P2}^{-1})(s) &= \nabla^1(\mathcal{D}_{P2}^{-1}(s)) - \mathcal{D}_{P2}^{-1}ds,\\ \widetilde{\nabla}^2(\mathcal{D}_{P2}^{-1})(s) &= \nabla^2(\mathcal{D}_{P2}^{-1}(s)) - \mathcal{D}_{P2}^{-1}ds.\end{aligned}$$

Hence from (57)

$$\begin{aligned}\widetilde{\nabla}^1(\mathcal{D}_{P2}^{-1})(s) - \widetilde{\nabla}^2(\mathcal{D}_{P2}^{-1})(s) &= (\nabla^1 - \nabla^2)(\mathcal{D}_{P2}^{-1}(s))\\ &= (\mathcal{M}_\phi(P^1dP^1 - P^2dP^2)\gamma\mathcal{D}_{P2}^{-1})(s)\,.\end{aligned}$$

And so

$$\mathcal{D}\widetilde{\nabla}^1\mathcal{D}_{P2}^{-1} - \mathcal{D}\widetilde{\nabla}^2\mathcal{D}_{P2}^{-1} = \mathcal{D}\mathcal{M}_\phi(P^1dP^1 - P^2dP^2)\gamma\mathcal{D}_{P2}^{-1}\,.\tag{61}$$

We have

$$\begin{aligned}\theta_1(s) &= -\mathrm{Tr}\,\{\widetilde{\Delta}_{P1}^{-s}\mathcal{D}\widetilde{\nabla}^1\mathcal{D}_{P1}^{-1}\}\\ &= -\mathrm{Tr}\,\{\widetilde{\Delta}_{P1}^{-s}\mathcal{D}\widetilde{\nabla}^1(\mathcal{D}_{P2}^{-1} - \mathcal{K}(P)\gamma\mathcal{D}_{P2}^{-1})\}\\ &= -\mathrm{Tr}\,\{\widetilde{\Delta}_{P1}^{-s}\mathcal{D}\widetilde{\nabla}^1\mathcal{D}_{P2}^{-1}\} + \mathrm{Tr}\,\{\widetilde{\Delta}_{P1}^{-s}\mathcal{D}\widetilde{\nabla}^1(\mathcal{K}(P)\gamma\mathcal{D}_{P2}^{-1})\}\end{aligned}\tag{62}$$

since the terms are trace class. From (61) we have that the first term of (62) is

$$\begin{aligned}&- \mathrm{Tr}\,\{\widetilde{\Delta}_{P1}^{-s}\mathcal{D}\widetilde{\nabla}^2\mathcal{D}_{P2}^{-1}\} - \mathrm{Tr}\,\{\widetilde{\Delta}_{P1}^{-s}\mathcal{D}\mathcal{M}_\phi(P^1dP^1 - P^2dP^2)\gamma\mathcal{D}_{P2}^{-1}\}\\ &= -\mathrm{Tr}\,\{\widetilde{\Delta}_{P2}^{-s}\mathcal{D}\widetilde{\nabla}^2\mathcal{D}_{P2}^{-1}\} - \mathrm{Tr}\,\{(\widetilde{\Delta}_{P1}^{-s} - \widetilde{\Delta}_{P2}^{-s})\mathcal{D}\widetilde{\nabla}^2\mathcal{D}_{P2}^{-1}\}\\ &\quad - \mathrm{Tr}\,\{\widetilde{\Delta}_{P1}^{-s}\mathcal{D}\mathcal{M}_\phi(P^1dP^1 - P^2dP^2)\gamma\mathcal{D}_{P2}^{-1}\},\end{aligned}\tag{63}$$

recalling that $\widetilde{\Delta}_{P1}^{-s} - \widetilde{\Delta}_{P2}^{-s}$ is trace class. Substituting (63) into (62) completes the proof. □

Proposition 10 Let $\omega_{\mathbf{z}}^1, \omega_{\mathbf{z}}^2$ be the **z**-function connection forms associated to the Grassmann sections $\mathbb{P}^1, \mathbb{P}^2$. Then

$$\omega_{\mathbf{z}}^1 - \omega_{\mathbf{z}}^2 = \mathrm{Tr}\left\{\mathcal{D}\widetilde{\nabla}^1(\mathcal{K}(P^1)\gamma\mathcal{D}_{P^2}^{-1})\right\} + \mathrm{Tr}\left\{\mathcal{D}\mathcal{M}_\phi(P^2dP^2 - P^1dP^1)\gamma\mathcal{D}_{P^2}^{-1}\right\}. \quad (64)$$

Proof For the first term on the right-side of (59) we have

$$\begin{aligned}\Theta_1(s) &:= \mathrm{Tr}\left\{\widetilde{\Delta}_{P^1}^{-s}\mathcal{D}\widetilde{\nabla}^1(\mathcal{K}(P^1)\gamma\mathcal{D}_{P^2}^{-1})\right\} \\ &= \mathrm{Tr}\left\{\widetilde{\Delta}_{P^1}^{-s-1}\widetilde{\Delta}_{P_1}\mathcal{D}\widetilde{\nabla}^1(\mathcal{K}(P^1)\gamma\mathcal{D}_{P^2}^{-1})\right\}.\end{aligned} \quad (65)$$

But $\widetilde{\Delta}_{P^1}^{-s-1}$ is norm continuous for $\mathrm{Re}(s) > -1$ and $\widetilde{\Delta}_{P_1}\mathcal{D}\widetilde{\nabla}^1(\mathcal{K}(P^1)\gamma\mathcal{D}_{P^2}^{-1})$ is a smoothing and so trace class operator. $\Theta_1(s)$ is therefore holomorphic for $\mathrm{Re}(s) > -1$ and this allows us to go down to $s = 0$ in (65). Thus, $\Theta_1(s)$ is regular at at $s = 0$ and is given there by

$$\Theta_1(0) = \mathrm{Tr}\left\{\mathcal{D}\widetilde{\nabla}^1(\mathcal{K}(P^1)\gamma\mathcal{D}_{P^2}^{-1})\right\}. \quad (66)$$

Similarly, $\Theta_2(s) := -\mathrm{Tr}\left\{\widetilde{\Delta}_{P^1}^{-s}\mathcal{D}\mathcal{M}_\phi(P^1dP^1 - P^2dP^2)\gamma\mathcal{D}_{P^2}^{-1}\right\}$ is regular at $s = 0$ and given there by

$$\begin{aligned}\Theta_2(0) &= -\mathrm{Tr}\left\{\mathcal{D}\mathcal{M}_\phi(P^1dP^1 - P^2dP^2)\gamma\mathcal{D}_{P^2}^{-1}\right\} \\ &= \mathrm{Tr}\left\{\mathcal{D}\mathcal{M}_\phi(P^2dP^2 - P^1dP^1)\gamma\mathcal{D}_{P^2}^{-1}\right\}.\end{aligned} \quad (67)$$

For $\mathrm{Re}(s) >> 0$ the remaining term is $(s\Theta_3(s))|_{s=0}^{\mathrm{mer}}$, where

$$\Theta_3(s) := -\mathrm{Tr}\left\{(\widetilde{\Delta}_{P^1}^{-s} - \widetilde{\Delta}_{P^2}^{-s})\mathcal{D}\widetilde{\nabla}^2\mathcal{D}_{P^2}^{-1}\right\},$$

which vanishes by a similar argument using (60).

Thus $\theta_1(s) - \theta_2(s) = \Theta_1(s) + \Theta_1(s) + \Theta_1(s)$ is holomorphic for $\mathrm{Re}(s) > 0$ with a meromorphic continuation to all of $\mathbb{C}$, and we have

$$\begin{aligned}\omega_{\mathbf{z}}^1 - \omega_{\mathbf{z}}^2 &= \frac{d}{ds}_{|s=0}\{s(\theta_1(s) - \theta_2(s))\} \\ &= \Theta_1(0) + \Theta_2(0) + \frac{d}{ds}_{|s=0}(s\Theta_3(s)|^{\mathrm{mer}}).\end{aligned} \quad (68)$$

By (66), (67), the Proposition is proved. □

Since $P^1 dP^1 - P^2 dP^2$ is finite-rank and $\gamma\mathcal{D}_{P^2}^{-1}$ is bounded we have using (30)

$$\begin{aligned}
&\mathrm{Tr}\left\{\mathcal{D}\mathcal{M}_\phi(P^2dP^2 - P^1dP^1)\gamma\mathcal{D}_{P^2}^{-1}\right\} \\
&= \mathrm{Tr}\left\{\gamma\mathcal{D}_{P^2}^{-1}\mathcal{D}\mathcal{M}_\phi(P^2dP^2 - P^1dP^1)\right\} \\
&= \mathrm{Tr}\left\{\gamma(I - \mathcal{K}(P^2)\gamma)\mathcal{M}_\phi(P^2dP^2 - P^1dP^1)\right\} \\
&= \mathrm{tr}\left\{(I_{2n} - P_K S(P^2)^{-1}P^2)\gamma\mathcal{M}_\phi(P^2dP^2 - P^1dP^1)\right\} \\
&= \mathrm{tr}\left\{(I_{2n} - P_K S(P^2)^{-1}P^2)(P^2dP^2 - P^1dP^1)\right\} \\
&= \mathrm{tr}(P^2dP^2) - \mathrm{tr}(P^1dP^1) \\
&\quad - \mathrm{tr}\left\{P_K S(P^2)^{-1}P^2(P^2dP^2 - P^1dP^1)P_K\right\} \qquad (69)
\end{aligned}$$

where $P_K := P(\mathcal{D})$, I_{2n} denotes the identity on $\mathbb{C}^{2n}$ and we use (55). Note that the trace $\mathrm{tr} = \mathrm{tr}_{\mathbb{C}^{2n}}$ in the third line equals the trace $\mathrm{tr} = \mathrm{tr}_{W_2^\perp}$ over range $(P^2)^\perp$ since $P^2(I - P_K S(P^2)^{-1}P_K)) = 0$.

For the first term $\Theta_1(0)$ in (64) one has to work a little harder. To study the trace $\mathrm{Tr}\left\{\mathcal{D}\widetilde{\nabla}^1(\mathcal{K}(P^1)\gamma\mathcal{D}_{P^2}^{-1}\right\}$, we consider the operator $\mathcal{K}(P^1)\gamma\mathcal{D}_{P^2}^{-1}\colon \mathcal{H}_b \longrightarrow \mathrm{Ker}(\mathcal{D}_b)$ as the composition

$$\mathcal{K}(P^1)\gamma\mathcal{D}_{P^2}^{-1} = \mathcal{K}(P^1)P^1 \circ P^1\gamma\mathcal{D}_{P^2}^{-1}\colon \mathcal{H}_b \longrightarrow W_b^1 \longrightarrow \mathrm{Ker}(\mathcal{D}_b) \subset \mathcal{H}_b\,,$$

where $W_b^1 = \mathrm{range}\,(P_b^1)$ is the fibre of the bundle $\mathcal{W}^1$ at $b \in B$. The bundles $\mathcal{H}$, $\mathcal{W}^1$ have the connections $\nabla^{\mathrm{triv}} = d$ and $\nabla^{\mathcal{W}^1}$, while the bundle $\mathrm{Ker}(\mathbb{D})$ in $\Theta_1(0)$ has the induced connection $\nabla^1_{|\mathrm{Ker}(\mathbb{D})}$. Let $\nabla^{(\mathcal{W}^1,\mathrm{ker})}$, $\nabla^{\mathcal{H},\mathcal{W}^1}$ denote the induced connections on $\mathrm{Hom}(\mathcal{W}^1, \mathrm{Ker}(\mathbb{D}))$ and $\mathrm{Hom}(\mathcal{H}, \mathcal{W}^1)$. Then we have[3]

$$\begin{aligned}
&\widetilde{\nabla}^1(\mathcal{K}(P^1)\gamma\mathcal{D}_{P^2}^{-1}) = \widetilde{\nabla}^1((\mathcal{K}(P^1)P^1 \circ P^1\gamma\mathcal{D}_{P^2}^{-1}) \\
&= \nabla^{(\mathcal{W}^1,\mathrm{ker})}(\mathcal{K}(P^1)P^1).\ P^1\gamma\mathcal{D}_{P^2}^{-1} + \mathcal{K}(P^1)P^1.\nabla^{\mathcal{H},\mathcal{W}^1}(P^1\gamma\mathcal{D}_{P^2}^{-1})\,. \qquad (70)
\end{aligned}$$

Now since $\mathcal{K}$ has range in $\mathrm{Ker}(\mathcal{D})$, the second term in (70) is killed by $\mathcal{D}$ and so

$$\begin{aligned}
&\mathrm{Tr}\left\{\mathcal{D}\widetilde{\nabla}^1(\mathcal{K}(P^1)\gamma\mathcal{D}_{P^2}^{-1}\right\} = \mathrm{Tr}\left\{\mathcal{D}\nabla^{(\mathcal{W}^1,\mathrm{ker})}(\mathcal{K}(P^1)P^1).P^1\gamma\mathcal{D}_{P^2}^{-1}\right\} \\
&= \mathrm{tr}\left\{P^1\gamma\mathcal{D}_{P^2}^{-1}\mathcal{D}\nabla^{(\mathcal{W}^1,\mathrm{ker})}(\mathcal{K}(P^1)P^1)\right\} \\
&= \mathrm{tr}\left\{P^1(I_{2n} - P_K S(P^2)^{-1}P^2)\gamma\nabla^{(\mathcal{W}^1,\mathrm{ker})}(\mathcal{K}(P^1)P^1)\right\} \\
&= \mathrm{tr}\left\{(P^1 - S(P^1)S(P^2)^{-1}P^2)\gamma\nabla^{(\mathcal{W}^1,\mathrm{ker})}(\mathcal{K}(P^1)P^1)\right\}. \qquad (71)
\end{aligned}$$

[3] Note that for bundles ξ^i, $i = 1, 2, 3$, with connection inducing connections $\nabla^{i,j}$ on $\mathrm{Hom}(\xi^i, \xi^j)$ one has for respective sections A, B of $\mathrm{Hom}(\xi^2, \xi^3)$, $\mathrm{Hom}(\xi^1, \xi^2)$, $\nabla^{1,3}(AB) = \nabla^{1,2}(A)B + A\nabla^{2,3}(B)$, for any choice of connection ∇^2 on ξ^2.

But for $\xi \in C^\infty(B, \mathcal{W}^1)$ we have $P^1\xi = \xi$, pointwise, and $\nabla^{\mathcal{W}^1}(\xi) = P^1 d\xi$, and so

$$\begin{aligned}
&\nabla^{(\mathcal{W}^1,\mathrm{ker})}(\mathcal{K}(P^1)P^1)(\xi) = \nabla^{(\mathcal{W}^1,\mathrm{ker})}(\mathcal{K}S(P^1)^{-1}P^1)(\xi)\\
&= \nabla^1_{|\mathrm{Ker}(\mathbb{D})}(\mathcal{K}S(P^1)^{-1}P^1\xi) - (\mathcal{K}S(P^1)^{-1}P^1)\nabla^{\mathcal{W}^1}(\xi)\\
&= (d + \mathcal{M}_\phi P^1 dP^1\gamma)(\mathcal{K}S(P^1)^{-1}P^1\xi) - (\mathcal{K}S(P^1)^{-1}P^1 d(\xi)\\
&= d(\mathcal{K}).S(P^1)^{-1}P^1\xi + \mathcal{K}d(S(P^1)^{-1}P^1)\xi\\
&\quad + \mathcal{M}_\phi(P^1dP^1)P_K S(P^1)^{-1}P^1\xi\,. \qquad (72)
\end{aligned}$$

Since $\gamma\mathcal{K} = P_K$ we have $(P^1 - S(P^1)S(P^2)^{-1}P^2)\gamma\mathcal{K}d(S(P^2)^{-1}P^1 = 0$ (or we could use $\mathcal{D}\mathcal{K} = 0$ in the previous step), so

$$\begin{aligned}
\gamma\nabla^{(\mathcal{W}^1,\mathrm{ker})}(\mathcal{K}(P^1)P^1) &= \gamma d(\mathcal{K}).S(P^1)^{-1}P^1P^1dP^1P_KS(P^1)^{-1}P^1\\
&= dP_K.S(P^1)^{-1}P^1 + P^1dP^1P_KS(P^1)^{-1}P^1\,,
\end{aligned}$$

and we now have from (71)

$$\begin{aligned}
&\mathrm{Tr}\left\{\mathcal{D}\widetilde{\nabla}^1(\mathcal{K}(P^1)\gamma\mathcal{D}^{-1}_{P^2}\right\}\\
&= \mathrm{tr}_{W^1}\big\{(P^1 - S(P^1)S(P^2)^{-1}P^2)(dP_KP_KS(P^1)^{-1}P^1\\
&\quad + P^1dP^1P_KS(P^1)^{-1}P^1)\big\}\\
&= \mathrm{tr}_{W^1}\big\{P^1dP_KP_KS(P^1)^{-1}P^1 + P^1dP^1P_KS(P^1)^{-1}P^1\big\}\\
&\quad - \mathrm{tr}_{W^1}\big\{S(P^1)S(P^2)^{-1}P^2dP_KS(P^1)^{-1}P^1\\
&\quad - S(P^1)S(P^2)^{-1}P^2P^1dP^1P_KS(P^1)^{-1}P^1\big\}\\
&= \mathrm{tr}_K\big\{P_KS(P^1)^{-1}P^1(P^1dP_K + P^1dP^1)P_K\big\}\\
&\quad - \mathrm{tr}_K\big\{P_KS(P^2)^{-1}P^2dP_KP_K\big\} - \mathrm{tr}_K\big\{P_KS(P^2)^{-1}P^2P^1dP^1P_K\big\}. \qquad (73)
\end{aligned}$$

We consider next the canonical connection $\nabla^{C,\mathbb{P}}$ on DET $(\mathbb{D}, \mathbb{P})$, defined by the 1-form $\omega_C := \mathrm{tr}\big(S(P)^{-1}\nabla^{\mathcal{K},\mathcal{W}}S(P)\big)$ over $U \subset B$.

Lemma 3 Let $P_K = P(\mathcal{D}_b)$, $P = P_b$. One has

$$\omega_C = \mathrm{tr}_K\big\{P_KS(P)^{-1}(PdP + PdP_k)P_K\big\}. \qquad (74)$$

Proof Observe first that $P_K\xi = \xi$ for $\xi \in \Omega^0(B;\mathcal{K})$ and so $dP_K.\xi + P_kd\xi = d\xi$. Hence

$$Pd\xi - PP_Kd\xi = PdP_k\xi. \qquad (75)$$

But

$$\begin{aligned}
\nabla^{\mathcal{K},\mathcal{W}}(S(P))(\xi) &= \nabla^{\mathcal{W}}(S(P)(\xi)) - S(P)\nabla^{\mathcal{K}}\xi\\
&= Pd(P\xi) - PP_Kd(P_K\xi)\\
&= PdP.\xi + Pd\xi - PP_KdP_K.P_K\xi - PP_Kd\xi\\
&= PdP.\xi + PdP_k.\xi = PdP.P_K\xi + PdP_k.P_K\xi,
\end{aligned}$$

using (75) and since $P_KdP_KP_K = 0$. Equation (74) now follows. □

From (73) and (74)

$$
\begin{aligned}
&\mathrm{Tr}\,\{\mathcal{D}\widetilde{\nabla}^1(\mathcal{K}(P^1)\gamma\mathcal{D}_{P^2}^{-1}\} \\
&= \omega_C^1 - \mathrm{tr}_K\{P_K S(P^2)^{-1}P^2 dP_K P_K\} \\
&\quad - \mathrm{tr}_K\{P_K S(P^2)^{-1}P^2 P^1 dP^1 P_K\} \\
&= \omega_C^1 - \omega_C^2 + \mathrm{tr}_K\{P_K S(P^2)^{-1}P^2 dP^2 P_K\} \\
&\quad - \mathrm{tr}_K\{P_K S(P^2)^{-1}P^2 P^1 dP^1 P_K\} \\
&= \omega_C^1 - \omega_C^2 + \mathrm{tr}_K\{P_K S(P^2)^{-1}(P^2 dP^2 - P^2 P^1 dP^1)P_K\}\,.
\end{aligned}
\tag{76}
$$

Putting together (64), (69) and (76), we have proved that the $\mathbf{z}$ and C connection forms are related over U by

$$
\omega_{\mathbf{z}}^1 - \omega_{\mathbf{z}}^2 = \omega_C^1 - \omega_C^2 + \mathrm{tr}(P^2 dP^2) - \mathrm{tr}(P^1 dP^1)\,. \tag{77}
$$

We have from (77)

$$
\begin{aligned}
\Omega_{\mathbf{z}}^1 - \Omega_{\mathbf{z}}^2 &= \Omega_C^1 - \Omega_C^2 + d\,\mathrm{tr}(P^2 dP^2) - d\,\mathrm{tr}(P^1 dP^1) \\
&= \Omega_C^1 - \Omega_C^2 + \mathrm{tr}(dP^2 \wedge dP^2) - \mathrm{tr}(dP^1 \wedge dP^1)\,,
\end{aligned}
$$

and by the symmetry of the trace $\mathrm{tr}(dP^i \wedge dP^i) = 0$. This completes the proof of Theorem 2.

References

1. Atiyah, M.F., Patodi, V.K., Singer, I.M.: Spectral asymmetry and Riemannian geometry. III. Math. Proc. Camb. Phil. Soc. **79**, 71–99 (1976)
2. Booss-Bavnbek, B., Morchio, G., Scott, S.G., Wojciechowski, K.P.: Determinants, Manifolds with Boundary and Dirac Operators. Fundam. Theor. Phys. **94**, 423–432 (1998)
3. Bismut, J.M., Freed, D.: The analysis of elliptic families:(I) Metrics and connections on determinant bundles. Commun. Math. Phys. **106**, 159–176 (1986)
4. Booss-Bavnbek, B., Wojciechowski, K.P.: Elliptic Boundary Problems for Dirac Operators. Birkhäuser, Boston (1993)
5. Lesch, M., Tolksdorf, J.: On the determinant of one-dimensional elliptic boundary value problems. Comm. Math. Phys. **193**, 643–660 (1998)
6. Quillen, D.G.: Determinants of Cauchy-Riemann operators over a Riemann surface. Funk. Anal. I Ego Prilozhenya **19**, 37–41 (1985)
7. Scott, S.G.: Zeta determinants on manifolds with boundary. J. Funct. Anal. **192**, 112–185 (2002)
8. Scott, S.G., Wojciechowski, K.P.: The ζ–Determinant and Quillen's determinant for a Dirac operator on a manifold with boundary. Geom. Funct. Anal. **10**(2000), 1202–1236 (2000)
9. Seeley, R.T.: Complex powers of an elliptic operator. AMS Proc. Symp. Pure Math. X. AMS, Providence, pp. 288–307 (1967)

Local Gauss Law and Local Gauge Symmetries in QFT

Franco Strocchi

Contents

Abstract Local gauge symmetries reduce to the identity on the observables, as well as on the physical states (apart from reflexes of the local gauge group topology) and therefore their use in Quantum Field Theory (QFT) asks for a justification of their strategic role. They play an intermediate role in deriving the validity of Local Gauss Laws *on the physical states* (for the currents which generate the related global gauge group); conversely, we show that local gauge symmetries arise whenever a vacuum representation of a *local field algebra* $\mathcal{F}$ is used for the description/construction of physical states satisfying Local Gauss Laws, just as global compact gauge groups arise for the description of localizable states labeled by superselected quantum numbers. The above relation suggests that the Gauss operator $\mathbf{G}$, which by locality cannot vanish in $\mathcal{F}$, provides an intrinsic characterization of the realizations of a gauge QFT in terms of a local field algebra $\mathcal{F}$ and of the related local gauge symmetries generated by $\mathbf{G}$.

1 Introduction

Since by definition a gauge symmetry reduces to the identity on the observables, its physical meaning has been questioned and debated from a foundational/philosophical point of view.(For the conceptual and philosophical discussions

F. Strocchi (✉)
Dipartimento di Fisica "E. Fermi", Università di Pisa, Pisa, Italy
e-mail: franco.strocchi@sns.it

A. Cintio, A. Michelangeli (Eds.), *Trails in Modern Theoretical and Mathematical Physics*, https://doi.org/10.1007/978-3-031-44988-8_13

on the empirical meaning of gauge symmetries, which appear to be relevant for our understanding of the physical world, see [1]). Clearly, the final description of a physical system should be written in a gauge invariant language and therefore gauge symmetry are bounded to play only an intermediate role.

The following physical principles help to properly set the problem of understanding the strategic role of gauge symmetries. The *operational/experimental description of a physical system* (not necessarily quantum!) makes reference to the following elements:

A) the set $\mathcal{A}$ of its measurable quantities, briefly called **observables**;
B) the measurement of the **time evolution** of the observables;
C) the set Σ of configurations or **states** in which the system may be prepared, according to well defined protocols of experimental preparations; the measurement of an observable A, when the system is in the state ω, is operationally defined by the **experimental expectation** $< A >_\omega$.

Given a state ω, the set of states which can be prepared starting from ω, through physically realizable operations, is denoted by Γ_ω and called the **phase** of ω. This means that the protocols of preparations of states belonging to different phases are not related by physically realizable operations. Clearly, by definition, different phases describe disjoint realizations of the system (or disjoint "worlds") which cannot communicate through realizable observable operations.

The above *operational framework* has an essentially unique *mathematical transcription*. General physical considerations indicate that the set of the observables generate a normed algebra, actually a C*-algebra with identity, called the **algebra of observables**, for simplicity still denoted by $\mathcal{A}$. The time evolution defines a one-parameter group of transformations (technically automorphisms) of $\mathcal{A}$: $t\colon \mathcal{A} \to \alpha_t(\mathcal{A})$.

By its experimental expectations, each state ω, defines a linear positive functional $\omega(A) \equiv < A >_\omega$, $\forall A \in \mathcal{A}$, and by the GNS theorem a representation π_ω of $\mathcal{A}$ (as operators $\pi_\omega(A)$) in a Hilbert space $\mathcal{H}_\omega$; ω is represented by a vector $\Psi_\omega \in \mathcal{H}_\omega$ and the experimental expectations are represented by the matrix elements $< A >_\omega = (\Psi_\omega, \pi_\omega(A)\, \Psi_\omega)$. (For a discussion of the general description of a physical system, see [2]).

The set of states $\pi_\omega(\mathcal{A})\, \Psi_\omega$ are dense in $\mathcal{H}_\omega$, from which one may obtain also the mixed states associated to π_ω. The set of all such states is called the folium associated to ω, which may be considered as the mathematical description of Γ_ω. The family of physical states on $\mathcal{A}$ is denoted by $\Sigma(\mathcal{A})$.

In the following, we shall consider infinitely extended systems, which generically display the occurrence of different phases and the related role of gauge symmetries (see below). For this kind of systems a very relevant property is the **local structure** of $\mathcal{A}$: each observable is identified by the experimental apparatus used for its measurement and, since the physically realizable operations, as well as the corresponding experimental apparatuses, are inevitably localized in space, the algebra of observables is generated by localized observables. This means that the algebras

$\mathcal{A}(V)$ of observables localized in bounded regions V, generate $\mathcal{A}$, $\mathcal{A} = \cup_V \mathcal{A}(V)$, (hereafter called local algebra).

Another relevant property is the stability of $\mathcal{A}$ under the group of **space translations** $\alpha_{\mathbf{a}}$, $\mathbf{a} \in \mathbf{R}^3$, with $\alpha_{\mathbf{a}}(A)$ denoting the **a**-translated of A. (For lattice systems the space translations are replaced by the lattice translations).

For a reasonable physical interpretation, the measurement of a localized observable A should not be influenced by measurements of observables at infinite space separations, i.e. the following condition, called **asymptotic abelianess**, must hold, (B a localized observable): $\lim_{|\mathbf{x}| \to \infty} [\, A, \alpha_{\mathbf{x}}(B)\,] = 0$.

A distinguished role is played by the *pure homogeneous states* ω_0, characterized by being *invariant under a subgroup* $\mathcal{T}$ *of translations*

$$\omega_0(\alpha_{\mathbf{a}}(A)) = \omega_0(A), \quad \forall A \in \mathcal{A} \tag{1}$$

and by satisfying the *cluster property*

$$\lim_{n \to \infty} \omega_0(A\, \alpha_{n\,\mathbf{a}}(B)) = \omega_0(A)\, \omega_0(B), \tag{2}$$

where $n\, \mathbf{a}$ denote the group parameters of $\mathcal{T}$. Typically, but not necessarily, ω_0 is a ground state. The set of such states ω_0 on $\mathcal{A}$ is denoted by $\Sigma_0(\mathcal{A})$.

A very important consequence is that:

ω_0 *is the unique* $\mathcal{T}$ *invariant state in the pure homogeneous phase* Γ_{ω_0}.

The interest of considering the phases defined by such states is twofold.

In relativistic quantum field theory, the vacuum state is invariant under space translations and the validity of the cluster property is a necessary condition for the possibility of defining the scattering matrix, which requires the factorization of expectations of infinitely (space) separated clusters (describing scattering processes). Thus, the phase defined by a vacuum state in quantum field theory is a homogeneous pure phase in the above sense.

Quite generally, in a homogeneous pure phase the macroscopic observables, defined by space averages of local observables, take sharp (classical) values in agreement with the characteristic property of the standard pure phases in thermodynamics.

In conclusion, general physical principles characterize the mathematical description of a physical system, in terms of the local algebra of observables $\mathcal{A}$, its time evolution $\alpha_t(\mathcal{A}), t \in \mathbf{R}$, and the family of states $\Sigma(\mathcal{A})$ in which the system may be prepared. The task of theoretical physics is to devise concrete effective strategies for implementing and controlling such a general structure by determining the representations of the observable algebra corresponding to the various (physically realizable) states.

In quantum field theory, as advocated by Wightman, the standard and well established strategy is to study the vacuum representations of the relevant field algebra, provided by the vacuum correlation functions. However, already in the case of a system of free particles, physically important particle states are not present in the vacuum representation of the observable algebra, called the vacuum sector.

A simple example is provided by the system of free massive Dirac fermions, since the observable algebra and therefore its vacuum representation have zero fermionic charge; a complete QFT description is obtained by introducing the local field algebra $\mathcal{F}$ generated by the fermionic fields, whose (unique) vacuum representation contains the representations of the observable sub-algebra $\mathcal{F}_{\text{obs}}$, labeled by the quantum number of the fermionic charge.

This is the standard strategy adopted in the formulation and control of QFT models. One studies the vacuum representation of a local field algebra $\mathcal{F}$ generated by local fields, which describe the degrees of freedom inferred and borrowed from the limit of vanishing interaction. Such a local field algebra contains a (suitably characterized) observable field sub-algebra $\mathcal{F}_{\text{obs}}$ and one looks for the representations of $\mathcal{F}_{\text{obs}}$ contained in a vacuum representation of $\mathcal{F}$.

Therefore, the assumed role of the field algebra $\mathcal{F}$ is to provide the building stones of the theory: the fields of $\mathcal{F}$ describe the relevant states, generate the observable algebra and allow for a direct and simple definition of the dynamics (through a Hamiltonian or Lagrangian which is a polynomial function of the fields of $\mathcal{F}$).

In conclusion, the difficult problem of determining the representations of the local observable algebra and its time evolution, beyond the vacuum sectors, i.e. the solution of the problem $(\mathcal{A}, \alpha_t(\mathcal{A}), \Sigma(\mathcal{A}))$, is attacked by finding the vacuum representations of a local field algebra $\mathcal{F} \supset \mathcal{A}$, i.e. by solving the relatively easier problem $(\mathcal{F}, \alpha_t(\mathcal{F}), \Sigma_0(\mathcal{F}))$.

Behind such a strategic choice there is the implicit assumption (extrapolated from the non-interacting case) of considering the states outside the vacuum sector which are local states, i.e. may be obtained by applying the local fields of $\mathcal{F}$ to the vacuum; this property allows for their localizability in the DHR sense (for general discussion of such an important property, see [3]): the state ω is localized in the double cone $\mathcal{O}$ if

$$\omega(A) = \omega_0(A), \quad \forall A \in \mathcal{A}(\mathcal{O}'), \tag{3}$$

where $\mathcal{O}'$ denote the set of points which are spacelike w.r.t O and $\mathcal{A}(\mathcal{O}')$ the algebra of observables localized in $\mathcal{O}'$.

The quantum numbers which characterize the inequivalent representations of $\mathcal{A}$ contained in the Hilbert space $\mathcal{H}_{\omega_0(\mathcal{F})}$ of a vacuum representation of $\mathcal{F}$, define superselection rules; they identify a global gauge group G, which defines non-trivial symmetries of $\mathcal{F}$ and it is represented by the states of $\mathcal{H}_{\omega_0(\mathcal{F})}$ (if $\omega_0(\mathcal{F})$ is invariant under G).

Under general assumptions (namely the absence of infinite degeneracy of particle types with equal mass and the completeness of the asymptotic states) the so derived global gauge group is compact. These deep results, which explain the *origin of compact gauge groups from general physical principles* were obtained in a series of papers by Doplicher, Haag and Roberts. (For a comprehensive account see [3], esp. Section IV.4, and references therein.)

In conclusion, the experimentally detectable existence of *localizable states labeled by superselected quantum numbers* imply that (under general assumptions) there is a local field algebra which generates them from the vacuum has the symmetry of a global compact gauge group; this explains why such gauge groups arise in the theory of infinitely extended systems, typically in quantum field theory, where locality plays a crucial role and vacuum representations of local field algebras are the object of the standard approach.

More subtle is the problem of justifying the role of local gauge symmetries on the basis of general physical principles, since they reduce to the identity both on the observables and on the (physical) states. There is no doubt that local gauge symmetries proved to be useful for the formulation of the quantum field theory of elementary particles, but an interesting question is whether their *raison d'être* may be traced back to general a priori arguments beyond their a posteriori successful use.

The standard justification of local gauge symmetries, in particular in the prototypical case of quantum electrodynamics, is tight to the introduction of redundant degrees of freedom through the vector potential, used either for describing a massless spin one particle (the photon) or for defining the electron-photon interaction [4].

These reasons do not appear to be rooted in general principles better than the origin of local gauge symmetries they should explain. Actually, the photon is equally well described by the electromagnetic (quantum) field $F_{\mu\nu}$, whose Lorentz transformation does not require a gauge transformation. As a matter of fact, already for the solution of the free field equations $\partial^{\mu} F_{\mu\nu} = 0$, the introduction of the quantum field A_{μ} as a Hilbert space operator is not free of problematic choices, since it cannot be done without giving up covariance and/or locality (see [5], Chapter 7, Appendix 8).

Moreover, a Lagrangian invariant under the group $\mathcal{G}$ of local gauge transformations, labeled by the set of infinitely differentiable gauge functions *localized in space*, does not define a deterministic time evolution of the field algebra and therefore $\mathcal{G}$ must be broken by a gauge fixing; such a breaking need not to be down to the identity, as might be inferred from the functional integral argument, but only up to the extent of restoring deterministic evolution (see [6]).

In the standard approach to gauge theories (particularly in the perturbative treatment) the identification and construction of the physical states is usually obtained by starting with a Lagrangian invariant under a group $\mathcal{G}$ of local gauge transformations and by adding a gauge fixing which preserves the invariance under the related global gauge group G and allows to use a local field algebra $\mathcal{F}$. Then, the (infinitesimal) transformations of G on $\mathcal{F}$ may be generated by local currents J^a_{μ}, ($a = 1, \ldots, n$, $n =$ the dimension of G).

As a consequence of the gauge fixing, the second Noether theorem does not apply and such currents do not satisfy a Local Gauss Law, i.e. they are not the divergence of antisymmetric tensors $F^a_{\nu\mu}$. Nevertheless, one may show that a characteristic property of the physical states constructed in a vacuum representation of $\mathcal{F}$ is that the currents J^a_{μ} satisfy Local Gauss laws on them [5, 6]; i.e. for any physical state Ψ obtained through a vacuum representation of $\mathcal{F}$

$$(\Psi, (J^a_{\mu} - \partial^{\nu} F^a_{\nu\mu}) \Psi) = 0, \tag{4}$$

(with $F^a_{\mu\nu}$ actually the field strength) or, in a manifestly gauge invariant way,

$$(\Psi, \sum_a (G^a_\mu)^* \, G^a_\nu \, \Psi) = 0, \quad G^a_\mu \equiv J^a_\mu - \partial^\nu F^a_{\nu\,\mu}. \tag{5}$$

Such a form of the Local Gauss Law is obviously satisfied in QED and may be easily checked to hold in the (local) BRST quantization of Yang–Mills theories, thanks to the nillpotency of the BRST charge. Actually, for the same reasons, any monomial of G^a_μ has vanishing expectation on the physical states. For brevity, such physical states shall be called *LGL states*. Clearly a prototypical example is provided by the charged states in QED.

Thus, one may argue that even if local gauge symmetries reduce to the identity on the physical states, they play an intermediate role in the construction of the representations of the observable algebra by guaranteeing that they are defined by physical states obeying LGL.

Conversely, given the existence of LGL states, one may investigate which gauge symmetries emerge for a local field algebra $\mathcal{F}$ which allows for their construction through its vacuum representation; we shall argue that in this way one obtains local gauge symmetries. This means that, in order to allow for the construction of LGL states, the local algebra $\mathcal{F}$ must contains fields with non-trivial transformation under a local gauge symmetry.

This offers a possible explanation of the emerge and strategic role of local gauge symmetries, *without ever mentioning the vector potential*, which, being generically determined up to a (local) gauge transformation, automatically and obviously brings with it the freedom of local gauge symmetries.

In this perspective local gauge symmetries are not introduced by an *a priori* ansatz or Local Gauge Principle, nor as a consequence of the introduction of redundant degrees of freedom through the vector potential, but automatically arise as symmetries of a local field algebra which allows for the realization of states satisfying Local Gauss Laws.

2 Local Gauss Laws and Local Gauge Symmetries

The aim of this Section is to argue that physical LGL states (e.g. the charged states in QED or unconfined quark states in QCD) lead to the emergence of local gauge symmetries if a local field algebra is used for their construction.

In the abelian case (QED) the characterization of the physical LGL states is simply provided by the electrically charged states, labeled by the superselected electric charge (giving rise to a $U(1)$ global gauge group) and satisfying the LGL given by the Maxwell equations.

Less obvious is the non-abelian case and we adopt the following characteristic properties of LGL states, extracted from their actual construction:

i) they carry superselected quantum numbers corresponding to a compact global gauge group G, as in the DHR case,
ii) in contrast with the DHR states, LGL states cannot be described by local states in a vacuum representation of an auxiliary local field algebra $\mathcal{F}$ which contains the field strengths $F^a_{\mu\nu}$, as well as the currents J^a_μ, which generate the infinitesimal transformations of $\mathcal{F}$ under G; for brevity, the corresponding charges shall be called *Gauss charges*, to emphasize that as a consequence of the LGL they are not localizable charges,
iii) in the abelian case of Quantum Electrodynamics (QED) they do not define local states on the algebra of the observables and therefore they are not localizable in the DHR sense.

Remark 1 Property ii) follows from the fulfillment of LGL, if the vacuum representation of $\mathcal{F}$ satisfies semipositivity or the relativistic spectral condition. In fact, (5) with $\Psi = \Psi_0$, implies that $G^a_\nu\, \Psi_0$ is a null vector and therefore, if semipositivity holds, $(\Psi, G^a_\nu\, \Psi_0) = 0$, $\forall \Psi \in \mathcal{H}_{\omega_0(\mathcal{F})}$. As a consequence of this last equation, given a local state $\Psi = F\, \Psi_0$, where $F \in \mathcal{F}$ transforms non-trivially under G,

$$\delta^a F = i \lim_{R\to\infty} [\, Q^a_R,\, F\,], \quad Q^a_R = J^a_0(f_R\, \alpha), \tag{6}$$

(with the standard notation for the smearing of J^a_0, see e.g. [5], Chapter 7, Section 2), by the locality of F one has

$$0 \neq (\delta^a F\, \Psi_0, \delta^a F \Psi_0) = i \lim_{R\to\infty} (\delta^a F \Psi_0, [\, J^a_0(f_R\alpha) - \partial^i F_{i\,0}(f_R\alpha),\; F\,]\Psi_0).$$

Then, if $F\Psi_0$, and therefore $\delta^a F\Psi_0$, satisfies LGL, the r.h.s. reduces to

$$i \lim_{R\to\infty} (F^*\, \delta^a F \Psi_0, G^a_0(f_R\alpha)\, \Psi_0) = 0,$$

leading to a contradiction.

The same conclusion is reached even if the vacuum correlation functions of the auxiliary local field algebra $\mathcal{F}$ do not satisfy semipositivity, but the relativistic spectral condition holds. In fact, as above, by locality for a local state $\Psi = F\, \Psi_0$ one has

$$\lim_{R\to\infty} < \Psi, [\, Q^a_R, F\,]\Psi_0 >= \lim_{R\to\infty} < \Psi, [\, G^a_R, F\,]\Psi_0 >,$$

the limit is reached for finite R and, if Ψ satisfies the LGL, the r.h.s reduces to $< \Psi_0, F^*\, F\, G^a_R\, \Psi_0 >$, R large enough.

Now, by the relativistic spectral condition

$$W(y - x, z - x) \equiv < \Psi_0, F^*(x)\, F(y)\, G^a_R(z)\, \Psi_0 >,$$

(where $F(y) \equiv U(y) F U(y)^{-1}$ denotes the y-translated of F, and similarly for the other operators), is the boundary value of a function $W(\zeta_1, \zeta_2)$ analytic in the tube $\mathcal{T}_2$. Such a function vanishes for ζ_1, ζ_2 real, $\zeta_2 = (\mathbf{z}_2, z_{0,2})$, $|\mathbf{z}_2|$ sufficiently large, since then $[\, F(y),\, G^a_R(z)\,] = 0$ and LGL applies. Then, by the edge of the wedge theorem $W(\zeta_1, \zeta_2) = 0$ and $F\, \Psi_0$ is chargeless, contradicting i).

Remark 2 In the QED case, the current, which generates the global gauge group $U(1)$, and the field strength $F_{\mu\nu}$ are observable fields; then, if a state ω is localized in the DHR sense, say in a double cone $\mathcal{O}$, for sufficiently large R, $\partial^i F_{0i}(f_R\alpha) \in \mathcal{F}_{\mathrm{obs}}(\mathcal{O}')$, so that, according to the DHR criterion, $\omega(\partial^i F_{0i}(f_R\alpha)) = \omega_0(\partial^i F_{0i}(f_R\alpha)) = 0$ and by the LGL ω is chargeless.

The embedding of the observable algebra into a larger algebra $\mathcal{F}$ for the description of LGL states, through a vacuum representation of $\mathcal{F}$, is not unique and the local gauge symmetries of $\mathcal{F}$ depend on $\mathcal{F}$. As we shall see, the generators of the infinitesimal local gauge transformations on $\mathcal{F}$ are provided by the Gauss operators G_0^a. A limiting case in QED is the choice of the Coulomb gauge field algebra $\mathcal{F}_C$, since in $\mathcal{F}_C$ the Gauss operator G_0 vanishes, $\mathcal{F}_C$ is non-local and the Coulomb gauge fixing excludes any local gauge symmetry of $\mathcal{F}_C$, (for the general structure of the Coulomb gauge see [7, 8]. For the necessary ultraviolet regularization, see, for a perturbative control, [10] and, for a general control which exploits the properties of the Feynman–Gupta–Bleuler (FGB) gauge, [11]; see also the discussion in [6] Section 2.3).

Actually, a crucial property for the strategy outlined above, leading to local gauge symmetries, is the locality of the field algebra; this choice, motivated by the no-interaction limit, is also suggested by technical reasons, since locality helps for the control of the dynamics (it is well known that the renormalizable gauges are local gauges). Also from a constructive point of view, the infinite volume limit is better handled for a local field algebra with a local dynamics.[1]

For these reasons, the auxiliary local field algebra $\mathcal{F}$ should not satisfy the LGL; otherwise, by locality, $\mathcal{F}$ would be pointwise invariant under the global gauge group G. In fact, if the LGL hold in the local algebra $\mathcal{F}$ one would have

$$\delta^a F = \lim_{R\to\infty} [\, J_0^a(f_R\alpha),\, F\,] = \lim_{R\to\infty} [\, \partial^i F^a_{i\,0}(f_R\alpha),\, F\,] = 0, \quad \forall F \in \mathcal{F},$$

and, consequently, $\mathcal{H}_{\omega_0(\mathcal{F})} = \overline{\mathcal{F}\,\Psi_0}$ would not contain states carrying non-trivial charges of G.

[1] For the problems arising for a non-local dynamics, see e.g. [9], Appendix A.
In the case of DHR states the existence of a local field algebra for their description is guaranteed by general principles; in the case of LGL states it may be motivated by the standard way of treating gauge field theory models, e.g. in the perturbative treatment of QED or more generally of the standard model. Some hint is provided, e.g. in QED, by giving a small mass μ to the photon, so that the charged states become DHR states and DRH analysis applies with the existence of a local field algebra $\mathcal{F}$ for their description. As shown by Blanchard and Seneor the vacuum correlations of $\mathcal{F}$ have a limit for $\mu \to 0$ (preserving locality) and the problem is reduced to the construction of the physical LGL states in terms of the vacuum representation of the local algebra $\mathcal{F}$. Actually, in Symanzik's treatment of the Proca theory, with the use of the Stuckelberg field B, in analogy with QED, (*Lectures on Lagrangian Field Theory*, DESY report T-71/1), the fields $\psi_g = \exp^{-ie[(-\Delta)^{-1}\partial_i A^i]}\,\psi$, $A_g^\mu = A^\mu - \partial^\mu[(-\Delta)^{-1}\partial_i A^i]$, with ψ, A_μ the Proca fields, commute with B, which in the limit $\mu \to 0$ generates the local gauge transformations, (with gauge parameters $\varepsilon(x)$ satisfying $\Box\varepsilon = 0$), and should yield the (Coulomb) physical charged states.

The different choices of the local algebra $\mathcal{F}$ are characterized by the way LGL fail, i.e. by the non-vanishing Gauss operators G_0^a. To this purpose, it is useful to remark that LGL correspond to a combination of hyperbolic evolution equations and constraint/elliptic equations for the field strengths $F^a_{\mu\nu}$:

$$\Box F^a_{\mu\nu} = \partial_\mu J^a_\nu - \partial_\nu J^a_\mu + C^a_{\mu\nu}, \quad \partial^i F^a_{i0} = J^a_0. \tag{7}$$

with $C^a_{\mu\nu}$ a bilinear function A^c_λ, $F^b_{\rho\sigma}$ and their first derivates. If, as required, LGL do not hold in $\mathcal{F}$, the above equations get modified by the non-vanishing Gauss operators G^a_μ, $\partial^\mu G^a_\mu = 0$.

a) Time Independent Local Gauge Symmetries

As remarked above, a non-trivial representation of the global gauge group G by a local field algebra $\mathcal{F}$ requires that $G^a_0 \neq 0$, since, by locality the infinitesimal transformations of $\mathcal{F}$ by G are given by

$$\begin{aligned} \delta^a F &= \lim_{R\to\infty} [\, J^a_0(f_R\alpha),\, F\,] = \lim_{R\to\infty} [\, J^a_0(f_R\alpha) - \partial^i F^a_{i0}(f_R\alpha),\, F\,] \\ &= \lim_{R\to\infty} [\, G^a_0(f_R\alpha),\, F\,]. \end{aligned} \tag{8}$$

A possible realization of such a framework, which, in a certain sense, minimizes the needed violation of the LGL in $\mathcal{F}$, even at the expense of loosing manifest covariance, is given by $G^a_0 \neq 0$, $G^a_i = 0$, which plays the role of a gauge fixing; then, the continuity equation for G^a_μ requires that $G^a_0(\mathbf{x}, t)$ is time independent. As a consequence, (7) are replaced by

$$\Box F^a_{ij} = \partial_i J^a_j - \partial_j J^a_i + C^a_{ij}, \quad \partial^0 F^a_{0i} = -\partial^j F^a_{ji} + J^a_i, \tag{9}$$

with the equal time constraint

$$\partial^i F^a_{i0} = J^a_0 - G^a_0. \tag{10}$$

This is the choice adopted by the temporal gauge; in fact, (9), (10) are the evolution equations of the fields $F^a_{\mu\nu}$ in the temporal gauge.

Equation (5) imply that for any physical state Ψ, $(G^a_0\,\Psi,\, G^a_0\,\Psi) = 0$, i.e., if positivity holds, $G^a_0\,\Psi = 0$.

One should remark that a mathematical subtlety occurs in the standard local and positive temporal gauge, namely the local field algebra $\mathcal{F}$ represented by a vacuum state is generated by the exponentials of the standard (non-observable) fields, with algebraic relations corresponding to those of their formal generators. Then, the condition which selects the physical states should rather read

$$V^a(\Lambda)\Psi = \Psi, \quad \Lambda \in \mathcal{D}(\mathbf{R}^3),$$

where the unitary operator $V^a(\Lambda)$ is formally the unitary exponential of $G^a_0(\Lambda)$.

Furthermore, since $\mathcal{F}$ is not covariant under relativistic transformations the relativistic spectral condition is not satisfied by the vacuum correlation function of $\mathcal{F}$, the Reeh–Schlieder theorem does not apply and $G_0^a \Psi_0 = 0$, or better $(V^a(\Lambda) - \mathbf{1})\, \Psi_0 = 0$ does not imply that the local operator G_0^a or better $(V^a(\Lambda) - \mathbf{1})$ vanishes.[2]

For simplicity, in the following discussion, we shall sometimes use the formal generators of the exponential fields, the more accurate mathematical discussion being easy to obtain.

The operator G_0^a plays the role of a static charge density which is not seen by the physical states and compensates the vanishing flux of $F_{i\,0}^a$ at (space) infinity on the local states of $\mathcal{H}_{\omega_0(\mathcal{F})}$.

The non-vanishing G_0^a modifies the equal time constraint implied by the LGL on the local states; the vanishing of G_0^a on the physical states requires that the physical states of $\mathcal{H}_{\omega_0(\mathcal{F})}$, carrying a non-trivial charge of G, are non-local limits of local states.

Now, for any test function $\Lambda(\mathbf{x}) \in \mathcal{D}(\mathbf{R}^3)$, the operator $G_0^a(\Lambda, t)$ generates a *time independent derivation* on $\mathcal{F}$

$$\delta^{a,\Lambda} F \equiv i[\, G_0^a(\Lambda, t),\, F\,] = i[\, G_0^a(\Lambda, h),\, F\,], \quad h \in \mathcal{D}(\mathbf{R}), \quad \int dt\, h(t) = 1. \tag{11}$$

This is the basic property of the derivations generated by local (covariant) currents, corresponding to infinitesimal symmetry transformations, where the time independence is a consequence of current conservation and locality.

If the auxiliary local field algebra $\mathcal{F}$ contains the formal exponentials of G_0^a, (as in the standard temporal gauge [5, 6]), the following local transformations are defined on $\mathcal{F}$:

$$\beta^{\Lambda}(F) = V^a(\Lambda)\, F\, V^a(\Lambda)^*. \tag{12}$$

Moreover, since the subspace of physical state vectors must be pointwise invariant under the application of the observable operators, these operators should commute with $V^a(\Lambda)$, so that the above transformation defines a *local time independent gauge transformation.*

As a matter of fact, in the temporal gauge, where the local algebra $\mathcal{F}$ is generated by canonical fields, the derivation (11) corresponds to the standard time independent gauge transformations, with local gauge parameter $\Lambda(\mathbf{x})$.

The time independence of the generators G_0^a implies that the time evolution of $\mathcal{F}$ commutes with such local gauge transformations; thus, the Hamiltonian, as a function of the fields of $\mathcal{F}$, must be gauge invariant, a property which in the standard approach corresponds to the requirement of minimal coupling.

Thus, by the above arguments, the strategic role of local gauge symmetries may be traced back to the realization of LGL states through a vacuum representation of a local field algebra $\mathcal{F}$, where the necessarily *non-vanishing Gauss operators generate local gauge symmetries.*

---

[2] For a more detailed discussion see [5] Chapter 8, Section 2.1.

b) Local Gauge Symmetries in QED

Another distinguished example of local gauge symmetries, arising according to the pattern discussed above, is provided by Quantum Electrodynamics.[3]

This example is obtained by requiring that the time evolution of the observable electromagnetic field as operator in $\mathcal{F}$ is not modified by the non-vanishing Gauss operator G_μ, i.e. that

$$\Box F_{\mu\nu} = \partial_\mu J_\nu - \partial_\nu J_\mu, \tag{13}$$

the only effect of $G_\mu \neq 0$ being a modification of the equal time constraint

$$\partial^i F_{i\,0} = J_0 - G_0. \tag{14}$$

This implies that $\partial_\mu G_\nu - \partial_\nu G_\mu = 0$ and in a *Lorentz-covariant local field algebra* this is obtained by $G_\mu = \partial_\mu \mathcal{L}$, with $\mathcal{L}(x)$ a scalar field of $\mathcal{F}$; then, the continuity equation obeyed by G_μ implies $\Box\mathcal{L}(x) = 0$.

Such a choice of the local field algebra corresponds to the Feynman–Gupta–Bleuler (FGB) gauge, albeit in a more general context, since no reference is made to the vector potential; in fact, (13), (14) coincide with the equations for $F_{\mu\nu}$ in that gauge.

The relativistic covariance of $\mathcal{F}$ lead to the validity of the Reeh–Schlieder theorem and therefore positivity cannot hold, since otherwise (5) implies $G_\mu\, \Psi_0 = 0$; hence, by the Reeh–Schlieder theorem the local operator G_μ vanishes, $\mathcal{F}$ commutes with the charge and $\mathcal{H}_{\omega_0(F)}$ does not contain charged states.

A way out of this difficulty is to guarantee the validity of (5) on the vacuum state by a non-local condition; since $\mathcal{L}(x)$ is a free field, its negative energy part $\mathcal{L}(x)^-$ is well defined, it is a non-local operator and the equation

$$\partial_\mu \mathcal{L}(x)^-\, \Psi = 0 \tag{15}$$

implies that Ψ satisfies (5); then, such a condition may be chosen for selecting the physical states.

As in the temporal gauge discussed above for the general case of a compact gauge group G, one may show that in QED the Gauss operator generates a time independent derivation on the local field algebra $\mathcal{F}$.

To this purpose, one considers infinitely differentiable functions $\Lambda(x)$ satisfying

$$\Box\Lambda(x) = 0, \quad \Lambda(\mathbf{x}, 0),\ \partial_0\Lambda(\mathbf{x}, 0) \in \mathcal{D}(\mathbf{R}^3), \tag{16}$$

---

[3] For a general discussion of the occurrence of local gauge symmetries in QED, in a C*-algebra setting, see F. Ciolli, G. Rizzi, E. Vasselli, QED Representation for the Net of Causal Loops, Rev. Math. Phys., **27**, 1550012 (2013); D. Buchholz, F. Ciolli, G. Rizzi, E. Vasselli, The universal algebra of the electromagnetic field. III. Static charges and emergence of gauge fields, arXiv: 2111.01538 [math-ph]. When the present note was in reparation a very important result was obtained by the same authors, for QED in the presence of external charges [12].

and the operator

$$G(\Lambda)_R(x_0) \equiv \int d^3x\, f_R(\mathbf{x})\, (\Lambda(x) \overset{\leftrightarrow}{\partial_0} \mathcal{L}(x)), \tag{17}$$

with $(A \overset{\leftrightarrow}{\partial_0} B) \equiv (A\partial_0 B - \partial_0 A\, B)$.

Since $\Lambda(x)$ and $\mathcal{L}(x)$ satisfy the free wave equation, by locality the following commutator is independent of time, i.e.

$$i\partial_0 \lim_{R\to\infty} [G(\Lambda)_R(x_0),\, F] = 0, \quad F \in \mathcal{F}. \tag{18}$$

In fact,

$$\begin{aligned}\partial_0 \lim_{R\to\infty} [G(\Lambda)_R(x_0) &= \int d^3x\, f_R(\mathbf{x})(\Lambda(x)\Delta\mathcal{L}(x) - \Delta\Lambda(x)\mathcal{L}(x)) \\ &= \int d^3x\, \partial_i f_R(\mathbf{x})(\Lambda(x)\partial_i\mathcal{L}(x) - \partial_i\Lambda(x)\mathcal{L}(x)).\end{aligned}$$

Then, since supp $\partial_i f_R(\mathbf{x}) \subset (R \leq |\mathbf{x}| \leq R(1+\varepsilon))$, by locality the commutator (18) vanishes. In conclusion, one has a time independent derivation on $\mathcal{F}$, labeled by the infinitely differentiable functions $\Lambda(x)$ of compact support in space

$$\delta^\Lambda F \equiv \int dx_0\, \alpha(x_0) i \lim_{R\to\infty} [G(\Lambda)_R(x_0),\, F\,]. \tag{19}$$

The stability of the subspace of physical states under application of observable operators is guaranteed if the observables commute with $\mathcal{L}(x)^-$, (and therefore with $\mathcal{L}(x)$). Then the above derivation has the meaning of an infinitesimal local gauge transformation with gauge function $\Lambda(x)$.

Indeed, if the algebra $\mathcal{F}$ is generated by local canonical fields, as it is the case of the FGB realization, one gets the standard (infinitesimal) local gauge symmetries of the FGB gauge.

As discussed above in point a), the time independence of the derivation (19) implies that, as a function of the local fields of $\mathcal{F}$, the Hamiltonian should be invariant under (infinitesimal) local gauge transformations with gauge parameter $\Lambda(x)$, (satisfying $\Box\Lambda = 0$).

3 Conclusion

In conclusion, a possible physical/empirical explanation of gauge quantum field theories is the existence of (particle) *states which carry the quantum numbers of a (global) compact gauge group G and satisfy Local Gauss Laws*, for the conserved currents associated to G. The prototype is clearly Quantum Electrodynamics where the electric charge (the generator of the global compact $U(1)$ gauge group) labels the (physical) charged states, which satisfy the Local Gauss Law corresponding to the Maxwell equations. Quite generally, the description of such LGL states through a vacuum representation of a ***local** field algebra* $\mathcal{F}$ leads to the emergence of ***local** gauge symmetries* for $\mathcal{F}$, which commute with the time evolution of $\mathcal{F}$.

Thus, whereas states carrying localizable superselected charges lead to global compact gauge groups for the local field algebra which obtains them from the vacuum, the realization of states carrying Gauss charges through a vacuum representation of a local field algebra $\mathcal{F}$ implies a local gauge symmetry for $\mathcal{F}$, i.e. such $\mathcal{F}$ must contains fields which transform non-trivially under local gauge transformations. Therefore, the local extension of the global gauge group G (related to the superselected quantum numbers) needs not to be *a priori* assumed with no compelling physical motivations (as in the standard characterization/definition of gauge theories), but it is required by the *physical* existence of states carrying Gauss charges and their construction through the vacuum representation of a local field algebra.

In our opinion, the role of the Gauss operator as the generator of the local gauge symmetries suggests a classification of the possible local field algebras used for realizing the LGL states, better than the gauge fixings which typically involve the vector potential.

Acknowledgements I am indebted to Francesco Serra for a stimulating discussion.

References

1. Brading, K., Castellani, E. (eds.): Symmetries in Physics: Philosophical Reflections. Cambridge University Press (2003). L. Felline, A. Ledda, F. Paoli and E Rossanese eds., *New Directions in Logic and the Philosophy of Science*, SILFS 3, College Publications 2016.
2. Strocchi, F.: The Physical Principles of Quantum mechanics. A critical review. Eur. Phys. J. Plus **127**, 12 (2012)
3. Haag, R.: Local Quantum Physics. Springer (1996). Chapters III, IV
4. Weinberg, S.: The Quantum Theory of Fields, Vol. I. Cambridge University Press (1995). Section 5.9
5. Strocchi, F.: An introduction to non-perturbative foundations of quantum field theory. Oxford University Press (2013). 2016
6. Strocchi, F.: Symmetry Breaking in the Standard Model. A non-perturbative outlook. Scuola Normale Superiore, Pisa (2019). Chapter 2
7. Dirac, P.A.M.: Canad. Jour. Phys. **33**, 650 (1955)
8. Symanzik, K.: Lectures on Lagrangian Field theory. DESY report T-71/1. (1971)
9. Strocchi, F.: Symmetry Breaking, 3rd edn. Springer (2021)
10. Steinmann, O.: Perturbative Quantum Eelctrodynamics and Axiomatic Field Theory. Springer (2000)
11. Buchholz, D., Doplicher, S., Morchio, G., Roberts, J.E., Strocchi, F.: Ann. Phys. **290**, 53 (2001)
12. Buchholz, D., Ciolli, F., Rizzi, G., Vasselli, E.: Gauss's law, the manifestations of gauge fields, and their impact on local observables (2022). https://doi.org/10.48550/arXiv.2212.11009. Published in the present volume

Part III
Remembering Gianni

"For the sake of culture"

Riccardo Adami

I retain a precise memory of when I first saw Gianni Morchio: it was in November 1992, when I was a third-year student in Physics at the University of Pisa. He was in charge of the exercise sessions of the course of "Metodi Matematici della Fisica", held by Giampaolo Cicogna. He entered the room timidly, left his books and notes on the desk, and suddenly started speaking and writing.

It became crystal clear that his style was defintely new to us: we were not used to look at things from such a height. Gianni talked to us as he would to his colleagues: it was for him the only possible way to talk about Mathematics and Phyiscs: with the deepest respect for science as well as for the audience.

As a student, I greatly appreciated the way Giampaolo Cicogna introduced the analysis of Hilbert spaces: fast but clear, so that everything seemed to be simple and natural. Extremely pedagogical and effective, so how could one ask for more?

And then Gianni showed up and destabilized all my criteria of judgement. His method was different from Cicogna's: for him, everything had to be understood immediately in full depth, far-reaching connections had to be drawn from the very beginning, and all unnecessary tools, like heavy notation, collateral properties, pedantic corollaries, could be avoided, and they always were. For such things, there were books. Lectures were a matter of humans, and humans should not get lost in minor technicalities. Gianni used to get right to the point, leaving to us the task of fixing details. But, in spite of his concision, he used no shortcuts nor tricks. Important results and techniques were never easy nor natural, they could became as such only at the expenses of an enormous work to be carried out after the lectures. A hard work of reconstructing concepts and writing them down in perfectly rigrous mathematical terms.

Were he another teacher, I would have given up on his course, as I did for most of the university courses I was supposed to attend. But it proved impossible for me to elude the strength of Gianni's personality. It was a gentle strength: he never got

R. Adami (✉)
Dipartimento di Scienze Matematiche "G.L. Lagrange", Politecnico di Torino, Torino, Italy
e-mail: Riccardo.Adami@Polito.it

A. Cintio, A. Michelangeli (Eds.), *Trails in Modern Theoretical and Mathematical Physics*, https://doi.org/10.1007/978-3-031-44988-8_14

upset nor angry, he used to smile and stare at the students with mildness and kindness. Even when lectures seemed to be too dense, burdensome or abstract, trying to decipher them was more than a good student's duty: it was a moral categorical imperative.

The only concession he made were his legendary pauses. He used to stop writing, quite unexpectedly, and to say these precise words: *A scopo cultura*, namely "for the sake of culture". And then, he opened a world of unprecedented connections: a deep meditation on Kolmogorov's axioms of probability, an unforeseen but strict connection between Kant's philosophy and Quantum Mechanics, the idea of generality developed during the French Revolution, the influence of China on the present world ... At the end of such digressions, he started writing again. Such interruptions made mandatory not to skip a single moment of his lectures, and the categorical imperative to meditate on them became even more stringent.

So, some of us started meeting during the afternoons with the only purpose of getting through the notes we had taken at his morning lecture. These afternoons stand among the most stimulating and worthwhile intellectual experiences I ever lived. Reconstructing the line of Gianni's lectures, that were sometimes too ambitious to be entirely grasped "live", made us aware of the depth and of the originality of his vision. Another feature of this way of thinking emerged overwhelmingly and sometimes unexpectedly: the absolute elegance of his reasoning, much beyond our reach during the lectures, and also beyond the level of textbooks, sometimes already advanced, that we used at the time. Years later, I discovered how such elegance was recognized and appreciated worldwide. It was during an edition of the School on Mathematical Methods of Quantum Mechanics in Bressanone, when Jürg Fröhlich interrupted a lecturer giving a talk on Quantum Electrodynamics by quoting "one of those incredibly beautiful papers by Morchio and Strocchi", literaly.

We coined a specific vocabulary for the afternoons dedicated to the reconstruction of Gianni's lectures: in Italian, the verb "smorchiare" refers to an agricultural practice, but for us it meant to manage to get to the same intellctual height where Gianni lived. The word arose unconsciously as a crasis between Gianni's family name, and the verb "scollinare" that means to overcross a hill. In my experience, no hill had ever been so tough, but no view so awesome as the one from its top.

At that time, Gianni was also in charge of the course of Quantum Mechanics. It was the only course of Mathematical Physics available for the Degree in Physics in Pisa, and it was considered as a challenge: every year the number of students attending the course oscillated between one and three, and not all of them were able to take the final exam. Nowadays, a department could not afford to offer a course like that: too few students, so the ratio between cost and benefit of the course would not fit some parameters predetermined by the present bureaucratic obtusity.

In fact, should one accept such a ratio as a meaningful parameter to judge a course, it would be honest to keep into account, as a benefit, the number of specialists for which that course proved crucial for their scientific life. Under this light, I have no knowledge of university courses for which the ratio between cost and benefit is as low as for Gianni's. Nowadays, a significant fraction of Italian researchers of my generation on the mathematics of Quantum Mechanics come from such a course: for instance Alessandro Pizzo, Michele Correggi, Alessandro Michelangeli, Giuseppe De Nittis, myself, and many others who were able to enlarge their expertise, like Andrea Cintio and Fabio Acerbi, and others who, despite significant contributions to scientific research, found other fields where to spend their education, like Lucattilio Tenuta and Emanuele Costa.

I had the privilege to be the first to experience a path that later became classical: first Gianni's course, then Degree in Pisa, and finally Ph.D. with Gianfausto Dell'Antonio in Rome or Trieste. It was almost an obliged path, in those times when PhD students on mathematical approach of Quantum Mechanics were as rare as four-leaf clovers. The point is that twenty years ago courses like that of Gianni were exceptional for Italy. Nowadays they are hosted in almost every Department of Mathematics, as the unavoidable result of the inesorable importance of this field of research. Italian university suffered from a heavy retard, if not in research, surely in the teaching of such topics. Gianni's course was one of the exceptions, and thanks to these exceptions Italian research was later able to recover the delay.

I remember the sliding doors effect unchained in my life by one lecture of Gianni's. It was on the theory of self-adjointness. I was deeply unsatisfied with the notion of hermitian operators supplied in previous courses. Already simple models of quantum systems on the half-line or on finite intervals showed that the concept of hermiticity was not well suited to guarantee the well-posedness of the dynamics. But it sounded heretical to draw questions on that matters. I always received vague answers and foggy sentences, even from books: "these are subtle questions of domain", I read once. I was explicitly discouraged to pursue such issues, and this resulted for me in a deep frustration. Gianni showed that my unease was not only legitmate, it was even definitely right! And the answer was already there since decades, provided that someone knew it and aimed at disseminating it. That lecture changed my life.

Furthermore, even though it seems now unbelievable, in those years most Italian universities showed a strong closure towards the emerging field of Quantum Information. Not only there were no courses on such subject, but as students we were warned against losing time and energy in thinking of foundations of Quantum Mechanics, so that our prevailing feeling towards that newborn realm of science was mistrust. Gianni did not directly speak of Quantum Information nor Communcation, but he was fond of philosophy of Quantum Mechanics and gave me hints, like reading the books by Bernard d'Espagnat, topics to study in deep, like Bell's inequalities and Kochen–Specher's Theorem, and I know that in that time he gave theses on Nelson's stochastic mechanics as well as on Bohmian Mechanics, that he criticized. My mistake was to mix the two levels of mathematics and foundations, so I left Gianni's course convinced that every conceptual aporie of Quantum Mechanics could be healed by the rigorous use of Mathematics. Of course it was a wrong conclusion, but to understand this I had to spend twenty years more in doing research. It was not Gianni's fault, it was my limit.

My regret is not to have been able to tell Gianni how lucky I had been to meet him. I last met him one evening in Pisa, in 2005. I was Lizzanello Postdoctoral Fellow at the Centro di Ricerca Matematica Ennio De Giorgi, enthousiastic about the recent rigorous derivations of effective one-body nonlinear equations, like Hartree's, Hartree–Fock's, or Gross–Pitaevskii's. He patiently listened to me, trying to overwhelm him with BBGKY hierarchies and estimates in Schatten spaces. He smiled, and then he said: "Good. But do not exaggerate with fireworks". I greatly appreciated this suggestion, as a philosophical motto: quite hard for me, since I really like fireworks.

This was his last *a scopo cultura*, that I will never forget.

Remembering Gianni

Fabio Bagarello

When I think to my youth, I still consider a real privilege the possibility of spending some time with Gianni. I suspect he was not aware of how much he knew Mathematics and Physics, and how simple it was for him to understand things that are very complicated for most of the human beings, and restate these things in some much simpler form.

To me, Gianni was half of an single entity: Franco (Strocchi) and Gianni: Franco&Gianni. I met Franco at SISSA as a doctoral student. When I had to choose a supervisor for my thesis, I had no doubt: no one was as clear as him when teaching. Every topic during my first two years in Trieste were new, and difficult, and I still remember how hard I had to study to reach a reasonable level of knowledge. Well, Franco was the only one I was able to understand from the beginning to the end of his lectures, despite of my poor background. And this is why I asked him to be my supervisor. And Franco introduced me Gianni, his own *other half*.

It was immediately clear that Gianni was a special person, different from many others I had met before. First of all, he was far from being interested in his academic career, a very rare attitude in the Italian panorama. In my eyes, he was mainly (if not only) focused on a sort of personal *Holy Trinity*: Family, Politics and Science. And career is not Science, as we all know. Sometimes they are connected, sometimes they are not. In Gianni's case, they were not, since he was not interested in improving his academic position.

Gianni was much more interested in discussing physics with Franco, and with each student or colleague. And there was always something to learn when chatting with him: once I was working on the fractional quantum Hall effect, and in particular on the possibility of getting a wave function for the system with strong localization properties. It took few weeks to me (with the help of Franco, in Trieste) to deduce what we thought was an interesting theorem on this problem. When I

F. Bagarello (✉)
Dipartimento di Ingegneria, Università di Palermo, Palermo, Italy
Sezione di Catania, Istituto Nazionale di Fisica Nucleare, Catania, Italy
e-mail: fabio.bagarello@unipa.it

A. Cintio, A. Michelangeli (Eds.), *Trails in Modern Theoretical and Mathematical Physics*, https://doi.org/10.1007/978-3-031-44988-8_15

went to Pisa to discuss this result with Gianni, just after few seconds my report on what he had done, Gianni started laughing. It was unexpected, and I was a bit angry, and I asked why he was laughing. He replied that it was quite easy to produce a counterexample for our result, ... and he did! Then, in few seconds, he destroyed my hard work of several days. After a first shock, this was maybe one of the key moments for my scientific growth: I realized that imagination is more important than computations! And Gianni is still, in my mind, the perfect example of *the man full of ideas*, happy to share and discuss his suggestions with anyone interested.

This brings me to another interesting episode that helps having an idea of Gianni's pure interest for Science. We (Franco, Gianni and myself) were working on a problem in abstract differential equations, in connection with a spin chain, and we (mostly Gianni and Franco, to be honest) proved an existence and uniqueness result for a special class of operator-valued differential equations. I had to study operator algebras, and topology to face with this problem and to become an active member of the team. In the same period I started working with a colleague (and friend) of mine in Palermo, Camillo Trapani, on a different, but related, problem in statistical mechanics and I used with Camillo some of the techniques I was learning from Gianni and Franco. Eventually my paper with Camillo appeared before the one with Gianni, and Franco was quite unhappy with this: "Fabio, this way what you did with Gianni looks like a sequel of what you did with Camillo! And this is not correct!!". Actually, it was not really like this: there is, and there cannot be, any barrier between what you learn when working with somebody and what you can use, or learn, when working with somebody else. But I understand Franco considered himself to be responsible of this specific "unpleasant time ordering between papers". However, and this is how Gianni considered the whole situation, Gianni just replied: "I don't see the problem! Fabio learned something, and he can (and must!) use as he likes what he is learning from us, and from others scientists."

Unfortunately, after my Ph.D. I had not many other possibilities to interact with Gianni, but I am quite happy of the few years we spent working on the same problems, discussing Physics, and I am sure Gianni's attitude toward Science, but also toward his family, are now part of my personal history, and contributed to make of me what I am now. Grazie Gianni.

With Gianni Morchio at the Interface Between Mathematics and Physics

Bernhelm Booß-Bavnbek

I met Gianni Morchio around 1994 in Pisa. My collaborator Krzysztof P. Wojciechowski and I came to work with him and Franco Strocchi on a fascinating problem, the invariance of the Fermi determinant under global chiral transformations. Gianni posessed a rich mathematical culture. He had a very particular way of combining factual thinking with far reaching scientific aims and visions. That and his generosity and friendliness impressed everybody. We became friends, also on the family level, and met again and again for the rest of his life.

How We Met

Gianni and I did not meet by chance. "Gott würfelt nicht" ("God does not play dice"), as Einstein once held against Bohr. We met following a suggestion by the Polish physicists and polymath Andrzej Trautman. Here is what happened.

Trautman Was There

Gianni and I were admirers, and I still am an admirer and follower of Trautman. In 1993, Krzystof P. Wojciechowski and I had published our monograph on "Elliptic boundary problems for Dirac operators" [3]. Later that year I had given a talk on our main results at a Clifford algebra conference in Belgium, where I met Trautman. A few months later in Trieste, Franco Strocchi and Gianni discussed with Trautman a 1985 paper [7] by M. Ninomiya and C.I. Tan pointing to a peculiar case of seemingly incompatibility between

1. the conventional wisdom on *chiral symmetry* and *gauge invariance* in quantum chromodynamics (QCD),

[1] Actually in a letter of 4 Dec. 1926 to Max Born, Einstein wrote: "Quantum mechanics is certainly imposing. But an inner voice tells me that it is not yet the real thing. The theory says a lot, but does not really bring us any closer to the secret of the 'old one'. I, at any rate, am convinced that He [God] does not play dice." [4, page 91]

B. Booß-Bavnbek (✉)
Department of Sciences and Environment, Roskilde University, Roskilde, Denmark
e-mail: booss@ruc.dk

A. Cintio, A. Michelangeli (Eds.), *Trails in Modern Theoretical and Mathematical Physics*, https://doi.org/10.1007/978-3-031-44988-8_16

2. the appearance of a non-necessarily vanishing integer, a *topological term* in the conservation equation of the chiral current $\partial^\mu j_\mu = -1/16\pi^2 \tilde{F}F$, and
3. the wanted *invariance of the fermion determinant* under global (i.e., rigid) chiral transformations.

The arguments by Ninomiya and Tan were based on the calculation of the index of the (spectral, global, pseudo-differential) Atiyah–Patodi–Singer (APS) boundary value problem for the partial Dirac operator D_A^+, i.e., the one component of the full Dirac operator D_A over the four-dimensional disc $\mathbb{B}^4$ of radius R. The full Dirac operator acts on sections of the Clifford bundle of Euclidean spinors with coefficients determined by a chosen connection A acting on the trivial bundle. Then it is natural to assume that in a collar neighbourhood of the boundary of $\mathbb{B}^4$, the connection takes the form $d + h^{-1}dh$, where $h\colon S^3 \to \mathrm{SU}(2)$ is a smooth map. Determining the tangential operator B_h of the partial Dirac operator D_A^+ and imposing on it the corresponding APS boundary condition $\mathcal{P}(D_A^+)$, one finds

$$n_+ - n_- = \operatorname{index}(D^+_{A,\mathcal{P}(D_A^+)}) = \deg(h) \neq_{\text{in general}} 0.$$

Devastating Result: No Errors Found

When I arrived in Pisa and met all three, Morchio, Strocchi, and Trautman, they were concerned, but only a little bit: all three, as I remember, were convinced that the concept of the fermion determinant was sound and its invariance under global chiral transformations could not be doubted. But they acknowledged at once that the supposed lack of chiral symmetry was fatal.

Formally, 0-eigenvalues (i.e., a non-trivial kernel of $D_{A,\mathcal{P}(D_A)}$) are a nuisance since they make the determinant vanish. Therefore physicists came up with a regularization which takes care of this problem in the case $n_+ := \dim \ker D^+_\Pi = \dim \ker D^-_{\Gamma(\mathrm{Id}-\Pi)\Gamma^{-1}} = n_-$, where Γ denotes the Green's form. However, in the case $n_+ \neq n_-$ the imaginary part of the determinant appears, unnatural in the case of an operator with symmetric spectrum.

With Trautman's help [11] and a reference he passed on to us [9], I checked Ninomiya's and Tan's arguments. I could not find an error and, even worse, I could not understand the placidity of my new physicists friends. They had invited me; I had traveled from Copenhagen to Pisa; and then we were together and one single mathematical result diminished x years of their work and the aspirations of many physicists, at a single stroke.

They were neither depressed nor tired. On the contrary, they were excited and wide-awake. Their attitude, see also [6], was something like "Man hat schon Pferde (vor der Apotheke) kotzen sehen" ("You never know, anything can happen"). Gianni explained to me later, roughly recording: "Look, there you have the difference between mathematics and physics. In mathematics, the validity of a single equation can be decisive for a wide field. After Euclid had shown the irrationality of $\sqrt{2}$, and Lindemann the transcendence of π, Abel had given an algebraic equation of fifth degree with a solution that can not be expressed in elementary algebraic terms, and D'Alembert the factorization of any real polynomial in monomials and binomials etc. Mathematics as a field of research has always changed. Physics is different. Seldom do we attribute decisive importance to a single observation or a single argument. Usually, we cover our results by multiple and preferably mutually independent approaches. *You* are trained to draw the most far reaching conclusions from minimal assumptions. Contrary to you, *we* wear both belts and suspenders."

Our Main Result

Jointly and with help by Wojciechowski over phone, in writing, and in direct discussion we found that there are other pseudodifferential projectors beside the APS-projection that

(i) defined a self-adjoint Fredholm extension of the full Dirac operator with finite dimensional kernel, consisting only of smooth spinors;
(ii) with gauge-invariant domain;
(iii) with a symmetric spectrum, i.e., in physics terminology that the boundary condition is γ_5 invariant.

Finally, for one of them, the *Calderón projector*, we proved the optimal result $n_+ = n_- = 0$.

I thought my job with Gianni was done. We wrote a nice report on that problem and our lucky solution [2], and a follow-up paper devoted to the calculation of the determinant. Suddenly, Franco and Gianni had new gripes, as Gianni formulated it in [12, p. 431]:

> From the point of view of physics this solution [our suggestion with the Calderón projector] is not completely satisfying because, unlike the Atiyah–Patodi–Singer condition, which depends only on the boundary data, the Calderón projection varies with change of the operator inside of the manifold. Therefore some alternative choices [of the boundary projection] have to be discussed.

Gianni proposed a joint work on the variation of the ζ-determinant under the change of the boundary conditions. However, our own understanding of the ζ-determinant was evolving, see [10]. We were not yet ready. Then Wojciechowski felt ill, and now Gianni has passed away. Alas, for now nothing more to report. Surely, other people will push Gianni's vision forward!

Gianni's Way of Commenting

To Wojciechowski and me, there was something solemn around having published two papers with these outstanding physicists, Morchio and Strocchi. In the slipstream of the Manhattan Project, claiming a deep connection to physics helped many mathematicians in their applications for funds or positions. In most instances it would suffice to pay lip service. Since the 1960s, when the race to the Moon became a dominant topic in science and technology financing, buffoons claimed to prove an over-representation of functional analysis NSF funds for mathematics due to the high frequency of the terms "space" and "research" in their applications.

When I met Gianni, I was not active on that playground. I was mostly interested in certain teaching questions, such as: How much physics do I have to teach in a class on PDEs? How much mathematics would Gianni have to teach in a class on electrodynamics or quantum field theory? And our teaching, how interchangeable would it be? But foremost, what is the role of mathematical arguments in physics?

Various Views on the Interface Between Mathematics and Physics

As I remember Gianni, he was never dogmatic. He admitted, on one side, (I) the *negation* role of mathematical arguments, the refutation, the falsification: "no way, our idea does not work", as we had (almost) experienced on the issues that had brought us together, and others that led us to doubt on the physical meaning of the Fermionic determinant. Historically, it seems that William Harvey (1578–1657), the discoverer of the closed circulation of blood, was one of the pioneers of this new way of using rigorous mathematical arguments (elementary arithmetical calculations in his case) to falsify common scientific wisdom—see Box 1!

Falsification, Gianni agreed, was and is one of the main roads of the application of formal mathematical reasoning in science. However, at the same time Gianni recognized (II) the *heuristic* power of mathematical calculations, evident already with Harvey: refusing elder concepts as unfounded and erroneous invited to the invention of new concepts and suggested new experiments. In Harvey's case the search for capillaries.

Box 1. Use of mathematical arguments. Harvey's falsification road

Between 1616 and 1628 William Harvey refused the common wisdom in medicine since Galen of Pergamon (131–201 CE), who taught: All blood flows from the liver and the heart to all parts of the body where it is consumed. In 1628 in [5], on the back of an envelope, (as one would later say), Harvey just solved the following elementary arithmetic problem:

- Heart capacity > 1.5 ounces
- Blood expelled per heart pump > 1/8 of heart capacity, i.e. > 1/6 ounce
- Number of heartbeats > 1000 per half an hour
- Sum1: More than 10 pounds 6 ounces arterial blood are produced in half an hour
- Sum2: More than sum1 × 48 = 540 pounds of arterial blood are produced and consumed in a day
- Consequence: There must be capillaries.
- Marcello Malpighi's light microscope histology confirmation (1661).

For Gianni, the mathematical results on the ζ-function regularization of the determinant were an even better case of the heuristic value of rigorous mathematical arguments, i.e., the partial justification of the crude infrared and ultraviolet approximations (chopping the eigenvalues near 0 and ∞)—see my attempt of an accessible presentation of the basic ideas in [1].

Gianni and I discussed also on the visibility and invisibility of the mathematical and physics content in engineering devices and that slick design can conceal essential differences between the math content.

A simple case is the similarity between a CT engine and an MRI engine. The principles of computed tomography scans are mathematically and physically fully understood since the discovery of Cavalieri's Principle in the 17th century and the discovery of X-rays in the 19th century. The use of magnetic spin resonance imaging has other challenges. Admittedly, in atomic physics and quantum mechanics the physical principles are basically understood; moreover, in modern analytical chemistry, and hence for small probes only, the implementation in devices has been perfected for years. The delicacies of time length, magnetic field strength, and image resolution seem, however, to keep challenging mathematically. (For advances towards a coherent mathematical theory of MRI see [8].) Thus, CT scans provide a type III of mathematics-physics interface, namely the perfect engineering design that is fully understood also theoretically, while MRI provides a type IV, namely the design of instruments that do function without being fully understood in all their details and side aspects.

How Much Physics Understanding May We Demand from Mathematicians? It was during a warm and bright midsummer week in Djursholm near Stockholm at the *Mittag-Leffler Institut* (MLI) in 2016 that we met last time. I was initiator and had, jointly with Matthias Lesch (Bonn), George Marinescu (Köln), Nicolai Reshetikhin (Berkeley), and Boris Vertman (Oldenburg), organized a conference on *Interfaces between Geometric Analysis and Mathematical Physics* at the MLI.

On the front page of our program I had set two quotes to emphasize the conference's focus on interfaces and intelligibility for a mixed audience and to support the preparation of talks devoted to that aim:

> David Mumford: The thing that leaps to mind is something about the suicidal tendency in math to get more and more technical and never to think about explaining one's ideas to mathematicians in other fields of math (let alone other scientists or even the general public). The field has a strange psychology linked to the fear of being thought dumb if you don't know everything.
>
> Norbert Wiener: ... there has been a tendency, visible here and there, to give up the search for a great stroke or a great aperçu and to be content with a sort of mathematical embroidery ... This reinforces the tendency toward the thin and the bodiless change, which is one of the besetting sins of the pure mathematics of the present time and often burgeons into mountains of triteness and bad taste.

I had asked Gianni to be the first speaker with a talk under the programmatic title *Topology and symmetries in quantum mechanics on manifolds* he himself had chosen for our conference. He did not feel comfortable with his opening role. After all, he and Holger Bech Nielsen from Copenhagen were the two sole senior physicists participating in the conference. But he ensured me: "I have appreciated very much the quotations that you proposed for the spirit of the conference."

Gianni was accompanied by his wife Antonella. Their children were too old at that time to accompany their parents to Scandinavia. But for him and Antonella it was the first time they were in Sweden—and in gorgeous weather!

Two days later, we took the afternoon off and strolled together through the historic center of Stockholm. Then he came back to my quotes with his typical engaging smile—and a serious admonition: "As I wrote to you, I have appreciated your quotes, but do not push your fellow mathematicians. That will only increase the mountain of papers with physics terminology but without physics meaning. You may deplore the abstract turn of mathematics, but it will not help to return to times where math and physics were united. Now, we belong to different fields of research, different working traditions, different quality criteria. You need not accept the schism and, hopefully, keep on creating platforms where the two communities may meet like in this conference. Yet, do not push your fellow mathematicians ..."

Gianni and I never discussed politics. I do not know why we avoided that topic. Perhaps because I had the impression that Italian politics and the perpetual reshuffling of positions and social strata belong to the very, very few truly non-intelligible subjects in the world. Gianni on his side may have felt uneasy to address the interdiction, purely based on political reasons, from continuing working as mathematician, as it was experienced in my home country. Reading now the biographic contribution *Chi ha compagni non morirà* (Part I), it strikes me how deep-rooted Marx's thinking about the past, the present, and the future must have been in Gianni's thoughts about physics and mathematics, as well, when he ridiculed my *romantic yearning* for *more fully developed individuals* and my recalling of *a merely local connection resting on blood ties, or on primeval, natural or master-servant relations* (such as between mathematics and physics, i.e., Marx' *Bluturenge*), and at the same time shared my discontent with *this complete emptiness*, i.e., the common disregard for humans' mutual relations and for the necessity of constant bridging the gap, beyond lip service and occasional touch.

For comparison with my preceding quote from Gianni, I cite some lines on *alienation* by Karl Marx, *Grundrisse*, p. 77–78, where Marx himself points to analogies with our perception of science at the end of the quote:

> In earlier stages of development the single individual seems to be developed more fully, because he has not yet worked out his relationships in their fullness, or erected them as independent social powers and relations opposite himself. It is as ridiculous to yearn for a return to that original fullness as it is to believe that with this complete emptiness history has come to a standstill. The bourgeois viewpoint has never advanced beyond this antithesis between itself and this romantic viewpoint, and therefore the latter will accompany it as legitimate antithesis up to its blessed end. (The relation of the individual to science may be taken as an example here.)[2]

There Was No Gianni Without Antonella: Memories with a Generous Host Couple

From our first meeting in Pisa I recall that we all were tired after some hours of explaining things to each other in different terminology. Then Gianni had his smile and announced: dinner is served in one hour. Trautman and I could shortly return to our hotel, and then we met at Gianni's and Antonella's comfortable flat, also in

[2] English translation from https://www.marxists.org/archive/marx/works/1857/grundrisse/ch03.htm.

picturesque downtown. From then on we met almost all evenings at Antonella's dining table. Nobody knows how she could be fresh after a days teaching in senior high school, initiate intriguing conversations—and serve the most delicious food for us. Usually, Franco's wife Anna, an art historian and valuator, was with us, and later Sussi as well, my late wife, a military historian. She accompanied me on all my later travels to Pisa and we felt always welcome in Antonella's and Gianni's home.

Talking about math was not forbidden. However, as I remember, we spoke mostly about Antonella's school—and about history, history of art and Italian history. These evenings will remain unforgettable for me.

References

1. Booß-Bavnbek, B.: The Determinant of Elliptic Boundary Problems for Dirac Operators. Reader on The Scott–Wojciechowski Theorem (1999). http://thiele.ruc.dk/~booss/ell/determinant.pdf. 203 pages, mostly corrected, mostly completed, compiled, edited, revised, augmented
2. Booß-Bavnbek, B., Morchio, G., Strocchi, F., Wojciechowski, K.P.: Grassmannian and chiral anomaly. J. Geom. Phys. **22**, 219–244 (1997)
3. Booß-Bavnbek, B., Wojciechowski, K.P.: Elliptic boundary problems for Dirac operators. Mathematics: Theory & Applications. Birkhäuser Boston, Inc, Boston, MA (1993). https://doi.org/10.1007/978-1-4612-0337-7
4. Einstein, A., Born, M., Born, H.: The Born-Einstein Letters: Correspondence Between Albert Einstein and Max and Hedwig Born from 1916-1955, with Foreword by Bertrand Russell, Introduction by Werner Heisenberg, and Commentaries by Max Born. Macmillan (1971). ISBN 978-3-88682-005-4. Translated into English by Irene Born from the German edition, Munich, 1969, Nymphenburger Verlagshandlung
5. Harvey, W.: The Circulation of the Blood and Other Writings. Everyman: Orion Publishing Group, London (1993). English edition of "Exercitatio anatomica de motu cordis et sanguinis in animalibus" of 1628 and other original articles in Latin (trans: Franklin KJ). Introduction by Dr. Andrew Wear
6. Morchio, G., Strocchi, F.: Boundary terms, long range effects, and chiral symmetry breaking. In: Mitter, H., Schweifer, W. (eds.) Fields and Particles, Proceedings Schladming, Austria. Springer, Berlin, Heidelberg, New York (1990)
7. Ninomiya, M., Tan, C.-I.: Axial anomaly and index theorem for manifolds with boundary. Nucl. Phys. B. **257**, 199–225 (1985)
8. Schempp, W.J.: Magnetic Resonance Imaging: Mathematical Foundations and Applications. Wiley, New York, Chichester, Weinheim (1998)
9. Schmidt, J.R., Bincer, A.M.: Chiral asymmetry and the Atiyah-Patodi-Singer index for the Dirac operator on a four-dimensional ball. Phys. Rev. D **3**(35), 3995–4000 (1987)
10. Scott, S., Wojciechowski, K.P.: The ζ-determinant and Quillen determinant for a Dirac operator on a manifold with boundary. Geom. Funct. Anal. **10**(5), 1202–1236 (2000)
11. Trautman, A.: The Dirac operator on hypersurfaces. Acta Phys. Polon. B **26**, 1283–1310 (1995)
12. Wojciechowski, K.P., Scott, S.G., Morchio, G., Booß-Bavnbek, B.: Determinants, manifolds with boundary and Dirac operators. In: D., V. (ed.) Clifford Algebras and Their Apllication in Mathematical Physics, pp. 423–432. Kluwer Academic Publishers (1998)

Memories of Gianni

Costantino Budroni

I first met Gianni when I was an undergraduate student at the University of Pisa. I attended his lectures on algebraic methods in quantum mechanics. Gianni was an extraordinary teacher, as many remarked. From his lectures, which included topics such as the Bell and Kochen–Specker theorems, I was particularly amazed by the discovery that basic questions about quantum mechanics were still unanswered. Encountering these problems, presented from Gianni's perspective of doubting and questioning everything, was a pivotal moment for me. After taking his course, I asked him to be the supervisor of my MSc thesis, which ended up being on the problem of classical representability for partial Boolean structures in quantum mechanics, i.e., commutative subalgebras of projectors and the associated probability measures.

During the writing of my thesis, but also in the several months I spent in Pisa after my graduation, we would meet once a week, sometimes even more often, and discuss for hours. Gianni was an inexhaustible source of profound observations, always able to point at the root of a problem and formulate it in a clear and precise way. After these meetings, I would come back home with several pages of notes handwritten by Gianni during our discussions, and I would go through them in the following hours and days, trying to reconstruct all that we said and all the details I may have missed. After I left Pisa, we remained in contact, and I periodically came to visit him. Gianni was always eager to discuss with collaborators and students alike. One could talk or write to him, asking for his opinion or advice about anything, and he would always have some acute observation, as if he already reasoned about that problem for a long time.

Going back to our email exchange, I'm still astonished today by the depth of his remarks, which may spread across the most diverse topics. I could mention several examples, but I'll just pick one, which is, in my opinion, representative of the

C. Budroni (✉)
Dipartimento di Fisica "E. Fermi", Università di Pisa, Pisa, Italy
e-mail: costantino.budroni@unipi.it

A. Cintio, A. Michelangeli (Eds.), *Trails in Modern Theoretical and Mathematical Physics*, https://doi.org/10.1007/978-3-031-44988-8_17

combination of Gianni's intellectual integrity and his awareness of (and approach to) the social issues.

In one of our last email exchange, in which I was expressing my dissatisfaction with some of the dynamics and practices of the academic system (in particular, in theoretical physics), Gianni wrote:

> Io credo che [...] invece ci sia un grandissimo bisogno di un pensiero 'scientifico' vero nel senso più largo [...] che miri alla diffusione di strumenti critici, di applicazione universale, e in questo io credo che l'astrazione (direi quella RESPONSABILE, che non inventa problemi, ma li RICONOSCE e li analizza) abbia un ruolo POSITIVO essenziale.

An approximate translation would be:

> I believe that [...] instead there is a great need for a true 'scientific' thought in the broadest sense [...] which aims at the dissemination of critical tools, of universal application, and in this I believe that abstraction (I would say, the RESPONSIBLE one, which does not invent problems, but RECOGNIZES them and analyzes them) has an essential POSITIVE role.

Meeting with Gianni had a significant impact on my scientific career and on my life in general, as it was the case for many of his students and collaborators. I rarely met scientists with such a capability of deep and clear thought, and its combination with his kindness and human qualities is almost unique. I'm extremely grateful for everything I learned from him, for his guidance and patience, and his dedication to the role of teacher and mentor. Grazie Gianni.

Morchio's Axe

Paolo Christillin

There are two anecdotes that I would like to add to what was said at Gianni's funeral and to the tribute paid to him back then, as on that occasion I didn't feel like sharing them, in order to avoid further emotion.

As at the physics department my office was in the same corridor as Gianni's, I happened to speak to him multiple times, and yet I had always had some sort of restraint for possible joint scientific collaborations.

It seemed to me that he knew everything and that he was way better than me!—even if I did appreciate the concreteness with which he didn't disdain to make numerical estimates, if necessary.

However, some time earlier I was contacted by a layman who, as typical, thought he was revolutionising physics and was looking for someone from the academy willing to listen to him. Naturally, I accepted to talk to that folk, and to my surprise I discovered that the only other person in the department who had offered to do so was Gianni.

So I dropped my previous reserve, and Gianni and I eventually started talking about physics. This way we discovered that we were two iconoclasts—the both of us, albeit in different ways: he, because he thought nothing was rigorous enough; I, because I reckoned that nothing was derived along a simply enough path.

This common attitude lead us to deal with General Relativity. I think we were both very proud of the result of this collaboration because it showed that results of General Relativity could be obtained in a much simpler way in another metric (Painleve–Gullstrand), using essentially only Special Relativity.

That turned out to be quite an onerous work, as I often used to tell him, jokingly: "*You see, Gianni, once there was Occam's razor, but now we have to face Morchio's axe*" (because his criticisms were always punctual and severe).

The reactions were obviously totally negative! How could we poor people even think to commit such an act of treason. With one exception: once, reaching the

P. Christillin (✉)
Dipartimento di Fisica "E. Fermi", Università di Pisa, Pisa, Italy
e-mail: christ.pao@gmail.com

A. Cintio, A. Michelangeli (Eds.), *Trails in Modern Theoretical and Mathematical Physics*, https://doi.org/10.1007/978-3-031-44988-8_18

printer to fetch a copy of the manuscript, I found a colleague who inadvertently had already taken it as part of his printouts, and when he realised that below the title one of the authors was Gianni he commented: but then it's serious stuff!

Instead, at the umpteenth specious and insulting review report, Gianni blurted out: *goddamn!* It was the only time I saw him over the top. This second episode unsettled even his Savoy restraint. In fact, at home I thought: today it is as if I had heard the Pope blaspheme.

Goodbye Gianni.

Years of Teaching Collaboration with Gianni

Giampaolo Cicogna

I begin my contribution (actually, a brief contribution, Gianni would deserve much more) with a personal memory going back several decades, when Gianni, a third-year student, came to take the exam of my course of '*Mathematical Methods of Physics*'.

Among the several thousand students I have examined in my long career, Gianni's is the oral exam I still remember well.

In fact, having checked that his written test was all correct, I had posed a rather challenging question to him. His answer was prompt, but with the use of technical terms far beyond the standard syllabus. I remember being almost annoyed by this and was led to think that he was the "know-it-all student" trying to impress the examiners by exhibiting big words he does not even know about. Then I tried to delve deeper into the subject and, to my surprise, I found that Gianni had perfectly mastered the topic, in which he moved with confident and correct language, even in depth, and with absolute naturalness (to be precise, the argument concerned Lie algebras, a topic that at that time was barely mentioned and only for a few simpler cases).

Then, even as a student, Gianni demonstrated his characteristic gifts that we have all come to appreciate: a deep knowledge and enormous culture concerning any subject of physics and mathematics, together with a disarming naturalness in discussing them.

Unfortunately, I never interacted with Gianni on matters of scientific research, but in return I established with him a legendary ironclad partnership of teaching collaboration.

When Gianni was assigned to my class on mathematical methods to run tutorials and exercise sessions, I was obviously well pleased, knowing his outstanding preparation, but I was also somewhat embarrassed, as that role formally placed him in a "subordinate" position with respect to the course's lecturer. On the other hand (and

G. Cicogna (✉)
Dipartimento di Fisica "E. Fermi", Università di Pisa, Pisa, Italy
e-mail: giampaolo.cicogna@unipi.it

A. Cintio, A. Michelangeli (Eds.), *Trails in Modern Theoretical and Mathematical Physics*, https://doi.org/10.1007/978-3-031-44988-8_19

this is another aspect of his character), Gianni never had ambitions for advancement in his academic career: this was not for some sort of exhibition of nonconformism or display of originality, but simply because he was perfectly happy as he was.

Our teaching collaboration lasted for several years until my retirement, and it was exceptionally happy and fruitful.

Gianni was always precise and precious in helping the students of my courses with illuminating examples and discussions, careful and scrupulous in the formulation of the exercises to be proposed to the students (I do not recall ever having presented assignments that were not well-balanced, instructive and meaningful—I trust none of my past students would contradict me!), invaluable in correcting papers and in oral examinations, where our understanding was perfect and resulted from the very beginning completely "spontaneous".

Other endowments I remember: Gianni's discretion and restraint in making judgments. I do not think I ever heard from him a rave or heavy critique of people or anything else.

And, in this regard, I want to end this contribution to Gianni's memory just as I began it, that is, with a personal recollection. I had assigned to a student, as a topic for his B.Sc. thesis, an introduction to Noether's Theorem; after carefully reading the pages written by the student under my supervision, and after listening attentively to the candidate's exposition, Gianni stood pondering for a few moments, also causing me some disquiet; finally, he simply said to me: "This thesis should be read by all third-year students from now onwards." In my career, I think this was the best judgment of my teaching activity.

Stirring of the Conscience

Andrea Cintio

I got to speak with Gianni for the first time ever in the afternoon of one day in Autumn. He had been running the exercise sessions of the one-year long course on mathematical methods for physics taught by Giampaolo Cicogna. I was preparing for the exam of that course and, particularly, I was studying the part of the program on group theory. I was well aware of the elegance of the theory, yet not able to grasp how its concepts could be applied. So I met with Gianni and together we went through topics still unclear to me about representation theory and characters of finite groups. I was impressed how Gianni's explanation was as explicit as a description of something in plain sight can be. I could appreciate his subtle ability in easily moving through the topic using a language that provided me with the vivid impression of "surgical" precision.

For their effective teaching collaboration, Cicogna–Morchio are an inseparable pair in my imagination. Together they formed one thing, but at the same time they were complementary. Giampaolo Cicogna was masterly in exposition, approaching a topic through the analysis of various examples that were frequently chosen as often appearing in the literature, hence meaningful. However, this was a course on *applicable* mathematics: the role of the physical applications herein was to illuminate the mathematics, not the other way around. The aim was essentially to lead students to the point where they can manage the quantum mathematical machinery in the books and papers of physics. Gianni, on the other hand, was used to digress in the course of an exercise, which was the occasion to trace connections to often quite difficult questions that were explored introducing advanced topics: hence, the original problem appeared in a different light and as a source of deeper insight. Just as a small example, I remember a lecture when Gianni talked about *Sobolev spaces*: the subject is of central importance for the study of differential equations but it sounded rather "exotic" to me back then. Gianni's approach would require a significant workload and, maybe, more than the needed mathematical knowledge

A. Cintio (✉)
Institute for the Chemical-Physical Processes, National Council of Research, Pisa, Italy
e-mail: andrea.cintio@pi.ipcf.cnr.it

A. Cintio, A. Michelangeli (Eds.), *Trails in Modern Theoretical and Mathematical Physics*, https://doi.org/10.1007/978-3-031-44988-8_20

for the solution of an exercise for the exam. However, an audience receptive to it was every student who both is interested in a conceptually clear and mathematically precise understanding of physical problems and tries to avoid *formally* applying a method only why it works. Gianni's exercise sessions were an amazing opportunity for a first acquaintance with some material that practicing (functional-)analysts and operator-theorists put into their kit-bag.

When I had to choose an elective course for the M.Sc. study "program in theoretical physics" I had no hesitation in picking the one Gianni was delivering to senior students interested in quantum mathematical physics. Unquestionably, I had not been able yet to realize how privileged I was: more than twenty-five years ago, a course like Gianni's on mathematical structures of a rigorous formulation of Quantum Mechanics was such a rare opportunity for a physics undergrad at an Italian university. What likely guided my choice was the impression formed during the previous class on mathematical methods: a good-natured and gentle person whose lectures were an enhancement of an ordinary exercise session to something with breadth and depth of approach to the field.

Gianni's course on quantum mechanics was a real challenge to get through, but more than worth the effort. The two-hour-long classes took place in the afternoon, twice a week. I have a clear memory of when Gianni used to start each class with a brief recap of key concepts from the previous class. He, then, began to write on the blackboard and filled it from one side to the other, many times in the course of the same class. Sometimes, he went back and wrote on a still blank small area in order to make something more precise, thus making the blackboard completely full. Gianni presented complete proofs of the theorems when it was not too onerous to do so. He often simply sketched the technical parts of an argument but nevertheless he always spent a bit of time for exploring the ideas behind the technicalities.

The course covered some mathematical structures required for the standard formulation of Quantum Mechanics, with an ample perspective that included coherent and conceptually rigorous pictures of both a relativistic field theory and a theory of a large (infinite) system. The main topics of the course were the *probabilistic interpretation* of the expectation values of operators in Hilbert space (*spectral theorem*), the theory of self-adjoint extensions of symmetric operators, the identification of observable algebras and the classification of the corresponding states, a discussion on how a (partial) probabilistic interpretation of Quantum Mechanics departs from a classical probabilistic one.

For my preparation for the exam I almost exclusively read through the priceless lecture notes prepared by Gianni (my own ones were unhelpful, I was never able to take adequate notes from class!). In them, the course topics were presented in a condensed manner. Only by having an in-depth and thorough understanding of the subject matter could one "distill" the course contents down to their core, as Gianni managed to do with his lecture notes. I surely spent quite an amount of time going through those pages, filling in routine arguments, trying to decipher all steps. It was only after many further readings that precise concepts started arising naturally to me, and I became able to distinguish fundamental points from side ones, making the former crystal clear to me. Then, a glow of satisfaction welled up inside me.

Similar experiences have been rare and implied invaluable boosts in my understanding. It was nearly the same as when I got through the third volume of the Landau–Lifshitz series: the first time I browsed through it I believed I understood everything but a much bigger world opened up to me only after much more work. This personal "short list" also includes Arnold's *Mathematical methods* and *Local Quantum Physics* by Haag.

My inclination towards rigorous physical thinking had started before meeting Gianni, but I tended to lean towards a misconception of the connection between mathematical formalism and physical theory. Captivated by the "intrinsic beauty" of a mathematical structure, I was spoilt by an overattention towards details, focussing with missionary zeal on theorems' proofs, collateral properties and the like, but loosing the physical insight. "Theory", meaning the study of the ideas underlying the subjects and the reasoning behind the techniques, is not synonymous of "logical rigor", but the latter without the former is like Popeye without his spinach.

While supervising me for the M.Sc. thesis, Gianni, as respectful and considerate of my views as he always was, often urged me to re-think such "unbalanced" approach of mine. Also, it is thanks to Gianni's determination that, after fumbling around far too long in a fog, I was able to kick-start the writing of the thesis and get the first ideas written down on paper, even when they were still anything but rigorous.

The aim of the thesis was to determine the energy spectrum associated with density waves, for large wavelength, in large quantum systems of nonrelativistic fermions, from a minimal set of assumptions on dynamics. Energy spectra are expressed in terms of time derivatives of commutators: I computed many many commutators, but that was not the whole story. A part of the thesis contains some results on the mechanism of gap generation associated to a spontaneous symmetry breaking for systems with long range interactions that Gianni and Franco Strocchi obtained adopting a somewhat different approach in previous articles.

I have an amount of recollections of when I used to visit him on a weekly basis at his office on the first floor of the "*Edificio C*" (Building C) in the course of my work on the final thesis, but without precise details of the specific circumstance. I remember Gianni sitting at the desk, willing to answer any question and in front of him a bunch of paper ready for experimenting with various ideas. While Gianni was engaged in solving a problem I got the distinct impression that any single term of the equations on the sheet gave a sort of "spoke" to him and Gianni got the message ... His mathematical manipulations looked as mere operations to me whereas, in Gianni's mind, I am sure they were a mirror of physically insightful steps, mechanisms, underlying connections. Gianni's penchant for "breaking down" a problem into more fundamental blocks was the expression, it seems to me, of his imperative to understand everything in the simplest possible way. This was reflected in the elegant articulation of his arguments.

During our meetings, Gianni patiently answered the numerous questions I used to pose to him—of course, I am the sole responsible if many of his answers were comprehended by my brain in a possibly distorted or incomplete form. Every time I left Gianni's office I was enriched by the ideas he shared with me, but in the presence of an intelligence, a sensibility of such profundity and originality I realized that I felt myself mostly filled with wonder and excitement.

After graduating, I maintained contacts with Gianni, only occasionally though. It was customary that former students stayed in touch with Gianni, even though they were far away from Pisa. After all, losing touch with a person such as Gianni would have been no mean feat. When I visited him, Gianni was curious ("*Tell me what you are doing now*") and left me with the feeling of a genuine desire to learn, every time. The questions about what I was doing were pertinent even if the subject was far different from Gianni's research work and I often got into trouble for answering back.

When Gianni was engaged in a discussion with a student, it struck me that he often talked about scientific (and not scientific) matter as he was speaking to someone on the same level as his. It was clear to me that Gianni had no intention of causing difficulties to anyone. Only later I realised that that attitude of his was in fact his actual belief in *reason* and *human consciousness*, in the ability of everyone in understanding.

One day he told me that in the present times there would be an urgent need for a reclamation of the views of the *Enlightenment*. The *postmodernity*, i.e., the current phase in social organization, represents a resurgence of long running counter-enlightenment ideas, such as that of "death of history", the criticism of *universalism* (as opposed to *relativism*), which had an influence on the *theory of Right* according to the conception of "human nature" of the Enlightenment, the denial of normative ideals, central for democratic action, the increasing public interest in anti-scientific beliefs and myths. Kant spoke of reasoned knowledge as a means of achieving genuine freedom and equality, that are now under threat of being weakened. I am not an expert in political theory, but that was, I believe, what Gianni actually meant.

I last heard from him during the pandemic. In an email he wished there was a "stirring of the conscience" after the pandemic and people would stage a serious protest about what did not work. If Gianni was still here he would be disappointed, as am I: not only people did not wake up but also they are in a "deep sleep".

Gianni Morchio and SISSA

Gianfausto Dell'Antonio

Gianni Morchio had a central role in the development of the research activities hosted at SISSA (the International School for Advanced Studies, Trieste) in the field of the mathematical methods for quantum mechanics, even without ever being a faculty member there.

He has been indeed the first "master" of many young students who entered SISSA after completing their M.Sc. in Pisa and subsequently wrote valuable Ph.D. theses for the doctoral degree in mathematical physics. Through his classes of quantum mechanics and of mathematical methods for physics they arrived at SISSA equipped with solid grounds with which they could successfully grapple with the problems they later worked on in the course of their doctoral researches.

To better understand all this, it is worth recalling that SISSA had started its activities as a post-graduate school ("Advanced School of Physics") after the M.Sc. degree, at a time when no proper doctoral programmes were run in Italy. It was born out of an idea of the physicist Paolo Budinich, inspired to the recent foundation of the ICTP (the International Centre for Theoretical Physics), an institution operated by the UNESCO. The goal was to convince the Italian government to establish, next to the ICTP, a post-graduate school on scientific disciplines, somewhat linked to the Italian academic system.

Initially SISSA was therefore a school connected to the ICTP, but belonging to the Italian academy through the university of Trieste. The faculty was then formed by a small group, very close-knit, of extraordinary lecturers and scientists (among which, Ambrosetti, Cellina, Sciama, Tosatti, ...). Classes were attended by researchers of the neighbouring ICTP, as well as by students and researchers of the university of Trieste; the contact with lecturers was direct, and straightforward to establish on every single day. This way SISSA soon became an autonomous in-

G. Dell'Antonio (✉)
Department of Mathematics, La Sapienza University of Rome, Rome, Italy
Division of Mathematics, Scuola Internazionale Superiore di Studi Avanzati, Trieste, Italy
e-mail: gianfa@sissa.it

A. Cintio, A. Michelangeli (Eds.), *Trails in Modern Theoretical and Mathematical Physics*, https://doi.org/10.1007/978-3-031-44988-8_21

stitution, granting doctoral degrees at the end of a study programme consisting of courses and research activities, culminating in the defence of the Ph.D. thesis.

I joined SISSA "by chance": in a fortuitous scientific meeting in Munich, Budinich asked me to visit SISSA for a few months and to deliver a graduate class on quantum mechanics. Since then, I kept teaching courses on the mathematics of quantum mechanics and quantum field theory; initially they were one-semester long, later re-sized to a two-semester-long class, when for some years I was detached full time to SISSA from La Sapienza (the university of Rome) in order to supervise more closely a number of doctoral theses, and again after my retirement from La Sapienza, where I had meanwhile become an emeritus professor.

SISSA was at that time a stimulating environment indeed. A few research groups (the future "Sectors" of SISSA) of notable scientific calibre were then active in particle physics, solid state physics, mathematical analysis, astrophysics, neuroscience: they shared small spaces, thereby naturally allowing for close interchange of ideas. Under the impetus of Budinich additional internal structures were meanwhile taking shape: the Interdisciplinary Laboratory and the Science Centre Immaginario Scientifico.

A few years later doctoral programmes were finally established in many Italian universities, thus making SISSA formally a doctoral school "as the others", albeit without internal Bachelor and Master programmes prior to the doctorate.

In fact, the latter feature turned out to be an advantage for SISSA in that initial phase. Indeed, at the beginning other universities' new doctoral schools happened to experience many difficulties and delays, both of organisation nature, and concerning the choice of the research lines and the admission procedures to access fellowship funds—clearly, on the long term the lack of internal B.Sc. and M.Sc. programmes may turn into a disadvantage, as it limits the scientific offer, may constitute an obstacle for the natural development of new research lines, and complicates the organisation of more advanced classes for Ph.D. candidates.

Thus, while other Italian doctoral schools were initially struggling with organisation, bureaucracy, choice of structures and curricula, SISSA was already well structured and operational, and this resulted into the circumstance that for some years excellent students coming from several Italian universities turned out as applicants to enter SISSA. Among them, the students from Pisa coming from "Morchio's school".

For a fistful of years in a row SISSA awarded the doctoral degree to a number of Gianni Morchio's former undergraduate students, including Riccardo Adami (now at the university of Turin), Alessandro Pizzo (now in Rome), Michele Correggi (now in Milan), Alessandro Michelangeli (now in Bonn), Giuseppe De Nittis (now in Santiago de Chile), as well Lucattilio Tenuta and Emanuele Costa (who then had a brilliant professional career outside of the academia), all supervised by me, and also Fabio Bagarello (now at the university of Palermo), Dario Pierotti (now in Milan), and Stefano Cavallaro, supervised by Franco Strocchi at SISSA and, externally, by Morchio himself, to which one should also add Paola Ruggiero (who got her Ph.D. in Statistical Physics at SISSA), and Luca Sciortino (who joined SISSA for the Master in Science Communication, one additional flagship structure of SISSA).

Of Riccardo Adami I have been also the supervisor for his M.Sc. thesis in Pisa (Gianni Morchio then acting as "internal supervisor"), on a topic of particular interest in quantum mechanics, the Bohm–Aharonov effect. This gave me the opportunity to get in contact with the department of physics of the university of Pisa and to interact with Gianni more closely, as until then I had much enjoyed his works, yet only getting to know him superficially. In fact, I already knew well a bunch of mathematicians in Pisa, but I had no previous link with the physicists.

I had already read and appreciated what Gianni and Franco Strocchi had written on gauge theories and the problems arising in particular from gauge invariance and from indefinite metric. The discussions that I then had with Gianni during my frequent visits in Pisa allowed me to deepen my knowledge of those topics and to acknowledge the accuracy of Gianni's reasonings and the depth of analysis of the associated mathematical formalism.

It then became clear to me how it comes that Riccardo Adami, as well as, later, the other students of Gianni who then wrote their doctoral thesis with me, did have such a solid and not only formal knowledge of the quantum mechanical formalism and mathematical structure, and did possess the capability of catching quickly essential aspects to concentrate the main focus on, for the study of the problems under investigation.

In the course of my visits in Pisa I could value Gianni's ability of identifying the core parts of a problem or of a theory and dealing with complex formalisms with the greatest confidence. I also attended a number of local scientific meetings organised by Gianni for students and researchers (at the "*Domus Galileiana*", as far as I remember) on particularly significant physical subjects. Always, during such conferences, Gianni used to draw attention to the crucial, fundamental elements of the physical and mathematical discussion.

What then followed were stimulating debates on "conceptual" aspects: they reflected Gianni's desire to involve young people in conceptual problems and not just in mathematical formulations. During these meetings it was underlined that for a complete analysis the "foundational" issues should not be overshadowed. In other words, I had a clean perception of the scientific environment which Gianni's students could move across, which accounted for their so refined preparation.

Just Give It a Go!

Giuseppe De Nittis

It was the end of 2004.

At that time, I was a disoriented and unmotivated student reaching the completion of the M.Sc. programme ('laurea specialistica') in theoretical physics at the University of Pisa. My grade point average was excellent, but I had taken more time than necessary and furthermore over the years my initial passion towards theoretical physics had given way to an exciting admiration towards the stringent rigour of mathematics.

I remember myself as a not-so-young and somewhat confused senior in search for a subject to write his M.Sc. thesis on, which cover a problem of interest in physics and be at the same time mathematically deep rooted. In the physics department quite an amount of professors had a profound knowledge of mathematics, but "*Professor Morchio knows much more*", as a fellow mate used to tell me to encourage me to fix a meeting with Gianni.

It was in such a position that one day I finally went to Gianni's office and there, after a brief conversation, my adventure in the world of mathematical physics actually began.

For one year, I worked under the careful supervision of Gianni on a problem about the *Return to Equilibrium for Large (i.e., Infinite) Systems*. I cut my teeth on classical literature (Reed–Simon, Bratteli–Robinson, etc.) and filled in my gaps in spectral theory and operator algebras. And, most important of all, I learned from Gianni how a mathematical physicist thinks and imagines. Gianni's explanations and reasoning were always of an impressive depth and a bewitching elegance. Under his guidance I started loving his work and admiring his *modus operandi*.

After a little over a year I completed my thesis and I was awarded the final degree with full marks. This was the last act of my academic career—at least, that was what I was convinced of!

G. De Nittis (✉)
Departamento de Matemática, Pontificia Universidad Católica de Chile, Macul, Chile
e-mail: gidenittis@mat.uc.cl

A. Cintio, A. Michelangeli (Eds.), *Trails in Modern Theoretical and Mathematical Physics*, https://doi.org/10.1007/978-3-031-44988-8_22

Indeed, having concluded my studies with a delay of a couple of years, and that making me somewhat "older" than my colleagues, I felt doubtful about the idea of pursuing a doctorate, which I was about to desist from. I had concocted other, definitely more modest plans. Or, I would have settled for a position as a maths teacher in high school.

Then, a fistful of days after defending my M.Sc. thesis, Gianni invited me to his office to talk about my future and there I expressed my hesitations and uncertainties. With his typical kindness, and with an initiative that looked to me both amiable and uncommon, he asked me to to make at least one attempt: "*just give a go to the admission at SISSA in Trieste, and then what is meant to happen will happen*", he said.

I accepted his invitation and trusted his words more than I trusted my own abilities. The entrance examination went unexpectedly well, as I was offered a doctoral scholarship at SISSA and there, against all my personal expectations, I eventually obtained my Ph.D. degree four years later, under the guidance of Gianfausto Dell'Antonio.

What followed is the (probably uninteresting) story of my academic career, consisting of some postdoctoral years, travels, conferences, papers, and finally a permanent academic position at the *Pontificia Universidad Católica de Chile*.

However, if today I am who I am, and if I achieved what I did achieve, the credit goes to that lucky and distant beginning, when that nice and shy man with tousled white hair addressed me and urged me to have faith in me and in my dreams.

Thank you for all this dear Gianni!

Reminescences of Gianni Morchio

Jürg Fröhlich

Gianni Morchio and I first met each other in Zurich, in 1971. His advisor *Franco Strocchi* had organized a stay for him at ETH Zurich, so that he would have a chance to profit from discussions and advice by *Klaus Hepp*. I was then one of Klaus' PhD students. Gianni and I shared an office in the theory institute on the Hönggerberg campus of ETH. This gave us the chance to engage in countless discussions of various problems in mathematical physics and to develop a very friendly relationship. We also discovered that we were both actively interested in matters of society and politics.[1] I liked Gianni's soft-spoken, friendly manners in combination with his clear, uncompromising views and opinions. Regrettably, we did not socialize much, as we were both busy working on our theses.

After our first encounter we did not see each other anymore, for several years. When I had completed my PhD degree, in the summer of 1972, I first spent a year at the University of Geneva and then moved to the United States where, after a year at Harvard, I got a research and teaching job at Princeton University. During my stay at Princeton, I got to know Franco, who was invited to visit Princeton in order to collaborate with the late *Arthur S. Wightman* on various problems in quantum electrodynamics (QED). This was the beginning of a friendship between Franco and myself that I have cherished ever since and that eventually also included our wives.

At the beginning of 1978, my family and I moved to France, where I had been offered a job at the IHES in Bures-sur-Yvette. At that time, Franco, Gianni and I had discovered our common interests in various problems of QED, in particular in the infrared problem[2] and some of its consequences (e.g., concerning the action of Lorentz boosts on the state space of QED). After some amount of long-distance

[1] Gianni was much more courageous and committed to political activism than I was, and he payed a price for his convictions.
[2] A problem that *Alessandro Pizzo* and I returned to and worked on in more recent years.

J. Fröhlich (✉)
Institute for Theoretical Physics, ETH Zürich, Zürich, Switzerland
e-mail: juerg@phys.ethz.ch

A. Cintio, A. Michelangeli (Eds.), *Trails in Modern Theoretical and Mathematical Physics*, https://doi.org/10.1007/978-3-031-44988-8_23

collaboration—at that time through the medium of hand-written letters—I had the chance to invite Gianni to visit the IHES in order to speed up and deepen our collaboration. Franco, Gianni and I successfully completed some joint papers on QED that, for all I know, are of good quality and are still of interest. We also wrote two papers on a manifestly gauge-invariant formulation of the Higgs mechanism. This work appears to have become quite popular in recent years. I found my collaboration with Gianni and Franco very enjoyable. For a variety of reasons—not least because our interests diverged somewhat—it did unfortunately not continue.

Gianni's visit at IHES gave us a chance to socialize and talk about matters other than physics. I discovered that Gianni was an excellent cook. For my family and myself he prepared what have probably been the most delicious gnocchis I have ever been served.

In 1982, I returned to ETH where I taught theoretical physics for almost three decades. I kept up some connection to Franco, who kindly invited me to visit Trieste and Pisa several times; but I lost sight of Gianni. When my retirement approached I decided that I could and should finally work on problems of physics that, while being truly fundamental, may never be solved to anybody's satisfaction. Among other such problems I started to think about how one might go ahead and complete quantum mechanics to a theory with a clear ontology and without a measurement problem. (A glimpse of some of our efforts in this direction is provided in my contribution with *Del Vecchio, Pizzo* and *Ranallo* to this volume.) My ideas on quantum mechanics have not gained much popularity, yet – actually, my situation reminds me sometimes of the one of the man from the countryside who tries to get access to the "Law", described in Kafka's short story entitled *"Before the Law"*—although, with some optimism, this will surely change with time. Among the few people who apparently decided that my ideas were of some interest were my friends in Italy, including Gianni, Franco and one of their former students, namely my close friend and collaborator Alessandro Pizzo. Gianni and Franco invited me to visit Pisa for discussions and to give a talk about my ideas. Anna Strocchi took us on a wonderful guided tour of some of the famous sites of art and architecture in Pisa, and Gianni and his wife invited my wife, the Strocchi's and me to spend an unforgettable evening at their apartment. I found out that Gianni owned an impressive collection of books including—not surprisingly—a good number of books on politics and political philosophy.

Sadly, those few days in Pisa were my last encounter with Gianni. I am glad that destiny has made me meet and befriend him.

Thoughts of Gianni

Jürgen Löffelholz

I met Gianni precisely on 18 September 1983 in Leipzig during a conference. After his evening lecture on "Local States in QED and Gauss Law" we had a short talk about the subjcct. We agreed next day to skip lunch and go for a short excursion with my car to the Monument to the Battle of the Nations. He showed up together with his future wife Antonella. That was yet six years prior the fall of the Berlin Wall and I did not have a passport for the West. But we were young and we somehow believed in the future. For the time being we exchanged ideas by postal mail.

In 1991 I received an invitation to the department of physics in Pisa. When the train stopped at *Pisa Centrale* (Pisa's main rail station) Gianni was waiting for me in the rain with his almost five years old son Iacopo at his side. We embraced each other and from that moment I had a true friend.

Gianni was almost the same age as me. We both had young families and hence we discussed about life. We tried to understand why East Germany and other countries did not really succeed with building an alternative with respect to capitalistic economy. And what now?

At the University of Pisa a *corso di lingua italiana per stranieri* (Italian language course for foreigners) was announced for the period February-April 1994. I had lost my job in Leipzig like about three thousand other scientists at the former Karl-Marx-Universität. So I could come again to Pisa.

Usually, we used to meet in Gianni's office or at the Scuola Normale Superiore together with Franco Strocchi. Our common interest was to analyze gauge QFT models within the functional integral formulation at imaginary time. In a further paper we analyzed a gauge model of QED with Theta-vacua given by a complex-valued Euclidean functional measure. We worked together over 15 years, which was wonderful and I will never forget. From the both of them I benefited very much in science and human relations, something I will never forget and I am infinitely grateful for.

J. Löffelholz (✉)
Angewandte Informatik, Fachhochschule Erfurt FHE, Erfurt, Germany
e-mail: jloeffholz@web.de

A. Cintio, A. Michelangeli (Eds.), *Trails in Modern Theoretical and Mathematical Physics*, https://doi.org/10.1007/978-3-031-44988-8_24

Gianni was very creative, not only in his scientific activity. Once he wrote a computer program in PASCAL for Iacopo based on game theory, this way perhaps enticing his son to later study Economics. He was also very proud of "Ceci" (his daughter Cecilia), who has now become a medical doctor. My impression was that Gianni also had a very fine wire to his wife Antonella, while often she preferred to remain in the background.

In multiple occasions I had the pleasure to be invited to their ranch near Fauglia (in the hinterland of Pisa) for relax. There already in the 1990's Gianni had managed to connect a washing machine installed inside to a water drain black hose winding in a spiral on the dash: of course also the pumping system was his own construction ...

Once, on a weekend, he asked me for help to repair the steering of the old FIAT 126, which turned out to be an actually subtle task. Yet, after some hours of work the car could move again as desired: however, in the meanwhile we had totally screwed out an amount of pieces without fitting them back into their original position, on which Gianni commented: "I suspect they were all useless" ...

Sometimes he let me use the old FIAT. Unfortunately, I could never understand where in Italy parking was allowed. So one day, returning to the place where I had left the car, I found that it wasn't there any more. Looking around, on a wall nearby I spotted a piece of paper—without doubt placed there by Gianni—with "ATTENZIONE, MULTA!" written on it ("watch out! a fine!"), a sketchy drawing of the FIAT, and an arrow pointing rightward with the words "100 metres". There, with great relief, I could retrieve the lost item ...

I knew about the illness of Gianni, the surgery and the fears within medical control after that. Years before I too was confronted with cancer, chemotherapy, etc. Hence, really I believed in his rehabilitation. Unfortunately, after summer 2021 I lost contact with him and only later I understood why ...

It is hard for me to realize that Gianni Morchio is no more here. He was such a kind person.

At the Physics Department with Gianni

Pietro Menotti

Despite having been a senior colleague of Gianni for many years at the Department of Physics in Pisa, I never wrote a paper in collaboration with him. Nevertheless I had a rather deep interaction with him and I will try to summarize below a brief history of it.

I became acquainted with Gianni Morchio in 1967. I was assistant professor at the Scuola Normale Superiore in Pisa where I gave a course on classical physics for the "normalisti" freshmen. They were a dozen of students which was the average number for a class at Scuola Normale.

There were smart people in this selected audience. Among them there was Gianni Morchio. He always sat in the back row and from the back row came the most challenging questions.

Then for a period I lost contact with him until his coming back from the University of Turin and at the same time my coming back to Pisa from the University of Naples. Gianni was assistant professor and recalling his great capabilities, I asked him to give the exercise classes for the course of physics at the School of Mathematics where I taught for two years.

Then I stated teaching quantum mechanics at the Institute of Physics and Gianni followed me again as an assistant. In these years we had intense collaboration in planning the course, preparing the exam papers and at the oral examinations. It goes without saying how much the students profited from him.

Later he started his course on Advanced Quantum Mechanics which became a cornerstone of the curriculum in theoretical physics in Pisa. It contained an introductory part on advanced functional analysis an then developed the algebraic formulation of quantum mechanics.

I was in the exam committee and I always let him to lead completely the exam.

A few times I appealed to him when stuck in my research work and I always obtained some crucial help.

P. Menotti (✉)
Dipartimento di Fisica "E. Fermi", Università di Pisa, Pisa, Italy
e-mail: pietro.menotti@unipi.it

A. Cintio, A. Michelangeli (Eds.), *Trails in Modern Theoretical and Mathematical Physics*, https://doi.org/10.1007/978-3-031-44988-8_25

Gianni had two main features: his absolute intellectual integrity and his gift, when confronted with a new problem, to see immediately the clue.

Later in the years we had discussions on general relativity in particular in connection with the work he was pursuing with Paolo Christillin.

Rarely we spoke about philosophy. Once he expressed to me his appreciation of Spinoza's "Ethics", which Gianni considered a good starting point for the understanding of our world. A few times he expressed his admiration for Descartes' urge for clear and distinct ideas; and actually Gianni had clear ideas. Also about Descartes he pointed out the need to subject continuously our reasonings to repetitions and checks as we can always be wrong. A counsel which I think all of us should follow.

Teachings and Legacy of a Theoretical Physicist

Alessandro Michelangeli

Having met Gianni as an undergrad in theoretical physics, let alone having participated in his courses or having been supervised by him during the MSc thesis, was one of the most stimulating opportunity and greatest privilege for several "chosen" ones, and so was surely for me.

Among the many many recollections that I have now of my interaction with Gianni, which itself sounds somewhat odd to me as I still imagine him at office like all the uncountable times I used to visit him over the years when passing by Pisa, my original hometown, a significant one is my participation in his one-year-long course of the last year of the study programme in theoretical physics.

Actually, we all youngsters had already had to get to know Gianni in one or two previous classes. The first was Giampaolo Cicogna's "*Mathematical Methods for Physics*" mandatory course, which Gianni was a crucial part of by running exercise sessions and co-hosting the exams. That was a one-year long class that in retrospect I still reckon to have been extraordinary and the very best of its type, as far as I could see in my career so far, including when I myself taught something very similar in other universities, for it managed to provide such a deep and rigorous insight on complex analysis methods, Hilbert space methods, Fourier and Laplace transforms, elements of harmonic analysis and PDE's, by keeping at the same time the most genuine "theoretical physicist" approach of understanding what each tool was devised for and how each technique naturally emerged in appropriate contexts. In particular, I remember Giampaolo Cicogna starting his course by introducing Gianni as "*my right-hand man, and in fact also my left-hand*" ...

In parallel to that, the other opportunity to encounter Gianni was in the "*Fundamentals of Theoretical Physics*" course (essentially, non-relativistic quantum mechanics), run at that time and for many years by Pietro Menotti (other parallel

A. Michelangeli (✉)
Department of Mathematics and Natural Sciences, Prince Mohammad Bin Fahd University, Al Khobar, Saudi Arabia
Hausdorff Center for Mathematics, Bonn, Germany
Trieste Institute for Theoretical Quantum Technology, Trieste, Germany
e-mail: amichelangeli@pmu.edu.sa

A. Cintio, A. Michelangeli (Eds.), *Trails in Modern Theoretical and Mathematical Physics*, https://doi.org/10.1007/978-3-031-44988-8_26

classes being run in those years alternatively by Riccardo Barbieri, Luigi E. Picasso, or Paolo Rossi), again with Gianni centrally involved in exercise sessions and final exams—a remarkably hard class, so much so that passing it would make one quite rightfully an "almost graduate". (At that time to obtain the final (Bachelor+Master unified) degree in theoretical physics the main remaining steps were then Adriano Di Giacomo's class on quantum field theory, Sergio Rosati's class on nuclear physics, a solid state physics class run by Ennio Arimondo or Salvatore Carusotto, plus a class on general relativity and a variety of elective courses.)

Yet, through such two mandatory courses there was often the tendency to consider Gianni some "super-human" being, not to say an alien, who simply *knew everything*, with an aura of mystery and admiration that possibly discouraged students from approaching him more systematically. This, together with the above-mentioned non-trivial workload needed to complete the degree, might explain why Gianni's elective class of the last year of the study programme had usually just a fistful of participants.

Nevertheless, we all students were made well aware that some "something real deep" would occur around Gianni and his course. I remember, in this respect, an old booklet compiled by previous students of physics in Pisa some years before I enrolled in, which was still circulating at my time and was meant as a short guide for freshmen, providing fair concise descriptions of all mandatory and elective courses of the study programme in terms of content, difficulty, and workload. Gianni's class was clearly indicated as a tough one, but above all I remember this emblematic comment: "*in the [penultimate year's] theoretical physics class you'd learn things, in [last year's] Morchio's class you'd understand them*" ...

So, indeed, in these recollections of mine about Gianni I find it natural to spend some words on what and how he taught us in that precious jewel which was his course for senior students in theoretical physics.

The title itself was emblematic: merely "*Quantum Mechanics*". That was probably due also to the circumstance that many other suitable titles were already booked in the physics department (including the above-mentioned one-year-long course on non-relativistic quantum mechanics) and that there was still the virtuous habit to give short, weighty titles to courses; but certainly it also reflected Gianni's all-embracing approach to knowledge, a key feature of his throughout his life, I believe.

The course's starting point required an already advanced acquaintance with all standard aspects of non-relativistic quantum mechanics, its formalism, the main one- and three-dimensional models and their spectral analysis, and furthermore evolution, scattering, and perturbation theory. From those bases, the goal was to revisit such a grand picture in its conceptual, mathematical, and physical pillars.

Recently, together with my friend and colleague Andrea Cintio we retrieved our own notes from that class, as well as Gianni's own concise, ultra-deep, impeccable lecture notes (quite the opposite than a static document: Gianni kept making little but fundamental updates every year), and in retrospect we really wondered how an even one-year-long class could encompass such a vastness and depth of materials that would be normally suited for PhD students or researchers in the field.

By moving from a self-consistent treatment of classical measure theory (so as to lay the rigorous grounds for the probabilistic interpretation of quantum mechanics), participants were then brought into a deep survey of the functional-analytic and operator-theoretic tools for the formalisation of quantum models, including self-adjoint extension theory for the definition of physically meaningful observables (thus covering a large part of the first two volumes of the Reed–Simon series, to have an idea), and eventually faced the general discussion of the completeness/incompleteness of quantum mechanics, locality/non-locality, hidden variables, Bell's inequalities, Bohm, Gleason's theorem, ...

Although I cannot diminish other greatly impacting classes for us novices of physics, it is fair to claim that that course represented our first major cultural experience.

I should like to stress that Gianni's class was deeply and eminently *physical*, even if pervaded by advance mathematics throughout. A perspective completely poles apart from, say, Franco Strocchi's beautiful book on the mathematical structure for quantum mechanics, which incidentally was taking shape in those very years and was instead explicitly designed for mathematicians.

Among the main lessons, Gianni taught us the meaning of and the difference between *formalism* (observables, states, dynamics) and *interpretation* (rules) of a physical theory, also providing a most instructive comparison between classical and quantum mechanics—funnily enough, it was precisely by presenting that line of reasoning later at the interview to enter the PhD programme at the SISSA that quite surely granted my admission, which I clearly owe to Gianni. Fortunately, another contributor to this volume is Luca Sciortino, who graduated with Gianni some time before me with a thesis on objectification, and has written more precise and deeper words in this respect.

Besides, Gianni lectured us on the perils of deducing physically wrong conclusions from heuristically sound, yet ambiguous or ill-defined physical models, a perspective that both the shut-up-and-calculate and the sophisticated, intuition-driven theoretical physicists often tend to overlook, quite possibly with disastrous outcomes. An enlightening example, which I happened to "recycle" multiple times in my own teaching, was that celebrated discussion on Landau's book about the fall of a three-dimensional quantum particle onto the centre of a central force of inverse power-law magnitude. At that time, and for generations before, it was morally mandatory for decent physics students in Pisa to digest the main volumes of the Landau–Lifshitz series, and learning that the "bible" could be incomplete was a kind of real shock and rise of awareness, I should say—the issue there was the spectral analysis of the "natural" (i.e., Friedrichs extension) self-adjoint realisation of a minimally defined Schrödinger operator with inverse power-law potential as opposed to other realisations with boundary conditions at the origin ensuring the lower semi-boundedness of the Hamiltonian (the particle "goes through") or its unboundedness from below (the particle "falls onto the centre").

In fact, while recollecting such teachings, I ask myself how many university study programmes in physics offer such a deep presentation of the self-adjoint extension theory in application to quantum mechanics!

In those days we used to joke among us students of physics, also to keep up with the pressure arising from the great challenge of the course first, and later of the thesis work with Gianni, by imagining fictitious phone calls by us to Gianni's home—as a matter of fact, Gianni used to share his private phone number with some of the students writing their thesis with him, as an additional form of availability during their work, although that never occurred to me as I was too shy for that. Such a privilege, linking common mortals to a demigod, had become kind of legendary, but even more legendary was the fact that occasionally it was Gianni's young son to pick up the phone (in fact, I even never met in person Gianni's children, although I am well aware that in the meanwhile they succeeded extremely well in Economics and in Medicine, respectively). So, given their father's scientific stature, it had become common to joke on the possible answer of the kid, like: "*Dad's not at home, but just tell me your problem,* I *would solve it right now . . .*".

Each early afternoon lecture with Gianni in that two-semester-long course was a genuine discovery, as I now remember with the eyes of the then-student fascinated by that sort of omniscient alien who used to bike every day to the physical institute, enter the lecture room with an apparently chaotic bunch of papers, and take full possession of every square inch of the blackboard from the first to the last minute . . .

In that course Gianni initiated me, as many before and after me, to the rigorous physical thinking and to the mathematical-physicist approach to theoretical physics, let alone to the comprehension of a grand picture made of first principles, advanced mathematical tools, rigorous conclusions, and driving physical insight. I owe to him my subsequent passion and commitment to quantum mechanics and its mathematical methods, which then resulted in my choice to write the MSc thesis under his supervision and in following his encouragement to proceed with my doctoral studies at SISSA, a path that was followed my many other students of his before and after me.

That year and that class was the beginning of a beautiful, discrete, respectful, acquaintance among me and him for the years to follow.

Of course, the MSc thesis represented the occasion that cemented such an acquaintance. Gianni has always used to challenge his students with a fully research-oriented work for the MSc thesis, which in my case consisted of examining certain conditions that allow to discriminate between equilibrium (i.e., thermal, say, KMS) states and merely steady states within the C^*-algebraic formulation of quantum statistical mechanics, moving from a very recent discussion of Ruelle, and a classical analysis by Haag and Trych–Pohlmeyer. It is unfortunate that the lack of time which we both experienced back then right after my graduation prevented us from writing even a short paper on it: in fact, we had identified certain mechanisms, that were probably worth being advertised, for constructing non-trivial steady non-KMS states. Yet, the analysis had a successful follow-up in the MSc thesis of Giuseppe De Nittis, who then became a fellow PhD student of mine at SISSA.

In the years that followed, both during my doctoral studies and in the subsequent development of my career, I kept regular contacts with Gianni, whom I used to meet whenever I returned to Pisa, the town where I was born and grew up.

That was the phase in which the "master" turned into a discrete mentor, who was constantly eager to know what was being discussed in the math-phys circles abroad and to share his thoughts on current research projects of his. There was never need of any courtesy pleasantries when I entered his office, for Gianni was immediately driving our conversation onto each and any topic he was most impatient to discuss.

It is in those years that to my great enjoyment our discussions encompassed broader and broader subjects, including social trends, national and international politics, academic politics, recent and future tendencies in science, in physics in particular, as well as virtuous or wicked practices in the academia and academic research.

Last time I met him he was so enthusiastic of his recent reading of Federico De Roberto's novel *I Viceré*, which by accident has always been one of my favourite books: I have fond memories of our last chats on the so distinctive characters portrayed in that novel, where no aspect of society appears to be free from corruption, and on the analogies we drew with our contemporary times.

I cannot but feel privileged to have had Gianni as a lecturer, supervisor, mentor, while also recognising and having benefited from his boundless knowledge, scientific rigour, physical comprehension, example, human generosity and kindness.

Whereas he left us way too early, I am glad to witness today his far-reaching legacy through his friends, collaborators, colleagues, mentees, and former students like me and many many others.

Remembering Gianni Morchio

Dario Pierotti

I first met Gianni when I was a Ph.D. student in Mathematical Physics. At that time, I already knew Franco Strocchi, who had encouraged me to investigate mathematical problems related to gauge invariance in QED.

So, when Franco suggested starting a collaboration with Gianni and himself based on their recent results on infrared and vacuum structure in local quantum field theory, I was excited to work with them but also somewhat anxious.

In particular, since Franco told me how outstanding Gianni was as a mathematical physicist, I figured he wasn't very interested in interacting with a young researcher with little experience.

On the contrary, right from our first meeting in Pisa, I felt totally at ease with Gianni due to his kindness and friendly attitude. At the same time, I was astonished by his impressive skills and the depth of his thought. Thus, I can say that having had the opportunity to work under the guidance of Franco and Gianni represents a fundamental step in my scientific experience.

In the several meetings that followed, besides the intense research activity on the rigorous mathematical analysis of two-dimensional QFT models (the massless scalar field, the Thirring model and the Schwinger model), I got to know Gianni better.

In particular, I appreciated his pleasant character and the extent of his interests and culture. I still remember many passionate discussions on historical, political, and social issues.

Hence, besides his invaluable scientific heritage, it seems increasingly important to me to recall Gianni's sensitivity and vision regarding the problems of inequalities and injustices, especially now that they are even farther from being solved.

D. Pierotti (✉)
Politecnico di Milano, Milano, Italy
e-mail: dario.pierotti@polimi.it

A. Cintio, A. Michelangeli (Eds.), *Trails in Modern Theoretical and Mathematical Physics*, https://doi.org/10.1007/978-3-031-44988-8_27

Unfortunately, many years have passed since I last saw Gianni, but the more time goes by, the more I realize how important it is to have known him.

Over the years, I have hardly met other people with such an extraordinary combination of scientific and human qualities.

Gianni Morchio as Model and Mentor

Paola Ruggiero

In the course of my third undergrad year in physics at the *Università di Pisa*, the moment came to look for a supervisor for the B.Sc. thesis, which usually was supposed to be three-month long but instead in my case ended up to be six months long (and all those months were definitely worth!). Among the professors I went to talk to there was Gianni.

I was immediately struck by his politeness, his enthusiasm and the passion in his words, besides the depth of the subjects he debated. It did not take long for me to choose him.

Those months were fundamental for me and for my growth, both personal and professional. I will never forget all the time Gianni dedicated to me: we used to meet at least once per week (sometimes even more) and we spent hours together (don't remember how many!), only interrupted by someone who at a certain point knocked on the door to "claim" his favourite collaborator: it was prof. Franco Strocchi. At that point I had to leave my place to him: *ubi maior*....

Something I never dared to confess to Gianni is that during those hours he was introducing me to the algebraic formulation of quantum mechanics, I recorded him. The reason is that what he wanted to tell me each time was simply too much information for me, and, on my side, I did not want to lose a single word. By recording him, I could enjoy our conversations more, while the following days were devoted to re-listen word by word everything I could have missed, or I wanted to understand better. And that was one of the most exciting parts of the learning: every time I discovered something more, and I felt like I was understanding better and better, until I was finally ready for a new meeting the week after!

Clearly, apart from physics, those recordings contained much more that Gianni let escape him, from political considerations to the philosophical ones, and some jokes.

P. Ruggiero (✉)
Department of Mathematics, King's College, London, UK
e-mail: paola.ruggiero@kcl.ac.uk

A. Cintio, A. Michelangeli (Eds.), *Trails in Modern Theoretical and Mathematical Physics*, https://doi.org/10.1007/978-3-031-44988-8_28

Among all the lessons I got from our meetings and conversations of those months, I want to mention the following one, using Gianni's own words: "*apprezzare l'ordine e le 'simmetrie' del mondo, 'rassicuranti' riguardo alla sua semplicità e alla sua perfezione, quando ci sono; e viceversa 'accettare' e saper cogliere la bellezza della complessità, laddove sia questa il denominatore comune.*" (the translation goes more or less like this: "*to value order and 'symmetry' in the world, which are 'reassuring' about its simplicity and its perfection, when such conditions are present; and instead, to be able to 'accept' and get the beauty of complexity, when this is the common denominator*").

As undergrads at university, we all look for models to imitate and for mentors to seek advice from. From those months on, Gianni has been for me model and mentor at the same time. During all these years, I looked at him as a source of inspiration, and I consulted him for each important work decision and in difficult personal moments. And every time, he replied with the tenderness and affection which characterised him.

While re-reading the e-mails we exchanged over the years, I realised they are all full of insights and it would be worth to spend a word on each of them. However, a recurring phrase in almost each of them, always after many considerations, towards the conclusion, is this one: "*dimmi, se vuoi, cosa fai e che cosa pensi*" ("*tell me, if you wish, what you do and think*"). A simple sentence, but it gives, I think, the measure of how much Gianni used to care about people and life in general.

Grazie di tutto Gianni, mi mancherai.

Quantum Theory, Objectification and Some Memories of Giovanni Morchio

Luca Sciortino

More than twenty-two years have passed by since, as a young student at the physics department at the University of Pisa, I met Giovanni Morchio (1948–2021) for the first time. After taking his courses on mathematical methods for physics and on quantum mechanics, from 2000 to 2002 I have worked with him on a M.Sc. thesis entitled "The Problem of Objectification in Quantum Theory: the Case of Pyramidal Molecules" [10].

Even after so much time, some of his statements, expressed during our research meetings, still flash through my mind from time to time. They are short, almost lapidary sentences, expressed in a very colloquial language, yet logically rigorous and dense of meaning, which have generated reflections, produced ideas and suggested ways of approaching problems throughout all my intellectual life, sometimes even in my research work in philosophy of science and in my activity as a science writer. I consider the ability of *Gianni* (as he was informally known) to be present in a person's mind and to fertilize it as the rare gift of a few teachers as well as of the great masters of thought.

One of the main goals of this chapter is to explore Gianni's approach to scientific problems and his ideas regarding certain conceptual issues that arise in quantum mechanics. I will pursue this goal by retracing the main stages of my research experience with him and by highlighting some personal memories. In so doing, I hope to achieve another non-secondary goal: to offer some glimpses of Gianni's combination of human qualities that has contributed to earning him great esteem and admiration among friends and colleagues.

Prior to starting working on my thesis, I had become passionate about the controversies surrounding the interpretation of quantum mechanics. These debates began around 1926, when Max Born (1882–1970) put forward the statistical interpretation of the theory's formalism, constructed between 1924 and 1926 by Erwin Schrödinger (1887–1961) and Werner Heisenberg (1901–1976). Eventually, most

L. Sciortino (✉)
eCampus University, Novedrate, Italy
e-mail: luca.sciortino@uniecampus.it

A. Cintio, A. Michelangeli (Eds.), *Trails in Modern Theoretical and Mathematical Physics*, https://doi.org/10.1007/978-3-031-44988-8_29

physicists agreed on an interpretation that can be traced back to the Copenhagen school and is called the *probabilistic interpretation of quantum mechanics*. However, a smaller number of physicists attempted to reformulate quantum mechanics as a classical theory trying to demonstrate that the formalism was incomplete and that the probabilities were "epistemic", i.e., they reflected our lack of knowledge, as happens in classical physics. What aroused my interest was the fact that this disagreement was rooted in some fundamental epistemological questions: do atomic objects exist independently of our observations? Is it possible to correctly understand their behavior? Can formalism incorporate philosophical biases into its structure? Indeed, Heisenberg wrote that "*the opponents of the Copenhagen interpretation all agree on one point. It would, in their view, be desirable to return to the reality concept of classical physics or, to use a more general philosophic term, to the ontology of materialism. They would prefer to come back to the idea of an objective real world whose smallest parts exist objectively in the same sense as stones or trees exist, independently of whether or not we observe them*" [5, 1958, p. 115].

In their lectures, many teachers preferred to avoid these conceptual problems by focusing on the technical aspects of the formalism. Gianni, on the other contrary, did not seem to shy away from these questions. If anything, he would delightedly discuss the ideas on quantum theory of the great minds of scientific and philosophical thought of the twentieth century. In his course called *Quantum Mechanics* I had learned that the works of John Von Neumann (1903-1957), Simon Kochen (1934), Ernst Specker (1920–2011) and John Bell (1928-1990) provide constraints to the possibility of a *classical* reformulation of quantum mechanics [1, 7, 12]. In particular, a global probabilistic interpretation is not possible, i.e., an interpretation such that there exists a measure and a common space in which to represent the collection of all possible observations as functions and whose points assign a defined value to each variable [2, 8]. However, alongside these answers, many questions regarding the possibility of obtaining the interpretation of quantum mechanics from weaker hypotheses still remained unsolved. It was necessary to better understand on which assumptions the probabilistic interpretation of quantum mechanics was based and whether or not these assumptions could be inferred from the formalism.

When I asked Gianni to supervise my thesis on these issues, to my surprise, he suggested that I take a course in the history of physics. Indeed, he considered history, among other things, as an extremely important subject for understanding the true nature of scientific theories. Afterwards, the first preliminary discussions with him convinced me even more that a crucial question concerned what is called *the problem of objectification*. Gianni had an original way of formulating this problem: "*it seems that the formalism of quantum mechanics does not contain certain facts, which must be assumed*", he would say smiling, as he did when he was aware of the depth of a statement. I reflected on that concise sentence for some time, then I realized that the problem of objectification, as formulated by Gianni, could be rephrased in these terms: in order to obtain the current interpretation of quantum theory, the fact that a result of a measurement has been achieved (for example through a very precise position of a display pointer of the equipment) must be assumed. In other words, given the difficulty of deducing from the time evolution the empirical fact that a display pointer indicates a very precise result following a measurement, everything seems to suggest that the formalism of quantum mechanics never represents the notion of fact or event. That is, it seems that the current interpretation of quantum theory needs a hypothesis of objectification, i.e., an ontological assumption that consists in assuming the reality of facts or events such as, for example, having obtained a precise value in a measurement.

But, is it really so? Gianni and I wanted to question this general conclusion. What we wished was to find cases in which it was possible to remove the hypothesis of objectification. In other words, we wanted to suggest reasons for removing or, at least, weakening that assumption. The central aim of the thesis became to ask whether the hypothesis of objectification, which is currently added to the formalism, is not, at least in one case, deducible from it and in particular from the dynamics of the temporal evolution. The case study we were looking for had to: 1) represent a situation similar to that in which a macroscopic system such as a measurement apparatus, following interaction with a system, assumes a well-defined state (which for example represents the fact that the result of a measurement is a particular value), i.e becomes objective; 2) represent a fact in Nature in which it is not clear whether its explanation can be derived from the formalism of quantum mechanics or whether a hypothesis of objectification is necessary for its explanation. Proving that such a fact can be deduced from the time evolution of quantum mechanics would have allowed us to conclude that there are cases in which the hypothesis of objectification is superfluous or, in other words, obtainable from the formalism.

With these ideas in mind, after an exploratory phase in which we inspected a number of examples which seemed particularly illuminating, we selected the case of chiral molecules of the type $XH3$ as our case study. For these molecules, the Schrödinger equation predicts that the two lowest-energy stationary states of the nucleus X are symmetric or antisymmetric under reflection with respect to the plane of the hydrogens. These states can be seen as superposition of localized states which, in fact, correspond to molecules with different chemical properties. In this case, one may ask whether quantum mechanics is compatible with the existence of molecules of one type or another. It was important to find an answer to this question because not only can a molecule of this kind be a good prototype of systems that are never observed in states of superposition (measurement apparatuses, pointers and macroscopic systems tout court), but it also has marked analogies with macroscopic systems. What we wanted to do was to schematize this selected case study in order to demonstrate that it is possible to deduce from the Schrödinger time evolution of a quantum-mechanical system, defined in the formalism, that typically certain chiral molecules of the type $XH3$ exhibit only localized states. This result would have allowed us to conclude that there exist situations in Nature for which the formalism of quantum theory implies that a particular physical system, despite having delocalized superposition states, assumes well-defined and localized states, in a certain sense similar to those of a display pointer.

At that point the goals of the thesis were much clearer and my work, under his supervision, began with an attempt to present an exposition of the probabilistic interpretation of quantum mechanics. This was supposed to be the content of the first part of my dissertation. I wanted to accomplish this task by putting myself in the best perspective to identify and expose the critical points of this interpretation. Gianni was very generous with his time right from the start, as many students acknowledge. He allowed me to call him at home in the evening, at around half past nine, when I needed help. And once a week, in his office, I would discuss the work done with him. I have many memories of that period. For example, one thing that impressed me was this: during discussions in his office on matters that were sometimes extremely technical and complicated, while we were in the process of working something out mathematically, the phone would ring. He would answer and speak for several minutes. At the end of the phone call, he would resume speaking from the exact point he had stopped, as if no one had ever interrupted him. His ability to refocus his attention on a particular task was impressive. Anyone would have had a hard time picking up again so quickly, after the phone call, the thread of the reasoning.

What Gianni did make me understand was the importance of distinguishing between physical formalism and its interpretation. "*A theory?*—he said once smiling—*it is a heap of symbols, some rules and an interpretation*". What he wanted me to notice was the fact that, unlike in classical physics, where it is intuitive to represent the position of a planet as a vector of Euclidean space, in quantum mechanics the empirical meaning of the symbols is not immediately intuitive—there is a clear split between mathematical entities and the objects they represent. As a consequence, the relationship between the symbols of quantum theory and the language we use to describe experiments is a priori problematic. Chris Isham has well expressed this point by saying that "*in classical physics, the 'realist' and 'instrumentalist' views of science fit together seamlessly, whereas in quantum physics they differ sharply, especially in their attitudes towards the idea of physical properties. That such a distinction can arise at all is closely tied to the different mathematical structures employed in the formulations of classical and quantum physics*" [6, p. 3].

On the basis of these considerations, my exposition of the interpretation of quantum theory proceeded through these steps: 1) I distinguished two languages, on the one hand an *observational language* with which we describe the experiments, and on the other hand a *theoretical language*, which includes the mathematical entities of quantum mechanics and classical physics; 2) I defined within this scheme what to interpret a *symbolic language* means (the probabilistic interpretation of quantum mechanics associates propositions of the observational language with the terms of the quantum theoretical language); 3) I clarified, using the two previous points, the differences that exist, in the constitutive substance, between the classical and the quantum language; 4) I defined in terms of points 1) and 2) the postulates of the interpretation of the theoretical language of classical physics and of the probabilistic interpretation of the theoretical language of quantum mechanics.

The idea of distinguishing and analyzing two languages, the theoretical language, which contains the mathematical entities of classical physics and quantum mechanics, and the observational language, which is used to communicate the experimental results, was suggested to me by reading the works of proponents of logical positivism such as Rudolf Carnap (1891–1970) and Hans Reichenbach (1891–1953) who, together with other philosophers of science from Vienna in the first half of the 1900s, reflected on the problems raised by the crisis of classical physics and the birth of quantum mechanics. However, it was the conversations with Gianni that convinced me of the need for an analysis of the two formalisms and of the very meaning of interpretation. According to Gianni, the problems of interpretation of quantum mechanics still required a work of reformulation which had necessarily to deal with problems of linguistic nature. In his mind, the interpretation of quantum mechanics could be better understood if one first comprehended how its theoretical language has been constructed, from what needs it has emerged and why it has that particular structure. A view of this kind, which considers the problem of language central to science, can be traced back to Wittgenstein. Not surprisingly, the *Tractatus Logico-Philosophicus* [13] was once the subject of our conversations. I still remember that, commenting on the statement 1.1 of Wittgenstein in that work, "*The world is the totality of facts, not of things*", Gianni said: "*You see ... what is taken for granted in philosophy, i.e., the claim that there are facts, is under discussion in physics*".

After providing the general characteristics of the observational and the theoretical language, I was able to explain that the formalism of classical physics has a very close relationship with the observational language by making clear that it is possible to identify the logical structure of the statements concerning a physical system with the structure of Boolean algebra of the subsets of a topological space that defines that system. I then discussed classical probability and stated the postulate that defines its interpretation. Afterwards, I explained that, when the topological spaces of classical physics are replaced with Hilbert spaces, then the logical structure of the observational statements acquires the property that in the classical case it acquired from the Boolean algebra structure of the subsets of a topological space. There is a profound link between statements that can be connected through logical connectives and the commutativity of the operators. At that point I introduced the postulate that interprets the notion of commutativity in quantum mechanics. Starting only from a class of states and operators in a Hilbert space, it is possible to associate a measure with certain functionals. In particular, the measurement of an interval can be interpreted as an output frequency of particular results of measurements performed on a collection of systems prepared in the same way.

Classical physics poses no particular problem in terms of the interpretation of its formalism. Not only can the theoretical language of classical physics be extended to the description of the measuring apparatus itself, but it also "contains" its interpretation, without requiring any external assumption—there are sequences of points whose frequencies of visit of the intervals are equal to their measure μ. For reasons analogous to those of the classic case, the interpretation of the measure μ of an interval in terms of relative frequency implies that in every single measure of a set of systems all prepared in the same way a fact occurs. What we elucidated was that it is not clear whether the theoretical language of quantum mechanics, due to its mathematical structure, allows the existence of specific cases in which the hypothesis of objectification is not necessary. Maybe, we were wondering, it is possible to construct a factual model whose description in terms of the theoretical language of quantum mechanics does not require an external hypothesis of objectification. Due to the non-commutative structure of the theoretical language of quantum mechanics, the construction of a Bernoulli system is observable-dependent. From the quantum description of the measurement process it is not possible to obtain the idea that an event has occurred, for example the fact that the measuring apparatus indicates a definite value. At first sight it seemed that one had to resign oneself to the idea that the theoretical language of quantum mechanics predicts the facts and that the hypothesis of objectification is always necessary. But, at this point, we asked ourselves whether the phenomenon of the localization of certain pyramidal molecules is a case in which the Schrödinger evolution implies the occurrence of an event and does not require an external objectification hypothesis.

The chirality of a molecule can be considered to all intents and purposes a fact described by the observational language. In our analysis, the problems of chirality were problems of localization since the chirality of a molecule corresponds to the localization of the wave function of its component nuclei. It must be noted, though, that molecules of type $XY3$ are not necessarily right or left-handed: in a multidimensional standard model for these molecules, the atom X is subject to a double-well potential and the Schrödinger equation has symmetric and antisymmetric eigenstates under reflection of the atom X with respect to the $Y3$ plane. In particular, the ground state and the first excited state are represented, respectively, by a symmetric and an antisymmetric wave function. These wave functions can be superimposed to give rise to two particular wave functions called *right* and *left*, which are respectively located at the left and right side of the x axis and which correspond to two localized configurations of the molecule. Among pyramidal molecules, ammonia exhibits configurations that are not localized and the difference in energy between the ground and the first excited state can be obtained by spectroscopic measurements. However, if the nitrogen atom is replaced by heavier atoms, such as phosphorus or arsenic, to form arsine and phosphine, then such molecules exhibit localized patterns.

In the language of Hilbert spaces, the problem is to explain whether, due to the effect of time evolution, a system that has a space $\mathbb{C}^2$ available *chooses* two localized states defined by *special directions*, which correspond to the location of the molecule. Since chirality is a fact describable by observational language and since it can be reduced to localization, the property of a system of being in a localized state is a fact, in the same way as a pointer of an apparatus pointing in a certain direction. In this sense, the problem of chiral molecules, reduced to the simplest problem which consists in asking whether, due to the effect of time evolution, a system always finds itself in two precise states of a dimensional space and not in their superpositions, is a good prototype of the general objectification problem.

To explain the localization of molecules heavier than ammonia we took into consideration two significant physical facts: first, the times of electromagnetic transition from a localized state to the fundamental state are much greater than the times an atom X takes to oscillate from side to side of the plane of the atoms Y; second, contrary to ammonia, for heavy molecules such as arsine and phosphine, these times are much greater than the formation time of a molecule and the observation time. Hence we conjectured that the reason why arsine and phosphine exhibit localized configurations is that during the formation of these molecules the trivalent atom ends up in a localized state, resulting in, for times which are long compared to our times of observation, a localized configuration of the molecule.

Thus, the question we asked ourselves is whether, due to the effect of time evolution, at a time $t_0 \ll t \ll \tau$, where τ is the transition time and t_0 is the molecule formation time, the atom X is in a localized state. To address this question, we first wrote the one-dimensional Hamiltonian of a non-relativistic quantum particle interacting with the electromagnetic field and subjected to the force field of the atoms Ys. This Hamiltonian consists of: the sum of the particle kinetic energy and the potential due to the Ys atoms (a symmetric double well potential, according to our assumption); a term that represents the interaction energy between the particle and the magnetic field, expressed through the creation and annihilation operators; finally, a term representing the energy of the radiation field, also expressed by means of the creation and annihilation operators.

On the basis of the experimental data on pyramidal molecules, we hypothesized that the gap ω between a level relative to an even eigenstate and to an odd eigenstate is very small compared to all the energies involved, a hypothesis which allowed us to state that $\tau \gg t_0$. We showed that as the potential barrier V tends to infinity, the ω gap tends to zero; then we proved two essential points: the first consists in the fact that the right-hand state, superposition of the even ground state and the first excited odd state, has norm $\mathcal{L}^1$ to the left of zero which is of the order of ω; the second consists in the fact that the eigenstates of the continuum of the Hamiltonian H_0, which represent an incoming wave from the left, have upper limit on the positive x axis of the order of ω. Afterwards, we defined the initial state of the particle and we showed that the formation time t_0 of the molecule can be identified with the passage time through the plane of the atoms Ys of the wave packet describing the atom X.

The term representing the interaction energy in the Hamiltonian of our problem has been considered as a perturbation. I showed how the time evolution operation expands perturbatively in the interaction representation, I described the Hilbert space of the problem and I introduced the scheme and the notation necessary to calculate the amplitudes of the evolved state at first order of the Hamiltonian of the perturbing interaction in the basis product between the symmetric states and the one-photon Fock states. Our goal was now to establish the phase relationships between the amplitudes of the vector resulting from the time evolution, at time $t_0 \ll t \ll 1/\omega$, of the initial state in the aforementioned basis. According to our conjectures, for times t such that $t_0 \ll t \ll 1/\omega$, i.e., for times in which the packet "has finished interacting" but which are still less than the oscillation times, the atom arriving from the left will end up in a left state. For times of the order of $1/\omega$, the oscillations between right and left will take place as predicted by quantum mechanics. From a mathematical point of view, we expressed this conjecture by making these points: the first is the fact that for $t \gg t_0$ the state evolved at time t has "almost reached" its limit and can be replaced by its limit vector (up to errors in the ratio between t and t_0); the second is that, in Schrödinger's representation, for times

t such that $t_0 \ll t \ll 1/\omega$, this vector has phase differences of the order of ωt_0 between the amplitudes in the product basis obtained by taking the symmetric and antisymmetric states and the one-photon Fock states; and, for times of the order of $1/\omega$, it oscillates between a right and a left state. In order to prove these two points I estimated some expressions involving the amplitudes of the evolved vector at time t. I calculated these amplitudes explicitly and discussed the properties of the functions that compose them, then I calculated their punctual limit as t tends to infinity and I estimated the difference between the amplitudes at time t and their punctual limit in terms of t_0/t. Finally, I estimated the differences between the amplitudes showing that the state is localized. Given the observed values of ω, our model was inapplicable for ammonia molecules and we could not conclude that they are in localized states. However, our model was applicable for pyramidal molecules heavier than ammonia such as arsine and phosphine: the fact that these molecules exhibit localized states can be explained by referring to the dynamics of their formation. We therefore concluded that Schrödinger's temporal evolution implies, in a simple case, *the transition from potentiality to actuality* without requiring a principle of objectification external to the theory.

After my graduation, I visited Gianni several times in his office. I recall with pleasure several conversations on the most disparate topics, including some on popular science books such as *The Road to Reality* by Roger Penrose [9] and *A Brief History of Time* by Stephen Hawking [4]. Commenting on the latter's popular science works, Gianni said, half-jokingly, that they tended to "*wrap some scientific problems in mysteries*". My view is that Gianni believed that scientific dissemination must be done in a clear and rigorous way. When this is not possible, then it is better to give up: for example, according to him, the problems of quantum mechanics were so connected with the interpretation of his formalism that in some cases they could not be clearly expressed in the ordinary language.

The content of those conversations as well as some points made in my thesis have been of fundamental importance for my subsequent work [11]. However, after the beginning my doctorate studies in the United Kingdom, I have had only sporadic meetings with him. The best memory I have of that period is this: I had to write a popular article on quantum mechanics for a magazine aimed at the general public. He gave me some valuable advice but, when I asked him if I could quote him in the article, he replied that he preferred not to be quoted: "*in my life I follow the adage attributed to Epicurus:* Λάθε βιώσαζ"—he answered smiling. With that adage, which means "live hidden", he alluded to that kind of life based on genuine values and away from the spotlight, as it is suggested in the 14th of Epicurus' "sovran maxims", as collected by Diogenes Laertius [3, p. 1297].

I believe that, beyond his outstanding scientific works, the most important thing Gianni has left us is his example of a life "well lived", mindful of others and focused on what is authentic and meaningful.

References

1. Bell, J.S.: On the Problem of Hidden Variables in Quantum Mechanics. Rev. Mod. Phys. **38**, 447 (1966)
2. D'Espagnat, B.: Conceptual Foundations of Quantum Mechanics. W.A. Benjamin (1971)
3. Laerzio, D.: Vite e dottrine dei più celebri filosofi. Bompiani (1987)
4. Hawking, S.: A Brief History of Time. Bantam Dell Publishing Group, London (1988)
5. Heisenberg, W.: Physics and Philosophy: The Revolution in Modern Science. George Allen & Unwin LTD, Northampton (1971)
6. Isham, C.J.: Quantum Theory: Mathematical and Structural Foundations. Imperial College Press, London (1995)
7. Kochen, S., Specker, E.: The Problem of Hidden Variables in Quantum Mechanics. J. Math. Mech. **17**, 59 (1967)
8. Mermin, M.D.: Hidden variables and the two theorems of John Bell. Rev. Mod. Phys. **65**, 803 (1993)
9. Penrose, R.: The Road to Reality: A Complete Guide to the Laws of the Universe. Jonathan Cape (2004)
10. Sciortino, L.: Il problema dell'oggettivazione nella teoria quantistica: il caso delle molecole piramidali. University of Pisa (2002). M.sc. thesis
11. Sciortino, L.: History of Rationalities: Ways of Thinking from Vico to Hacking and Beyond. Springer Nature, Switzerland (2023)
12. von Neumann, J.: Mathematical Foundations of Quantum Mechanics. Princeton University Press, Princeton (1955)
13. Wittgenstein, L.: The Tractatus Logicus-Philosophicus. Routledge, London (1921)

Collaboration and Friendship with Gianni

Franco Strocchi

Both my wife and I have been bound by a deep friendship with Gianni, also involving his whole family.[1] It is hard not to think with regret of the beautiful days spent together in Sardinia, of the walks on the Apuan Alps and in the Fauglia countryside, especially in the years when his sons Jacopo and Cecilia were children; all those are happy and poignant memories that are an important part of our lives.

I met Gianni when he was an outstanding student of the course I was lecturing at the University of Pisa and then in 1971 as the supervisor of his thesis. The result was a scientific collaboration that continued until a few months before his death.

I have no words to express my deep gratitude for what I have received from Gianni during our collaboration; his contributions and his role have always been crucial to achieving results. I still regret not having been able to get him the recognition from the scientific community and the right university position that he did deserve.

In the course of my activity I have had the opportunity to meet illustrious and prestigious personalities in the field of Mathematical and Theoretical Physics, but I can affirm, without hesitation, that Gianni was no less, for his lucidity in grasping the core of the problems on the fly and solving them with the simplicity of essential logic, without resorting to gratuitous technicalities.

In his academic research, Gianni was mainly interested in understanding and solving problems, without much interest in the related public recognition. On more than one occasion, despite my advice, the important results he had obtained have remained "unpublished".

[1] Original Italian version included in Appendix A.2.

F. Strocchi (✉)
Dipartimento di Fisica "E. Fermi", Università di Pisa, Pisa, Italy
e-mail: franco.strocchi@sns.it

A. Cintio, A. Michelangeli (Eds.), *Trails in Modern Theoretical and Mathematical Physics*, https://doi.org/10.1007/978-3-031-44988-8_30

Another great gift of Gianni was his willingness to make himself available to those who turned to him for help. I could personally witness his generosity in providing the solution of thesis problems to his undergraduates, leaving to them the scientific credit.

We were lucky enough to know and have as a friend an exceptional person, in terms of intelligence, generosity and great humanity and his loss has caused us incurable pain.

In Memory of Giovanni (Gianni) Morchio

Simone Zerella

A cold day of February 2005. I am strolling across the department of physics's corridors as a first-year PhD student. I have recently arrived in Pisa while still dazed by the change of city and life, and I am in search of doctoral courses to include in the first year plan of my graduate study programme.

While passing by the open door of a lecture room where a class has already started, I happen to notice a notification sheet posted at the entrance, laconically saying "course on mathematical methods for quantum mechanics".

Intrigued, I enter the room, apologising for the delay, and I am greeted by a shy and good-natured lecturer. I sit down and start listening to his lecture. The sensation I felt that day is the same that has always accompanied me during the years of study and research that I had the privilege of sharing with Gianni.

I preserve with particular affection several memories of that period of my life.

A first one is related to the not so smart impression I made when, in a lecture, Gianni introduced the fascinating topic of spontaneous Lorentz symmetry breaking in the charged sectors. That caught me quite incredulous and I then asked him to repeat that part of the discussion, somewhat convinced that he should be wrong. And yet Gianni, after hinting a smile and without reminding me at all my own inexperience, calmly unfolded again the main parts of the reasoning, additionally providing enlightening clarifications.

A second memory is about the final oral exam of Gianni's doctoral course: he actually questioned me as if I was an "ordinary" candidate, even though the two of us had already agreed that he would be my Ph.D. supervisor in the years to come, thereby giving me one of the many proofs of his intellectual probity.

One further recollection that I cherish concerns one of the many afternoons I spent in Gianni's office discussing physics and any other subject of sort. At some point, for a reason that I cannot remember now, the discussion switched from the

S. Zerella (✉)
Farnesina State Scientific High School, Roma, Italy
e-mail: simone.zerella@liceofarnesina.edu.it

A. Cintio, A. Michelangeli (Eds.), *Trails in Modern Theoretical and Mathematical Physics*, https://doi.org/10.1007/978-3-031-44988-8_31

infrared problem on to personal memories of Gianni's military service, which eventually aroused quite an amount of hilarity in the both of us.

Regarding the scientific side of Gianni's activity, I have nothing to add to what other contributors to this volume, more qualified than mine, are surveying in depth.

Regarding the human sides, instead, I believe I can write words from a privileged point of view. I do not think I ever met a more generous person than Gianni. Indeed, I did admire his consideration of those social classes that have become victims of the economic system we happened to live in, as much as his boundless mathematical culture.

In particular I can claim that from Gianni, beside all subtleties of the infrared problem, the usefulness of algebras in the formulation and synthetic understanding of field theories, of statistical mechanics, and of quantum mechanics on manifolds, I learnt the quest for a mathematical rigour which be never an end in itself, the predilection for reasoning in terms of concepts, and the scrupulous attention to expository clarity.

Those are gifts that I will always carry with me and that every day I try to return to my students. Too bad not to be able to talk to him any longer about all that. For sure, this would have made him happy.

Appendix—Original Sources in Italian

This appendix collects the original version in Italian of some of the pages of memory and tribute to Gianni which appear in their English translation in the main body of the volume.

It was the wish of the authors to keep track of their original words, with all the nuances of their native language, in order to fully express their praise and eulogy.

A.1 Chi ha compagni non morirà

(Original Italian version of the Chapter "*Chi ha compagni non morirà*" by Cornolti et al.)

In un email di buon anno Gianni ci scrisse:

> *Ho trovato nell'Archivio Marxista Internazionale una cosa bella e semplice. La massima e il motto credevo fossero i miei, ma sono stato preceduto.*
>
> *La vostra massima:* Humani nihil a me alienum puto.
>
> *Il vostro motto:* De omnibus dubitandum.

Sono le risposte di Marx ad un questionario che gli sottoposero le figlie: era allora costume in Inghilterra che le ragazze strappassero delle "confessioni" a genitori e amici.

Crediamo che qui ci sia tutto Gianni.

Humani nihil in tutta la sua generalità. Ogni singolo momento della storia umana suscitava in Gianni maledizione ed entusiasmo. Maledizione per le sofferenze e le devastazioni materiali e morali patite dalla stragrande parte dell'umanità nella sua storia; entusiasmo per tutto quello che, nonostante tutto, l'uomo aveva prodotto e poteva produrre e che avvicinava l'umanità al mondo sognato.

Gianni si riconosceva completamente in queste parole di Marx:

> *Si vedrà allora che da tempo il mondo custodisce il sogno di una cosa, del quale gli manca solo di prendere coscienza, per possederla veramente. Si dimostrerà che non si tratta di tirare una linea retta tra passato e futuro, bensì di portare a compimento i pensieri del*

A. Cintio, A. Michelangeli (Eds.), *Trails in Modern Theoretical and Mathematical Physics*, https://doi.org/10.1007/978-3-031-44988-8

> *passato. Si vedrà in ultimo che l'umanità non inizia un nuovo lavoro, ma porta a termine con coscienza il proprio antico lavoro.*[1]

Non una fede cieca, ma una fede fondata sulla ragione.

Non la ragione feticistica delle *magnifiche sorti progressive* dello sviluppo storico, o la razionalità strumentale di chi vede come unica possibilità la sua applicazione per ottenere i migliori vantaggi personali in una società che è un fatto e perciò accettatta come destino; non la posizione contemplativa di chi si vive separato dalla società nel mondo delle idee o della scienza; ma la ragione che è a fondamento dell'etica e del bene collettivo. La ragione come fondamento della passione e dell'attività politica. Per questo Gianni amava molto Spinoza e lo ha letto fino all'ultimo, così come amava il napoletano Antonio Labriola.

Gianni era incrollabilmente convinto che ogni uomo sia in grado di poter capire le cose più complesse. Era questa la principale eredità dell'Illuminismo. Bisogna aver visto la pazienza e la passione con cui illustrava categorie astratte di economia agli operai, con cui ha condiviso decenni di attività sindacale e politica, ma anche nella convinzione che il confronto con loro lo arricchisse, perché il pensiero, le conoscenze sono il prodotto della vita reale di uomini reali.

> *Grazie al semplice fatto che ogni nuova generazione trova davanti a sé le forze produttive* (ivi compresa la scienza) *acquisite dalla vecchia generazione, che servono come materia prima di una nuova produzione (in senso lato: materiale, artistica, scientifica, ecc.), si forma un contesto nella storia degli uomini, si forma una storia dell'umanità, che è tanto più storia dell'umanità quanto più le forze produttive degli uomini, e conseguentemente i loro rapporti sociali, sono cresciuti. La necessaria conseguenza è: la storia sociale degli uomini è sempre e soltanto la storia del loro sviluppo individuale, che ne siano coscienti o no. I loro rapporti materiali formano la base di tutti i loro rapporti. Questi rapporti materiali non sono altro che le forme necessarie, nelle quali si realizza la loro attività materiale e individuale.*[2]

De omnibus dubitandum: Gianni aveva la certezza che "*non possono esserci conoscenze tanto lontane da non potervi alfine pervenire, né conoscenze tante nascoste da non scoprirle*",[3] ma anche che l'oggetto del conoscere e lo stesso soggetto sono il prodotto dell'attività umana, momenti del processo sociale di produzione.

Perciò ogni verità deve essere considerara provvisoria, sottoposta continuamente a critica, sia nella verifica della sua capacità di reggere i nessi interni, sia nella sua rispondenza ai fatti storici o scientifici. La verità di una teoria sta innanzitutto nella sua capacità di stimolare le azioni umane e le successive ricerche, nella sua capacità, in definitiva, di condurre oltre se stessa.

Dubitare dunque di ogni proposizione o risultato acquisito è qui il contrario esatto del relativismo e dello scetticismo. Riflettere costantemente su ciò che viene considerato assodato è il principio direttivo di ogni sviluppo del pensiero.

Gianni era un divoratore di libri, alla ricerca incessante di fatti storici che potessero arricchire il materiale dell'esperienza con cui confrontarsi, di idee su cui ragionare. Ogni qualvolta arrivava ad una conclusione, che arricchiva la compren-

[1] Karl Marx (1818–1983), Lettera a Arnold Ruge, Settembre 1843.

[2] Karl Marx, Lettera a Pavel Vasilevič Annenkov, Dicembre 1846.

[3] Cartesio, *Discorso sul Metodo* (1637).

sione della storia e dei rapporti tra la classi, la comunicava con fervore, impaziente di condividerla e di metterla alla prova, di sottoporla alle obiezioni e alla critiche. Non di rado, però, nel momento stesso in cui ne parlava e l'oggettivava, sentivi l'inquietudine del pensiero ancora insoddisfatto di sé e di quanto raggiunto. E allora riprendeva il problema da un'altra prospettiva.

Era bellissimo vedere e seguire il pensiero nel suo farsi.

Ma Gianni non ha mai pensato in nessun caso il pensiero come fine a se stesso. Ciò che lo dirigeva in ogni momento della sua vita era la passione per la verità, convinto fino all'ultimo nervo del suo corpo che la ricerca della verità facesse tutt'uno con la lotta per una società futura come comunità di uomini liberi qual è possibile con i mezzi tecnici oggi a disposizione, per la trasformazione di quei rapporti sociali sotto i quali gli uomini soffrono e la loro anima è certo destinata a intristire.

La scissione tra uomo e società ha caratterizzato tutte le forme storiche della vita sociale. L'esistenza della società si è fondata sinora o sull'oppresione immediata, oppure è una cieca risultante di forze contrastanti; non è il risultato della spontaneità cosciente di uomini liberi.

Era inossidabile convinzione di Gianni che la verità è sospinta innanzi solo in quanto gli uomini che la possiedono parteggiano inflessibilmente per essa, la applicano e la impongono, agiscono in conformità ad essa. "*Il processo della conoscenza implica il reale agire e volere storico nella stessa misura in cui implica l'esperienza e la comprensione.*"[4] Perciò il processo della conoscenza è sempre il processo collettivo della lotta per la trasformazione dello stato delle cose esistenti.

Sul piano storico politico, in particolare, Gianni, insieme ai compagni con cui lavorava, era arrivato alla conclusione della necessità di un bilancio del movimento comunista e dei problemi che lo avevano messo in difficoltà. Un bilancio che doveva prendere in considerazione in modo unitario il processo che risale sul piano storico all'estensione del capitalismo fuori dall'Europa occidentale e degli USA, su quello politico alla crisi del movimento operaio tra gli ultimi anni della II Internazionale e la crisi della III, da fare senza scorciatoie sulla base e nel quadro delle tesi di Marx e dei loro sviluppi successivi. Non è qui il luogo per riassumere e discutere i risultati di questo bilancio. Quello che qui importa evidenziare è che la ricostruzione coerente della storia dell'ultimo secolo e mezzo dal punto di vista dei comunisti è stata la principale attività teorica di Gianni. Non per mera curiosità intellettuale, ma in quanto presupposto indispensabile per la lotta politica del movimento operaio.

Tutta la vita di Gianni è in fondo testimonianza di quanto dice Anatole France:

> *Noi non possediamo nulla in proprio tranne noi stessi; l'uomo dona veramente solo quando dona il proprio lavoro, la propria anima, la propria intelligenza, e questa magnifica offerta di tutto se stesso a tutti gli uomini arricchisce tanto il donatore quanto la comunità.*[5]

[4] Max Horkheimer (1895–1973), *Sul problema della verità* (1935).
[5] Anatole France (1844–1924), *Il signor Bergeret a Parigi* (1901).

E sottoscriverebbe ogni parola dell'*Angelus Novus* di Walter Benjamin:

La lotta di classe, che è sempre davanti agli occhi delle storico educato su Marx, è una lotta per le cose rozze e materiali, senza le quali non esistono quelle più fini e spirituali. Ma queste ultime sono presenti e vivono in questa lotta come fiducia, coraggio, umore, astuzia, impassibilità, e agiscono retroattivamente nella lontananza dei tempi. Esse rimetteranno in questione ogni vittoria che sia toccata nel tempo ai dominatori. Come i fiori volgono il capo verso il sole, così, in forza di un eliotropismo segreto, tutto ciò che è stato tende a volgersi verso il sole che sta salendo nel cielo della storia.[6]

Anche se tutto questo oggi ai molti appare impossibile e folle.

A noi resta il raro, e immenso, privilegio di averlo conosciuto.

Pompeo, Maurizio, Fulvio

A.2 Collaborazione e amicizia con Gianni

(Original Italian version of the Chapter "*Collaboration and Friendship with Gianni*" by F. Strocchi)

A Gianni sia io che mia moglie siamo stati legati da una profonda amicizia coinvolgendo anche tutta la sua famiglia. È difficile non pensare con rimpianto alle belle giornate passate insieme in Sardegna, nelle camminate sulle Apuane e nella sua campagna di Fauglia, sopratutto negli anni in cui i figli Iacopo e Cecilia erano bambini; sono ricordi felici e struggenti che fanno parte importante della nostra vita.

Ho conosciuto Gianni quando è stato studente eccezionale del corso che tenevo all'Università e poi nel 1971 come relatore della sua tesi. Ne è nata una collaborazione scientifica che è andata avanti fino a pochi mesi prima della sua morte.

Non ho parole per esprimere la mia profonda riconoscenza per quanto ho ricevuto da Gianni nel corso della nostra collaborazione; i suoi contributi e il suo ruolo sono sempre stati cruciali per il raggiungimento dei risultati. Mi resta il rammarico di non essere riuscito a procurargli il riconoscimento da parte della comunità scientifica e la giusta posizione universitaria che meritava.

Nel corso della mia attività ho avuto occasione di incontrare personaggi illustri e prestigiosi nel campo della Fisica Matematica e della Fisica Teorica, ma posso affermare, senza esitazione che Gianni non era da meno, per la lucidità di cogliere al volo il nocciolo del problema e risolverlo con la semplicità della logica essenziale, senza ricorso a tecnicismi gratuiti.

Nella ricerca Gianni era soprattutto interessato a capire e risolvere i problemi, senza molto interesse ai relativi riconoscimenti pubblici. In più di un'occasione, nonostante i miei consigli, gli importanti risultati che aveva ottenuto sono rimasti "non pubblicati".

[6] Walter Benjamin (1892–1940), *Tesi di filosofia della storia.*

Un'altra grande dote di Gianni è stata la sua disponibilità a mettersi a disposizione di chi si rivolgeva a lui per chieder aiuto. Ho potuto constatare di persona la sua generosità nel fornire la soluzione dei problemi di tesi ai suoi laureandi, lasciando a loro il relativo credito scientifico.

Abbiamo avuto la fortuna di conoscere e avere come amico una persona eccezionale, per intelligenza, generosità e grande umanità e la sua perdita ci ha causato un dolore insanabile.

Franco Strocchi

A.3 Il nostro babbo

(Original Italian version of the Chapter "*Our Dad*" by C. and I. Morchio)

Siamo stati molto indecisi se scrivere qualcosa per raccontare nostro padre, e il motivo è che era una persona decisamente convinta che nessuno è mai davvero in grado di giudicare, descrivere, raccontare le altre persone. Lo pensava sinceramente, e non per un disinteresse nei confronti degli altri, ma perché provava un profondo rispetto per le vite altrui, e in lui erano caratteristici il pudore e l'umiltà.

In tutti i suoi rapporti era la persona che ascoltava l'altro, diceva sempre che sentire i pareri di chi la pensa diversamente è una delle cose più interessanti al mondo.

Ci piacerebbe comunque trasmettere un'immagine del nostro babbo, che possa in qualche modo aggiungere un lato magari meno noto alle pagine che seguono.

La prima visione che ci viene in mente quando pensiamo a lui è il momento in cui tornava a casa dal lavoro, e ci sentivamo quasi come se ad accoglierci fosse lui e non viceversa: entrava in casa con un sorriso enorme, desideroso di sentire come era andata la nostra giornata e di scambiare con noi idee, opinioni e racconti spesso abbastanza complessi da metterci in seria difficoltà, anche se sentivamo la sua fiducia illimitata nell'intelligenza e nella capacità di capire, nostra e di tutti.

Del suo lavoro e della fisica parlava poco, e quando lo faceva era sempre e solo in modo divertente, con battute su come secondo le leggi della fisica l'acqua bolle, il gelato si compone, la frittata si gira.

Ne parlava poco anche perché condivideva con noi tante altre sue passioni: traduceva dal latino e dal greco senza vocabolario, conosceva a fondo la filosofia, sapeva a memoria interi canti della Divina Commedia e poesie varie, era un grande amante della musica classica, al punto da cantare arie delle sue opere preferite nei dialoghi della vita quotidiana, spesso per stemperare nervosismi o litigi.

Oggi che siamo adulti con le nostre vite fatte di lavoro, amici e affetti, fatichiamo ancora di più a capire dove trovasse le energie per fare tutto quello che faceva, con riserve inesauribili di entusiasmo. Forse il modo più bello di condividere l'affetto è trovare e dare tempo agli altri, e lui era sempre disponibile e aperto ad avere tempo per noi, sia da piccoli per giochi e fiabe inventate da lui, sia da grandi per confronti e discussioni di ogni genere.

Babbo ci ha insegnato che è importante avere passioni, conoscere le cose in profondità, pensare autonomamente, vivere in modo indipendente, lavorare con serietà, avere tanti interessi, rimanere aperti al mondo e non smettere mai di imparare. Ma soprattutto, con poche parole ma dando molto l'esempio, ci ha insegnato che è fondamentale essere persone integre, coerenti con le proprie idee e i propri valori.

Il suo esempio ci ha aiutato a trovare la nostra strada, nella consapevolezza che qualunque cosa avremmo fatto sarebbe andata bene, purché la facessimo con passione e integrità.

Anche se a parole dichiarava di non credere all'importanza delle emozioni, ci ha sempre fatto sentire la profondità dell'affetto che provava nei nostri confronti, facendoci sentire apprezzati e amati per quello che siamo.

Ci manca tantissimo, ma tutto ciò che è stato vive in noi.

Iacopo e Cecilia

Giovanni Morchio's List of Publications

1. Bracci, L., Morchio, G., Strocchi, F.: Wigner's theorem on symmetries in indefinite metric spaces. Comm. Math. Phys. **41**(11), 289–299 (1975).
2. Bracci, L., Morchio, G., Strocchi, F.: Description of symmetries in indefinite metric spaces. Lect. Notes Phys. **50**, 551–556 (1976). https://doi.org/10.1007/3-540-07789-8_56
3. Fröhlich, J., Morchio, G., Strocchi, F.: Charged sectors and scattering states in quantum electrodynamics. Annals Phys. **119**, 241–248 (1979). https://doi.org/10.1016/0003-4916(79)90187-8
4. Gatto, R., Morchio, G., Strocchi, F.: Natural Flavor Conservation in the Neutral Currents and the Determination of the Cabibbo Angle. Phys. Lett. B **80**, 265–268 (1979). https://doi.org/10.1016/0370-2693(79)90213-2
5. Gatto, R., Morchio, G., Strocchi, F.: Symmetries leading to flavor conservation in Higgs induced neutral currents and implications on the Cabibbo angle. Phys. Lett. B **83**, 348–350 (1979). https://doi.org/10.1016/0370-2693(79)91124-9
6. Fröhlich, J., Morchio, G., Strocchi, F.: Infrared problem and spontaneous symmetry breaking of the Lorentz group in QED. Phys. Lett. B **89**, 61–64 (1979). https://doi.org/10.1016/0370-2693(79)90076-5
7. Morchio, G., Strocchi, F.: Infrared singularities, vacuum structure, and pure phases in local quantum field theory. SNS 9/79
8. Barbieri, R., Morchio, G., Strocchi, F.: Neutrino masses and SO(10) hierarchical breaking. Lett. Nuovo Cim. **26**, 445 (1979). https://doi.org/10.1007/BF02817027
9. Morchio, G., Strocchi, F.: Infrared singularities, vacuum structure and pure phases in local quantum field theory. Ann. Inst. H. Poincaré Sect. A (N.S.) **33**(3), 251–282 (1980)
10. Gatto, R., Morchio, G., Sartori, G., Strocchi, F.: Natural flavor conservation in Higgs induced neutral currents and the quark mixing angles. Nucl. Phys. B **163**, 221–253 (1980). https://doi.org/10.1016/0550-3213(80)90399-5
11. Barbieri, R., Nanopoulos, D.V., Morchio, G., Strocchi, F.: Neutrino masses in grand unified theories. Phys. Lett. B **90**, 91–97 (1980). https://doi.org/10.1016/0370-2693(80)90058-1

A. Cintio, A. Michelangeli (Eds.), *Trails in Modern Theoretical and Mathematical Physics*, https://doi.org/10.1007/978-3-031-44988-8

12. Fröhlich, J., Morchio, G., Strocchi, F.: Higgs phenomenon without symmetry breaking order parameter. Phys. Lett. B **97**, 249–252 (1980). https://doi.org/10.1016/0370-2693(80)90594-8
13. Morchio, G., Strocchi, F.: Infrared Singularities and General Structure Properties of Local Gauge Quantum Field Theories. SNS-8-80
14. Morchio, G., Strocchi, F.: A natural non-perturbative mechanism for gauge hierarchies. Phys. Lett. B **104**, 277–281 (1981). https://doi.org/10.1016/0370-2693(81)90125-8
15. Fröhlich, J., Morchio, G., Strocchi, F.: Higgs phenomenon without symmetry breaking order parameter. Nucl. Phys. B **190**, 553–582 (1981). https://doi.org/10.1016/0550-3213(81)90448-X
16. Morchio, G., Strocchi, F.: A nonperturbative approach to the infrared problem in QED: construction of charged states. Nucl. Phys. B **211**, 471–508 (1983). https://doi.org/10.1016/0550-3213(84)90042-7. [Erratum: Nucl. Phys. B **232**, 547 (1984)]
17. Morchio, G., Strocchi, F.: A remark on spontaneous breaking of supersymmetry. Phys. Lett. B **118**, 340 (1982). https://doi.org/10.1016/0370-2693(82)90199-X
18. Morchio, G., Strocchi, F.: Spontaneous Symmetry Breaking and Energy Gap Generated by Variables at Infinity. Commun. Math. Phys. **99**, 153–175 (1985). https://doi.org/10.1007/BF01212279
19. Morchio, G., Strocchi, F.: Spontaneous symmetry breaking and energy gap generated by variables at infinity examples. IFUP TH 84/27
20. Morchio, G., Strocchi, F.: Infrared problem in QED and electric charge renormalization. Ann. Phys. **168**(1), 27–45 (1986). https://doi.org/10.1016/0003-4916(86)90108-9
21. Morchio, G., Strocchi, F.: Spontaneous breaking of the Galilei group and the plasmon energy gap. Annals Phys. **170**, 310–332 (1986). https://doi.org/10.1016/0003-4916(86)90095-3
22. Morchio, G., Strocchi, F.: Infrared problem, Higgs phenomenon, and long range interactions. NATO Sci. Ser. B **141**, 301–344 (1986)
23. Morchio, G., Strocchi, F.: Confinement of massless charged particles in QED in four dimensions and of charged particles in QED in three dimensions. Annals Phys. **172**, 267–279 (1986). https://doi.org/10.1016/0003-4916(86)90184-3
24. Morchio, G., Strocchi, F.: Long range dynamics and broken symmetries in gauge models. The Stückelberg-Kibble model. Commun. Math. Phys. **111**(4), 593–612 (1987). https://doi.org/10.1007/BF01219076
25. Morchio, G., Strocchi, F.: Long-range dynamics and broken symmetries in gauge models. The Schwinger model. J. Math. Phys. **28**, 1912–1919 (1987). https://doi.org/10.1063/1.527454
26. Morchio, G., Strocchi, F.: Mathematical structures for long-range dynamics and symmetry breaking. J. Math. Phys. **28**, 622–635 (1987). https://doi.org/10.1063/1.527649
27. Morchio, G., Strocchi, F.: Effective nonsymmetric Hamoltonians and Goldstone boson spectrum. Annals Phys. **185**, 241–269 (1988). https://doi.org/10.1016/0003-4916(88)90046-2. [Erratum: Annals Phys. **191**, 400 (1989)]

28. Morchio, G., Pierotti, D., Strocchi, F.: Infrared and vacuum structure in two-dimensional local quantum field theory models. The massless scalar field. J. Math. Phys. **31**(6), 1467–1477 (1990). https://doi.org/10.1063/1.528739
29. Morchio, G., Pierotti, D., Strocchi, F.: Infrared and vacuum structure in two-dimensional models of local quantum field theory. II. Fermion bosonization. J. Math. Phys. **33**, 777–790 (1992). https://doi.org/10.1063/1.529757
30. Morchio, G., Pierotti, D., Strocchi, F.: The Schwinger model revisited. Annals Phys. **188**, 217 (1988). https://doi.org/10.1016/0003-4916(88)90101-7
31. Morchio, G., Strocchi, F.: Removal of the infrared cutoff, seizing of the vacuum and symmetry breaking in many-body and in gauge theories. In: IXth International Congress on Mathematical Physics (Swansea, 1988). pp. 490–493. Adam Hilger, Bristol (1989) [SISSA-125/88/EP]
32. Morchio, G., Strocchi, F.: Infrared structures in QFT models and the θ angle problem. SISSA-116/89/EP
33. Morchio, G., Strocchi, F.: Boundary terms long range effects and chiral symmetry breaking. In: Mitter, H., Schweifer, W. (eds.) Fields and Particles. Proceedings Schladming, Springer (1990). https://doi.org/10.1007/978-3-642-76090-7_6
34. Bagarello, F., Morchio, G.: Dynamics of mean-field spin models from basic results in abstract differential equations. J. Statist. Phys. **66**(3–4), 849–866 (1992). https://doi.org/10.1007/BF01055705
35. Acerbi, F., Strocchi, F., Morchio, G.: Algebraic fermion bosonization. Lett. Math. Phys. **26**, 13–22 (1992). https://doi.org/10.1007/BF00420514
36. Acerbi, F., Morchio, G., Strocchi, F.: Infrared singular fields and nonregular representations of CCR algebras. J. Math. Phys. **34**, 899–914 (1993). https://doi.org/10.1063/1.530200
37. Bagarello, F., Morchio, G., Strocchi, F.: Quantum corrections to the Wigner crystal: A Hartree-Fock expansion. Phys. Rev. B **48**, 5306–5314 (1993). https://doi.org/10.1103/PhysRevB.48.5306
38. Acerbi, F., Strocchi, F., Morchio, G.: Theta vacua, charge confinement and charged sectors from nonregular representations of CCR algebras. Lett. Math. Phys. **27**, 1–11 (1993). https://doi.org/10.1007/BF00739583
39. Acerbi, F., Morchio, G., Strocchi, F.: Nonregular representations of CCR algebras and algebraic fermion bosonization. Rept. Math. Phys. **33**, 7–19 (1993). https://doi.org/10.1016/0034-4877(93)90036-E
40. Loffelholz, J., Morchio, G., Strocchi, F.: Spectral stochastic processes arising in quantum mechanical models with a non-L^2 ground state. Lett. Math. Phys. **35**, 251–262 (1995). https://doi.org/10.1007/BF00761297. [arXiv:hep-th/9410083 [hep-th]]
41. Loffelholz, J., Morchio, G., Strocchi, F.: The QED in (0+1)-dimension model and a possible dynamical solution of the strong CP problem. [arXiv:hep-th/9501118 [hep-th]]
42. Loeffelholz, J., Morchio, G., Strocchi, F.: A quantum mechanical gauge model and a possible dynamical solution of the strong CP problem. Annals Phys. **250**, 367–388 (1996). https://doi.org/10.1006/aphy.1996.0097

43. Morchio, G.: Charged fields, Higgs phenomenon and confinement: Lesson from soluble models. Ann. Inst. H. Poincare Phys. Theor. **64**, 461–478 (1996)
44. Mnatsakanova, M., Morchio, G., Strocchi, F., Vernov, Y.: Canonical commutation relations in a positive and an indefinite metric spaces. In: Quarks'96. Proceedings, 9th International Seminar, Yaroslavl, Russia, May 5–11, 1996. pp. 51–62 (1996)
45. Mnatsakanova, M., Morchio, G., Strocchi, F., Vernov, Y.: Irreducible representations of the Heisenberg algebra in Krein spaces. J. Math. Phys. **39**(5), 2969–2982 (1998). https://doi.org/10.1063/1.532431
46. Booss-Bavnbek, B., Morchio, G., Strocchi, F., Wojciechowski, K.P.: Grassmannian and chiral anomaly. J. Geom. Phys. **22**, 219–244 (1997). https://doi.org/10.1016/S0393-0440(96)00032-0
47. Buchholz, D., Doplicher, S., Morchio, G., Roberts, J.E., Strocchi, F.: A model for charges of electromagnetic type. In: Operator algebras and quantum field theory (Rome, 1996). pp. 647–660. International Press, Cambridge (1997). https://doi.org/10.1006/aphy.2001.6136. [arXiv:hep-th/9705089 [hep-th]]
48. Wojciechowski, K.P., Scott, S.G., Morchio, G., Booss-Bavnbek, B.: Determinants, Manifolds with Boundary and Dirac Operators. Fundam. Theor. Phys. **94**, 423–432 (1998). https://doi.org/10.1007/978-94-011-5036-1_32
49. Gentili, F., Morchio, G.: Irreversible weak limits of classical dynamical systems. J. Math. Phys. **40**(9), 4400–4418 (1999). https://doi.org/10.1063/1.532975
50. Cavallaro, S., Morchio, G., Strocchi, F.: A generalization of the Stone-von Neumann theorem to nonregular representations of the CCR-algebra. Lett. Math. Phys. **47**(4), 307–320 (1999). https://doi.org/10.1023/A:1007599222651
51. Morchio, G., Strocchi, F.: Representations of $*$-algebras in indefinite inner product spaces. In: Stochastic processes, physics and geometry: new interplays, II (Leipzig, 1999). pp. 491–503. CMS Conf. Proc., 29, American Mathematical Society, Providence (2000). [ISBN 0-8218-1960-7]
52. Buchholz, D., Doplicher, S., Morchio, G., Roberts, J.E., Strocchi, F.: Quantum delocalization of the electric charge. Annals Phys. **290**, 53–66 (2001). https://doi.org/10.1006/aphy.2001.6136. [arXiv:hep-th/0011015 [hep-th]]
53. Correggi, M., Morchio, G.: Quantum mechanics and stochastic mechanics for compatible observables at different times. Ann. Physics **296**(2), 371–389 (2002). https://doi.org/10.1006/aphy.2002.6236. [arXiv:quant-ph/0112021 [quant-ph]]
54. Buchholz, D., Doplicher, S., Morchio, G., Roberts, J.E., Strocchi, F.: Asymptotic Abelianness and braided tensor C*-categories. Prog. Math. **251**, 49–64 (2007). [arXiv:math-ph/0209038 [math-ph]]
55. Mnatsakanova, M., Morchio, G., Strocchi, F., Vernov, Y.: Representations of CCR algebras in Krein spaces of entire functions. Lett. Math. Phys. **65**(3), 159–172 (2003). https://doi.org/10.1023/B:MATH.0000010715.93852.84. [arXiv:math-ph/0211025 [math-ph]]

56. Loffelholz, J., Morchio, G., Strocchi, F.: Ground state and functional integral representations of the CCR algebra with free evolution. [arXiv:math-ph/0212037 [math-ph]]
57. Loffelholz, J., Morchio, G., Strocchi, F.: Mathematical structure of the temporal gauge in quantum electrodynamics. J. Math. Phys. **44**(11), 5095–5107 (2003). https://doi.org/10.1063/1.1603957. [arXiv:math-ph/0212039 [math-ph]]
58. Morchio, G., Strocchi, F.: Charge density and electric charge in quantum electrodynamics. J. Math. Phys. **44**, 5569–5587 (2003). https://doi.org/10.1063/1.1623928. [arXiv:hep-th/0301111 [hep-th]]
59. Morchio, G., Strocchi, F.: Localization and symmetries. J. Phys. A **40**, 3173–3188 (2007). https://doi.org/10.1088/1751-8113/40/12/S17. [arXiv:math-ph/0607015 [math-ph]]
60. Morchio, G., Strocchi, F.: Quantum mechanics on manifolds and topological effects. Lett. Math. Phys. **82**, 219–236 (2007). https://doi.org/10.1007/s11005-007-0188-5. [arXiv:0707.3357 [math-ph]]
61. Cintio, A., Morchio, G.: Sum rules and density waves spectrum for nonrelativistic fermions. J. Math. Phys. **50**(4), 042102 (2009). https://doi.org/10.1063/1.3100756. [arXiv:0804.0675 [cond-mat.other]]
62. Morchio, G., Strocchi, F.: The Lie-Rinehart universal Poisson algebra of classical and quantum mechanics. Lett. Math. Phys. **86**(2–3), 135–150 (2008). https://doi.org/10.1007/s11005-008-0280-5. [arXiv:0805.2870 [quant-ph]]
63. Morchio, G., Strocchi, F.: Chiral symmetry breaking and θ vacuum structure in QCD. Annals Phys. **324**, 2236–2254 (2009). https://doi.org/10.1016/j.aop.2009.07.005. [arXiv:0907.2522 [hep-th]]
64. Morchio, G.: A simple derivation of decoherence energies. J. Phys. A: Math. Theor. **44**(6), 068001 (2011). https://doi.org/10.1088/1751-8113/44/6/068001
65. Budroni, C., Morchio, G.: The extension problem for partial Boolean structures in quantum mechanics. J. Math. Phys. **51**, 122205 (2012). https://doi.org/10.1063/1.3523478. [arXiv:1010.4662 [math-ph]]
66. Budroni, C., Morchio, G.: Bell Inequalities as Constraints on Unmeasurable Correlations. Found Phys **42**, 544–554 (2012). https://doi.org/10.1007/s10701-012-9625-0. [arXiv:1103.3644 [quant-ph]]
67. Morchio, G., Strocchi, F.: Dynamics of Dollard asymptotic variables. Asymptotic fields in Coulomb scattering. Rev. Math. Phys. **28**(01), 1650001 (2016). https://doi.org/10.1142/S0129055X1650001X. [arXiv:1410.5612 [math-ph]]
68. Morchio, G., Strocchi, F.: The Infrared Problem in QED: A Lesson from a Model with Coulomb Interaction and Realistic Photon Emission. Annales Henri Poincare **17**(10), 2699–2739 (2016). https://doi.org/10.1007/s00023-016-0486-5. [arXiv:1410.7289 [hep-th]]
69. Christillin, P., Morchio, G.: Relativistic Newtonian gravitation. [arXiv:1707.05187 [gr-qc]]
70. Morchio, G., Strocchi, F.: Manifold topology, observables, and gauge group. J. Math. Phys. **62**(11), 112102 (2021). https://doi.org/10.1063/5.0048336. [arXiv:2102.09632 [quant-ph]]

Source

INSPIRE	https://inspirehep.net/authors/996877
MATHSCINET	https://mathscinet.ams.org/mathscinet/author?authorId=200311
ARXIV	https://arxiv.org/search/?searchtype=author&query=Morchio

GPSR Compliance

The European Union's (EU) General Product Safety Regulation (GPSR) is a set of rules that requires consumer products to be safe and our obligations to ensure this.

If you have any concerns about our products, you can contact us on ProductSafety@springernature.com

In case Publisher is established outside the EU, the EU authorized representative is:

Springer Nature Customer Service Center GmbH
Europaplatz 3
69115 Heidelberg, Germany

Batch number: 10370736

Printed by Printforce, the Netherlands